MW00634816

Structural DNA Nanotechnology

Written by the founder of the field, this is the first text of its kind, providing a definitive introduction to structural DNA nanotechnology. Readers will learn everything there is to know about the subject from the unique perspective of the leading expert in the field.

Topics covered range from origins and history, to design, experimental techniques, DNA nanomechanics devices, computing, and the uses of DNA nanotechnology in organizing other materials.

Clearly written, and benefiting from over 200 full-color illustrations, this accessible and easy-to-follow text is essential reading for anyone who wants to enter this rapidly growing field. It is ideal for advanced undergraduate and graduate students, as well as for researchers in a range of disciplines including nanotechnology, materials science, physics, biology, chemistry, computational science, and engineering.

NADRIAN C. SEEMAN is the founder of the field of structural DNA nanotechnology. He is currently the Margaret and Herman Sokol Professor of Chemistry at New York University, and is the recipient of a number of awards including the Sidhu Award, the Feynman Prize, the Emerging Technologies Award, the Rozenberg Tulip Award in DNA Computing, the World Technology Network Award in Biotechnology, the NYACS Nichols Medal, the SCC Frontiers of Science Award, the ISNSCE Nanoscience Prize, the Kavli Prize in Nanoscience, the Einstein Professorship of the Chinese Academy of Sciences, a Distinguished Alumnus Award from the University of Pittsburgh, and a Jagadish Chandra Bose Triennial Gold Medal from the Bose Institute, Kolkata. He is also a Thomson Reuters Citation Laureate and a Fellow of the American Crystallographic Association.

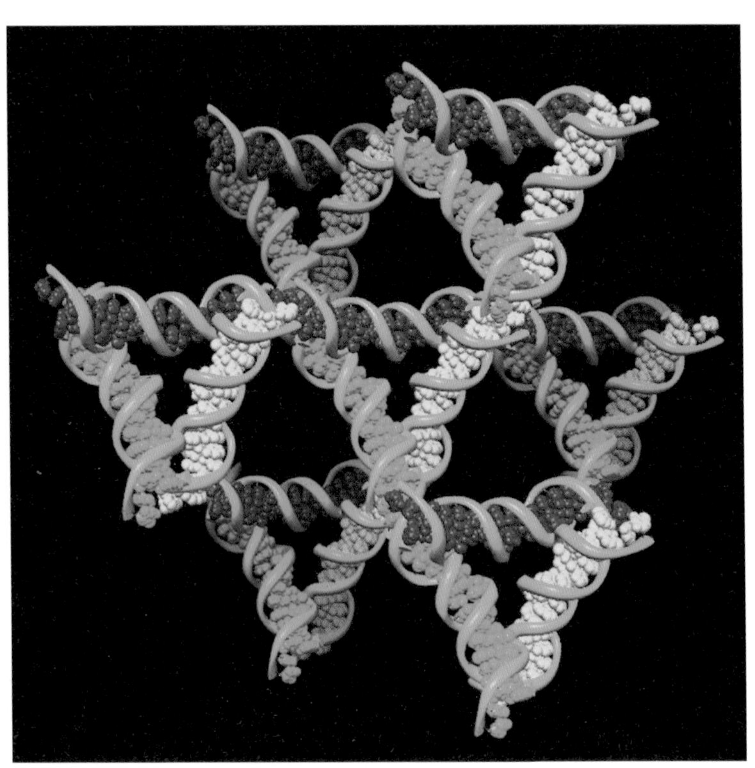

"The first of its kind, it will undoubtedly become the Bible for DNA self-assembly and nanoscale 3D printing. The visionary father of the field of structural DNA nanotechnology, Ned Seeman, lays out its principles lucidly and with superb graphics to match. For anyone curious about synthetic DNA technologies or in connecting these principles with current research, this is a must-have-must-read."

Yamuna Krishnan, University of Chicago

"Ned Seeman invented and pioneered structural DNA nanotechnology in the 1980s and he has been in the front line of the field since then. For many years he was alone in the field and it was considered as a mere curiosity by many scientists and ignored by most others. However, during the past 15 years the field has blossomed and today constitutes a unique approach to organize matter at the nanoscale by self-assembly. The book gives the best possible first-hand insight into this field and its amazing development."

Kurt Vesterager Gothelf, Aarhus University

"The book is an inspiring insight into the design and development of DNA motifs used as building blocks, molecular devices, and information processing tools. It is stimulating to both students and professionals with detailed introduction to blueprint composition and experimental strategies. These strategies have provided an exponential growth in the subject and established the field of DNA nanotechnology."

Natasha Jonoska, University of South Florida

"The pioneer of the field of structural DNA nanotechnology, Ned Seeman, presents the foundations, the state-of-the-art, and the stories leading to the development of this fascinating field that today allows researchers around the globe to control matter with sub-nanometer precision by means of self-assembly. Students in nanoscience-related fields will greatly benefit from this book, and for researchers planning to work in the fast growing field of DNA nanotechnology, it is a must."

Tim Liedl, Ludwig-Maximilians-Universität

"This is a wonderful book. It systematically covers all major aspects of DNA nanotechnology, a rapidly evolving research field. Though there are multiple books and reviews that cover the current topics of this field, this book is the only one that provides insights on how this field originated, developed, differentiated, and flourished. I enjoyed reading this book particularly because of its emphasis on structural bases of DNA molecules – quite often neglected by people now. I fully expect that this book will serve as a handy reference for practitioners in the field of DNA nanotechnology, as a textbook for graduate students and undergraduate students, and also as a historic book for people studying science history. For sure, this book will be the textbook for my graduate course, bionanotechnology, at Purdue University."

Chengde Mao, Purdue University

Structural DNA Nanotechnology

NADRIAN C. SEEMAN
New York University

CAMBRIDGE
UNIVERSITY PRESS

University Printing House, Cambridge CB2 8BS, United Kingdom

Cambridge University Press is part of the University of Cambridge.

It furthers the University's mission by disseminating knowledge in the pursuit of education, learning and research at the highest international levels of excellence.

www.cambridge.org
Information on this title: www.cambridge.org/ 9780521764483

First published 2015

Printed in the United Kingdom by Bell and Bain Ltd

A catalogue record for this publication is available from the British Library

Library of Congress Cataloguing in Publication data
Seeman, Nadrian C., 1945–
Structural DNA Nanotechnology / Nadrian C. Seeman, New York University.
pages cm
ISBN 978-0-521-76448-3
1. Nanobiotechnology. 2. DNA – Industrial applications.
3. DNA – Synthesis. I. Title.
TP248.25.N35S77 2015
620′.5–dc23
2015018281

ISBN 978-0-521-76448-3 Hardback

All illustrations are also available at www.cambridge.org/seeman

Contents

Preface *page* ix

1 **The origin of structural DNA nanotechnology** 1

2 **The design of DNA sequences for branched systems** 11

3 **Motif design based on reciprocal exchange** 28

4 **Single-stranded DNA topology and motif design** 44

5 **Experimental techniques** 64

6 **A short historical interlude: the search for robust DNA motifs** 88

7 **Combining DNA motifs into larger multi-component constructs** 97

8 **DNA nanomechanical devices** 130

9 **DNA origami and DNA bricks** 150

10 **Combining structure and motion** 172

11 **Self-replicating systems** 186

12 **Computing with DNA** 198

13 **Not just plain vanilla DNA nanotechnology: other pairings, other backbones** 213

14 **DNA nanotechnology organizing other materials** 231

Afterword 248

Index 251

Preface

I was approached by many publishers in the early years of this century to write a book about structural DNA nanotechnology. At the time, I was working on the central goal of my program, the control of the assembly of matter in three dimensions. I was rightfully afraid that I could get distracted from achieving that goal, so I turned them all down. Ultimately, in 2009, we published the 3D structure of a self-assembled DNA lattice, and I felt it was time to put my stamp on the field. In 2010, Cambridge University Press agreed to publish the book, I applied for and got a Guggenheim Fellowship, and I took my first sabbatical, to write it.

During the twentieth century, the field and what we were doing in my laboratory were kind of the same thing, but as the new millennium dawned, interest in DNA nanotechnology grew, and many laboratories were attracted to the field. The directions that the field has gone are not entirely reflective of my take on the issue of controlling structure with branched DNA motifs. I am interested in making lattices, not objects, but that is the main thrust of the field these days, largely owing to the popularity of DNA origami and DNA bricks. Both of those approaches are themselves consequences of the dropping price of DNA, a sort of Moore's law of DNA synthesis.

Thus, this book is heavily laden with the things that I do and that I have thought about since 1980. These include topology, sequence control, and other issues that are not thought about much today. I was about 2/3 of the way through the book at the 3/4 point of my sabbatical. At that point, I suffered an injury that kept me from finishing the book until my deadline approached at the end of 2014. The field has grown substantially from 2011 until now, but I never saw this monograph as a big review article containing the latest and greatest. Thus, the final two chapters are really just highlights of their topics, and the reader should not expect them to be even close to comprehensive.

Those most directly responsible for my getting to this point at all are the John Simon Guggenheim Foundation, whose financial support is gratefully appreciated, and my old friend Greg Petsko, who made his San Francisco house available to me during the writing of this volume. I have to thank Bruce Robinson for coming down the hall one day in late 1978 and asking me to build a model of a Holliday junction. Everything followed from that moment and the moment on a plane to Hawaii in 1979 when Greg mentioned a scheme to prepare hemoglobin intermediates. I should also thank Kathy McDonough for listening to me ramble on during a conversation when I had the idea for immobile junctions built from oligonucleotides. The late Malcolm Casadaban asked me if I could make immobile junctions with more than four arms; when I figured out that the answer was "yes," it wasn't too long until I wandered off to the Campus Pub and had the epiphany described in Chapter 1. Neville Kallenbach facilitated the earliest days of this work, both financially and collaboratively; if he hadn't brought me to NYU in 1988, I doubt much of this work would have happened. Jim Canary has supplied organic chemical knowledge when it was needed, as has Paramjit Arora. There have been others involved, but I can't list all my friends. However, my students and postdocs were essential to the development of the field, particularly Junghuei Chen, Chengde Mao, and Hao Yan. Nothing happens in my laboratory without the help of Ruojie Sha, and there is no way to express my appreciation to him adequately. Over the years, Natasha Jonoska, Erik Winfree, Paul Rothemund, and Bill Goddard have been invaluable colleagues. Special thanks are due to Paul Chaikin for pushing the self-replication project. I thank Hendrik Dietz for Figure A-3 and David Goodsell for Figure 7-25.

1

The origin of structural DNA nanotechnology

Everyone knows that DNA is the genetic material of all living organisms. Its double helical structure has become an icon for our age. The publication of its double helical structure by Watson and Crick in 1953 revolutionized biology.[1.1] Its most prominent applications today are in clinical diagnosis of genetic diseases and pathogenic organisms and in forensics. The key element of DNA is that it contains *information* – information in a form that is easy to understand, utilize, and manipulate. The central feature of this information is that it is linearly encrypted in the sequence of the DNA polymer. There are four different elements to this information, known as A, T, G, and C. We'll get into the chemical details of what those letters mean a little later. The important thing is that the molecule is in its most stable state (has the lowest free energy) when A on one strand is opposite T on the other strand, and when G on one strand is opposite C on the other strand. As Watson and Crick famously put it, it did not escape their attention that this complementary pairing leads immediately to a mechanism for replication: an A on the parental strand means you put a T on the daughter strand, and vice versa; similarly a G on the parental strand means you put a C on the daughter strand, and vice versa. It is important to realize that strands exhibiting this complementarity can be put together *in vitro*, a fact first noted by Alexander Rich and David Davies.[1.3] A key and often unvoiced aspect of this mechanism is that the helix axis is linear, not in the geometrical sense of being a straight line, but in the topological sense that it is not branched. This book is about what happens when the helix axis is branched and how we can use it to make new and interesting molecules and materials on the nanometer scale.

The chemical details of the classical structure of DNA are shown in Figure 1-1, and the backbone structure of DNA is shown in Figure 1-2. The double helical structure has many interesting features that need to be mentioned. First, the backbones are antiparallel. What do we mean by that? There is a chemical polarity

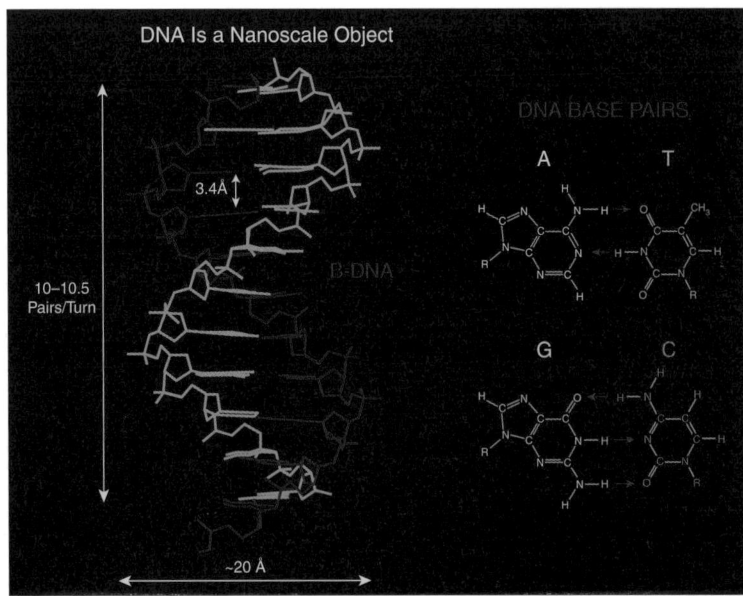

Figure 1-1 *The DNA molecule.* The double helical structure of DNA in the B form and its dimensions are indicated on the left. The details of the base pairing are shown on the right, where two hydrogen bonds are indicated between A and T, and three hydrogen bonds are indicated between G and C.

to the molecules, so that the strands have directionality. If you look at the left side of Figure 1-1, you will see pentagons every repeating unit (called a "residue") in both the gray strand and the red strand. They are pointing in opposite directions. For example, in the third step from the top of the helix, the pentagons are pretty much in the plane of the paper; the gray pentagon has a vertex pointing to the bottom of the page; by contrast, the pentagon opposite it on the red strand has a vertex pointing to the top of the page.

This arrangement is much like the opposite lanes in a two-lane highway with one lane going in each direction. Although the two strands of DNA are like the two opposite lanes of a highway, highways would not be very useful if they were like DNA: the road would go on forever, with no opportunity to turn off it. Branched DNA molecules can be thought of as intersections, where two roads meet. However, at the intersections in DNA, the lane is forced to turn and go off in a new direction, as shown in Figure 1-3. The lanes in Figure 1-3 are drawn in four independent colors, and the arrows of the same colors represent vehicles in the same lanes. Thus, the vertical red arrow pointing downward in the red lane is going south and the vertical cyan arrow opposite it pointing upward is going

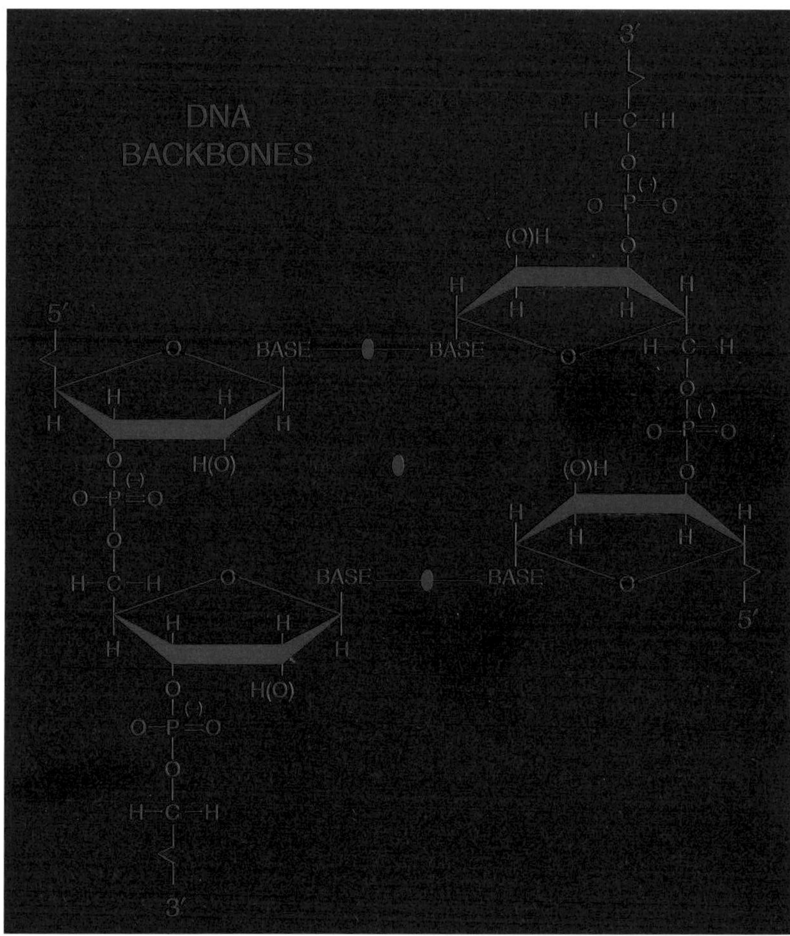

Figure 1-2 *DNA backbones*. The stereochemistry and directionality (5′ to 3′) are shown. The antiparallel nature of the backbone is evident from this representation. Note that there are two-fold axes indicated by little lens-shaped symbols. These are perpendicular to the helix axis and they relate only the backbones, and not the base pairs; there is one two-fold axis through each base pair plane, and a second one halfway in between them.

north. However, the red arrow makes a right turn at the intersection to go west, and after its right turn it is opposite the green arrow in the green lane going east. Similar relationships exist between the arrows in the other directions: all the lanes make right turns and change direction by 90°. The directions that they are going correspond to the movement of traffic in North America or in continental Europe, where vehicles drive on the right. Traffic in the UK, India, Japan, or

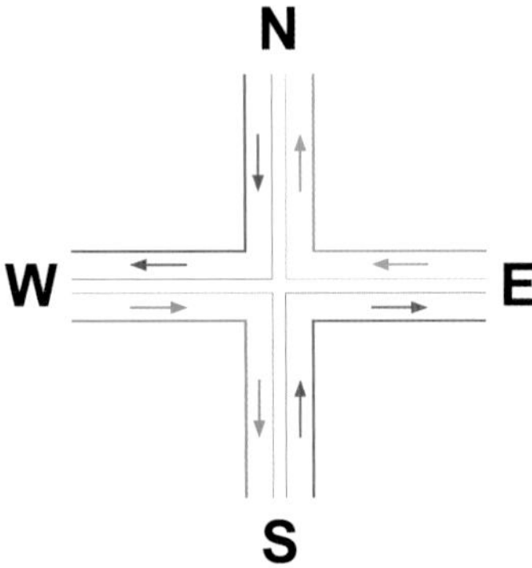

Figure 1-3 *A 4-way intersection.* Four different roads, coming from the north, west, south, and east, are shown, analogous to a 4-arm DNA branched junction. The flow of traffic through the intersection is color-coded, with the direction of traffic (in right-hand drive countries) indicated by arrows. Similar to DNA, each road has traffic going in two antiparallel directions. Reprinted with permission from Springer.[1.2]

Australia, where vehicles drive on the left, would be represented by a diagram that is the mirror image of Figure 1-3.

I started off talking about 4-way intersections, because that was what got me interested in branched DNA. The 4-way intersection or junction is analogous to an intermediate in genetic recombination, called the Holliday junction.[1.4] Although the possibility of branching had been mentioned earlier,[1.5,1.6] the Holliday junction was the earliest DNA junction that people thought about functionally, and in many respects it is still the most popular junction to make and with which to work, as we will see in later chapters. Of course, there are lots of kinds of intersections, not just those that consist of two roads crossing. It is possible to have three roads meet in a 3-way intersection,[1.7] or five or six,[1.8] or even more.[1.9] Generalizations of Figure 1-3 are shown in Figure 1-4, which illustrates 3-way and 6-way junctions.

Structural DNA technology got started one day in September 1980, when I went over to the pub on campus to think about nucleic acid junctions, which were analogous to 6-way intersections. Because they were easy to draw that way, I walked into the bar thinking of them as having snowflake-like six-fold symmetry. However, during my first beer, I suddenly thought about M.C. Escher's famous woodcut, *Depth*, which is shown in Figure 1-5. I realized that the fish in the drawing could be thought of as 6-arm junctions: there was no reason to draw 6-arm junctions as planar objects, like roadways usually are, but instead I could think of them in three dimensions. Furthermore, *Depth* doesn't show just a single fish, it shows a whole arrangement of fish. In fact, this

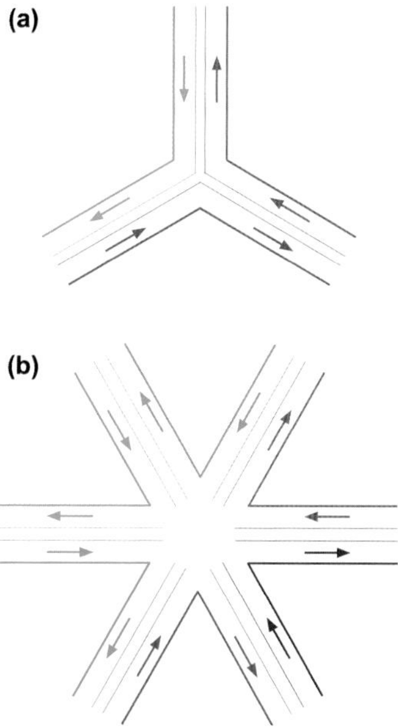

Figure 1-4 *Non-standard intersections.* (a) *A 3-way intersection.* (b) *A 6-way intersection.* In the same way that there are 3-way intersections and 6-way intersections, there are 3-arm DNA branched junctions and 6-arm DNA branched junctions. Junctions have been made with up to 12 arms. Reprinted with permission from Springer.[1.2]

arrangement is just like the arrangement of molecules in a molecular crystal: a periodic (repeating) pattern front to back, left to right, and top to bottom.

This was very meaningful to me, particularly at that time, because I was entering my fourth of five probationary years as an Assistant Professor. I had been hired to do crystallography on biological macromolecules, but had not by that time managed to crystallize any molecule of interest to myself or to others. Consequently, I was facing a fatal progression of "no crystals, no crystallography, no crystallographer." The fact that Escher had been able to organize these fish into a crystalline arrangement in a deliberate fashion made me think that I could organize nucleic acids the same way, and to get my crystals to form by a designed self-assembly process.

Of course, there was a catch! Escher was in complete control of every stroke in his artwork. As a natural scientist, I needed to work with a natural interaction to get branched DNA molecules to associate with each other and thereby form a crystalline lattice. The molecules themselves need to be connected to each other through some sort of chemical process. Fortunately, a process like this does exist, and has been used since the early 1970s to put simple double helices

Figure 1-5 *Escher's woodcut, Depth.* This image inspired the field of structural DNA nanotechnology. The fish are analogous to 6-arm junctions: starting from their centers, they have a head, a tail, a top fin, a bottom fin, a right fin, and a left fin. In addition, they are organized like the molecules in a molecular crystal, demonstrating periodicity front to back, left to right, and top to bottom.

together.[1.10] This process is based on a notion called "sticky ends." Sticky ends result when two strands of a double helical DNA molecule are not quite the same length. In the example shown in Figure 1-6, we see a red double helix and a blue double helix. Of course, the double helices don't look like helices here, because they have been unwound so that it is easy to display their sequences. Let's look at the two double helices at the top of the picture. The blue double helix has two strands, and the upper strand is four residues longer than the bottom strand. The overhang of four extra nucleotides is on the 3′ end, because it is near the arrowhead indicating the direction that the strand goes. The red double helix is similar, except that its overhang is on the 3′ end of the bottom strand. We always describe nucleic acid sequences from the 5′ end, even if we

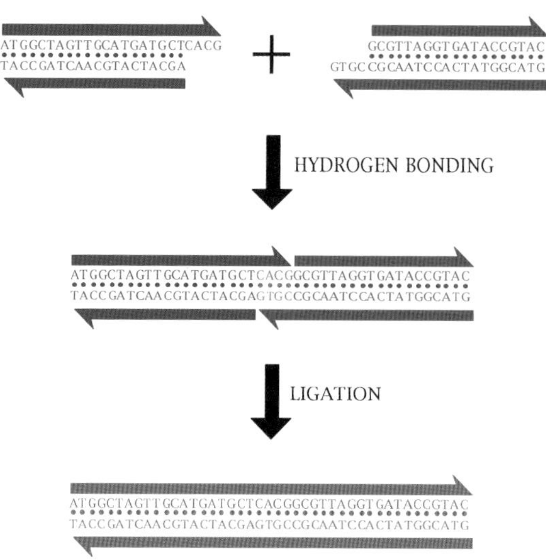

Figure 1-6 *Sticky-ended cohesion.* Two unwound double helices are shown at the top. Their strands are of slightly different lengths, creating overhangs that are called "sticky ends." If the sticky ends are complementary and conditions are proper, the two molecules can cohere, as shown in the middle. It is possible to join the ends covalently by enzymatic (or sometimes non-enzymatic) ligation.

have to read the picture backwards to do it. Given this rule, the sequence of the blue sticky end is CACG, and the sequence of the red sticky end is CGTG. The antiparallel nature of the double helix makes it a little tricky, because you have to read the sequence of one of the strands backwards to see it, but these two short strands are complementary to each other. Consequently, they can cohere with each other in just the same way that the two blue strands cohere or the two red strands cohere. This is evident in the middle part of the diagram, where the complementary bases are juxtaposed. The bases have been colored magenta, a color intermediate between red and blue, to indicate that they form a short segment of double helix all their own, part red, and part blue, just like the blue double helix and the red double helix. Under appropriate conditions (i.e., with sufficiently high concentrations of DNA and of cations, and at low enough temperatures), the structure in the middle of Figure 1-6 will cohere stably, and will form a single complex. If you want to make sure that the complex is really robust, it is straightforward to link the red strands to the blue strands covalently. This is done by ligating them enzymatically to form two total strands, rather than four. The ligated product is shown in the bottom panel of Figure 1-6, where there are now just two magenta strands, which are shown to be paired.

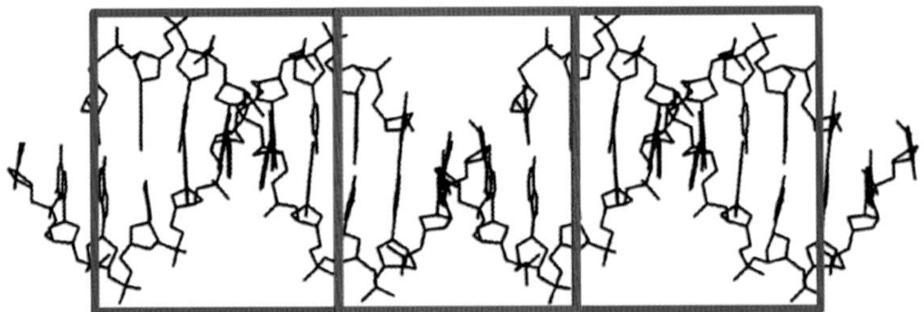

Figure 1-7 *A crystal structure showing sticky-ended cohesion.* This is the crystal structure of a self-complementary DNA decamer, but the dyad axis is displaced from the middle, creating a sticky end. Sticky-ended cohesion is shown in the red box in the middle, where the two gaps lacking phosphate groups are prominent. The important thing about this image is that the DNA in the red box has the same B-DNA structure as the DNA in the blue boxes, even though it is upside down, because it is a half-turn away. Thus, if one knows the coordinates of the DNA in the blue box on the right, one knows the coordinates of the DNA in the blue box on the left, even in solution. Reprinted by permission from Elsevier.[1.11]

The sticky-ended interaction is a very special affinity interaction. There are lots of affinity interactions in biology – for example, the binding of multi-subunit proteins to each other or the interactions of antigens and antibodies. However, all we know about those interactions is that there is affinity between the components of the complex. In every case, we need a crystal structure or some comparable analysis to know the details of which atoms are binding to which atoms.

Sticky-ended interactions are different in this regard, as shown in Figure 1-7. This drawing illustrates a crystal structure that forms an infinite DNA double helix in the horizontal direction.[1.11] It consists of two strands of DNA that are slightly offset from each other, so that the duplex is tailed in 2-base sticky ends. There are two gaps in the middle of the red box, places where one strand stops and the other strand hasn't started yet. These gaps delineate the position where the duplex is held together by sticky ends. The blue boxes outline roughly the same places in two successive unit cells. Their contents appears to be upside down from the material in the red box, but that is because they are a half-turn away. The key point, however, is that there is a great deal of similarity between the structure surrounding the place of sticky-ended cohesion and structures at the middle of the blue boxes (even if they are upside down). Consequently, if we know the coordinates of the atoms in the left blue box, we know the coordinates of the atoms in the right blue box. Thus, when two pieces of double helical DNA cohere, we know not merely that there is an affinity interaction but, on a predictive basis, what that interaction looks like structurally, without having to do the crystal structure. The details of the structure may not be so well determined a double helical turn or two away, but, at least locally, we know what the local product structure will be.

This is a very powerful property, unique, to my knowledge, in macromolecular recognition. If we have six nucleotides in our sticky ends, there are 4^6 unique sticky ends that can be made from the conventional bases: 2048 pairs (less 64 self-complementary hexamers, or a total of 1984) of different interactions that can be programmed uniquely. When we combine the notion of sticky ends with branched DNA that can contain at least 12 branches, we can imagine making huge numbers of stick-figure objects and lattices of DNA, where the vertices correspond to the points of branching and the edges consist of double helical DNA.

A simplified example of this concept is illustrated for a quadrilateral in Figure 1-8. At the left of the picture is a branched junction whose four arms meet at right angles, just like an intersection. Each of the four arms is tailed in a sticky end X, its complement X′, Y, and its complement Y′. What the diagram shows is that four of these junctions can stick together to form a quadrilateral. In addition, there are many unsatisfied sticky ends on the outside of the quadrilateral. Thus, more junction molecules could bind to the outside of the

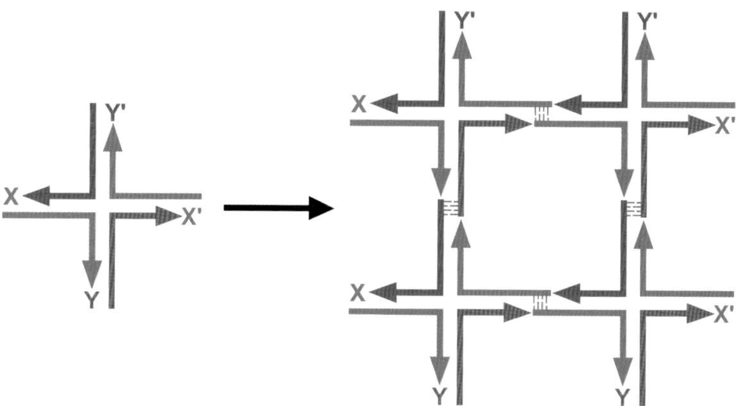

Figure 1-8 *The central concept of structural DNA nanotechnology: combining branched junctions into larger constructs.* The left side of the drawing contains a branched junction with four sticky ends, X, Y, and their complements, X′ and Y′. Four of these junctions have been combined to form a quadrilateral on the right. However, there are many sticky ends that are unsatisfied, so the construct could be extended, in principle, to form a lattice in 2D. In fact, since the ladder-like representation of DNA shown is not really appropriate to its double helical structure, it is possible to form 3D lattices from branched DNA molecules.

quadrilateral to form an infinite two-dimensional lattice. As we shall see, this system is not limited to two dimensions, but can in fact be extended to three dimensions, leading to infinite three-dimensional crystalline lattices. Figure 1-8 shows the fundamental notion of structural DNA nanotechnology: putting well-structured branched DNA molecules together to form lattices in two or three dimensions. There are many variations on this theme, going beyond lattices to devices and computation and other exciting topics. We shall explore them in the succeeding chapters of this book.

References

1.1 J.D. Watson, F.H.C. Crick, Molecular Structure of Nucleic Acids: A Structure for Deoxyribose Nucleic Acid. *Nature* **171**, 737–738 (1953).

1.2 N.C. Seeman, Another Important 60th Anniversary, *Advances in Polymer Science* **261**, 217–228 (2013).

1.3 A. Rich, D.R. Davies, A New Two-Stranded Helical Structure: Polyadenylic Acid and Polyuridylic Acid. *J. Am. Chem. Soc.* **78**, 3548–3549 (1956).

1.4 R. Holliday, A Mechanism for Gene Conversion in Fungi. *Genet. Res.* **5**, 282–304 (1964).

1.5 J.R. Platt, Possible Separation of Intertwined Nucleic Acid Chains by Transfer-Twist. *Proc. Nat. Acad. Sci. (USA)* **41**, 181–183 (1955).

1.6 A. Gierer, Model for DNA and Protein Interactions and the Function of the Operator. *Nature* **212**, 1480–1481 (1966).

1.7 R.-I. Ma, N.R. Kallenbach, R.D. Sheardy, M.L. Petrillo, N.C. Seeman, 3-Arm Nucleic Acid Junctions Are Flexible. *Nucl. Acids Res.* **14**, 9745–9753 (1986).

1.8 Y. Wang, J.E. Mueller, B. Kemper, N.C. Seeman, The Assembly and Characterization of 5-Arm and 6-Arm DNA Junctions. *Biochem.* **30**, 5667–5674 (1991).

1.9 X. Wang, N.C. Seeman, The Assembly and Characterization of 8-Arm and 12-Arm DNA Branched Junctions. *J. Am. Chem. Soc.* **129**, 8169–8176 (2007).

1.10 S.N. Cohen, A.C.Y. Chang, H.W. Boyer, R.B Helling, Construction of Biologically Functional Bacterial Plasmids *in Vitro*. *Proc. Nat. Acad. Sci. (USA)* **70**, 3240–3244 (1973).

1.11 H. Qiu, J.C. Dewan, N.C. Seeman, A DNA Decamer with a Sticky End: The Crystal Structure of d-CGACGATCGT. *J. Mol. Biol.* **267**, 881–898 (1997).

2

The design of DNA sequences for branched systems

Structural DNA nanotechnology rests on three pillars: (1) nucleic acid hybridization, (2) facile synthesis of designed DNA sequences, and (3) the ability to design branched DNA molecules. This chapter is primarily about the third topic, but before we get into it, we should briefly discuss the other two topics. The hybridization of DNA strands is taken for granted by virtually all investigators today, but this was not always so. When the first hybridization was done in 1956 by Rich and Davies (see Chapter 1), the result was treated with skepticism, typified by the comment, "You mean [the two strands hybridize] without an enzyme?"[2.1]

The first approaches to DNA nanotechnology entailed sequence design that attempted to minimize sequence symmetry in every way possible. Such sequences are not readily obtained from natural sources, so the synthesis of DNA molecules of arbitrary sequence is a *sine qua non* for DNA nanotechnology; the field would not exist without the phosphoramidite-based synthesis methodology developed by Caruthers and his colleagues.[2.2] Fortunately, DNA synthesis has existed for about as long as needed by DNA nanotechnology: synthesis within laboratories or centralized facilities has been around since the 1980s; today it is possible to order all the DNA components needed for DNA nanotechnology, so long as they are free of complex modifications, i.e., so-called "vanilla" DNA. In addition, the biotechnology enterprise has generated demand for many variants on the theme of DNA (e.g., biotinylated molecules), and these molecules are also readily synthesized or purchased.

The details of DNA base pairing. What about branched DNA? All of us know that A pairs with T and G pairs with C. That's how biology works. However, we are not talking about biology here. We are talking about making things out of DNA that do not form readily in biological systems. What problems arise in this case? What can go wrong, and why? Are there simple solutions to the issues that arise? To answer these questions we should examine the structure of DNA in more detail, and talk about the things DNA and its components can do, so as to be sure that we can get it to do what we want it to do.

When we talk about A pairing with T and G pairing with C, we are talking about hydrogen bonded interactions. When we look at Figure 1-1, we see the A–T base pair above the G–C base pair. The A and the G consist of a 5-membered ring fused to a 6-membered ring, and there are four nitrogen atoms in the nine atoms that make them up; such a group is called a purine. T and C consist of 6-membered rings containing two nitrogen atoms; these groups are called pyrimidines. If we look at the hydrogen bonding pairs, we see that there are two hydrogen bonds indicated between A and T, and there are three hydrogen bonds indicated between G and C. Given that hydrogen bonds stabilize interactions, this observation suggests that DNA sequences containing a higher percentage of G–C pairs will, for the most part, separate or "melt" at higher temperature than DNA sequences containing a lower percentage of G–C pairs.

In addition to their number, there is something else different about the hydrogen bonding between A and T and between G and C. We should recall that hydrogen bonds are directional interactions; indeed, those shown in Figure 1-1 are indicated as little red arrows, to emphasize this directionality. They are said to be donated from the hydrogen atom to the electron-rich atom at which the arrow points. If we focus on the N–H → N hydrogen bond at the center of each base pair, we see that in the A–T base pair the donation is from the T to the A, i.e., from the pyrimidine to the purine. By contrast, we see that the N–H → N hydrogen bond in the G–C base pair is donated from the G to the C, i.e., from the purine to the pyrimidine. The amino groups above these N–H → N hydrogen bonds donate to the carbonyl groups in the opposite direction. The third hydrogen bond in the G–C base pair can be thought of as a bonus. Thus, we see that there is information in the structures of the bases that selects the pairing partners.

However, there is not enough information in just these hydrogen bonding patterns. If all that were necessary for proper base pairing were to get the donation directions correct, then we would also have A–G and C–T pairs, because they would obey the same hydrogen bonding rules. Something else is needed. A–G pairing would involve two big molecules (purines) and C–T base pairing would involve two small molecules (pyrimidines), so further information involving the sizes of the molecules should be what we want. If we insist that each pair contain one big base and one little base in addition to following the hydrogen bonding rules, then we solve our problem. This notion is summarized in the table in Figure 2-1. The rows refer to the size of the bases (big or

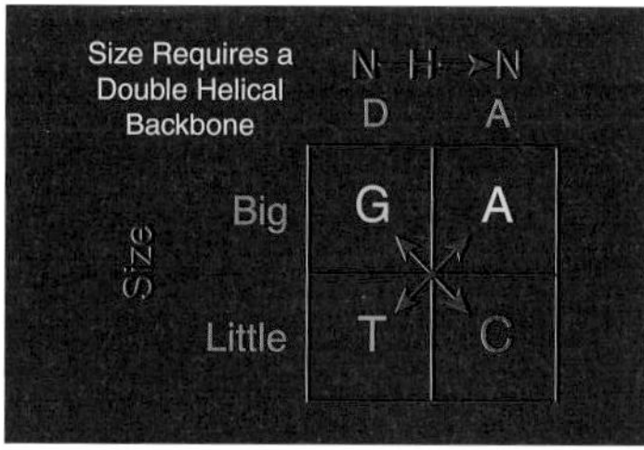

Figure 2-1 *Watson–Crick base pairing.* The two components of Watson–Crick base pairing are shown, the N–H → N hydrogen bonding direction and the size of the base. The big donor (G) pairs with the little acceptor (C), while the big acceptor (A) pairs with the little donor (T).

little), and the columns refer to the central N–H → N hydrogen bond (donor or acceptor). Thus, G is a big donor, and it pairs across the diagonal with C, a small acceptor; likewise A is a big acceptor, and it pairs across the other diagonal with T, a small donor.

The pairing schemes shown in Figure 1-1 seem to make a lot of sense, but they are only a few of the possibilities if the bases themselves are tossed into a pot individually or in pairs and are crystallized. Without a means of orienting the bases, many different paired hydrogen bonding arrangements are possible. In fact, it is well known that any base can pair with any other base, including itself, often in many different arrangements.[2,3] This is where the double helical backbone of DNA comes into play, in two different ways. First, it fixes the relative orientations of the bases, so that (to a good first approximation) only the hydrogen-bonding possibilities shown in Figure 1-1 are available. Second, it establishes the distance within which "big" and "little" become meaningful. It's easy for the reader to tell that two purines are "too big" and two pyrimidines are "too little" and that a purine and a pyrimidine yield the "Goldilocks pairing size." It's not so easy for a molecule to do this. However, in DNA two bases are fixed between the rails of a double helical backbone whose separation is exactly right for one pyrimidine and one purine to fit: two purines or two pyrimidines produce distortions of the backbone, which leads to increasing the free energy of the system. This point is seen readily by looking at the double helix in Figure 1-1.

The positioning of the two bases is set by a *cis* interaction. In other words, the two glycosyl bonds that attach the bases to the backbones are on the same side of the helix axis, rather than on opposite sides. The most important thing about the glycosyl bonds is that they are related to each other by a two-fold axis that is perpendicular to the helix axis. This axis is prominently drawn in Figure 1-2. Thus, an A–T base pair can fit in the same space as a T–A pair, and a G–C base

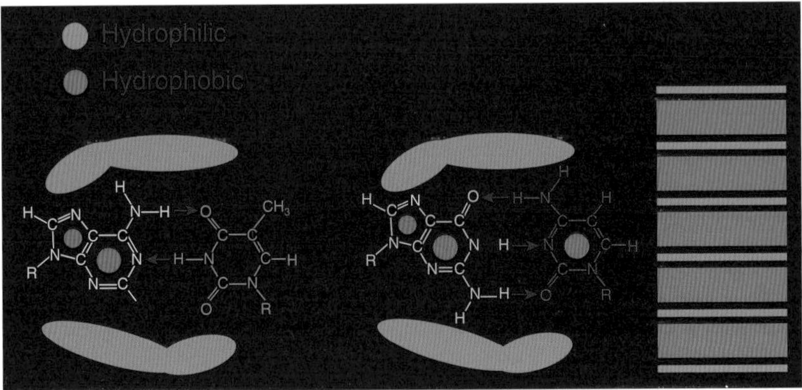

Figure 2-2 *The environment of the bases.* The A–T base pair is shown at the far left, and the G–C base pair is shown to its right. We are looking down on those pairs, and the tops of the bases are shown as orange circles, representing their hydrophobic character. Their edges are shown as being in blue regions, representing their hydrophilic nature. On the right is a view of the bases stacking. The hydrophobic parts of the bases (orange) form stacks, making a stack of the hydrophilic edges of the bases (blue lines).

pair can fit in the same space as a C–G pair. The angle between the glycosyl bonds is under 180°, meaning that if you look down on the base pair, there is a small angle between the glycosyl bonds, and a big angle. This leads to two differently sized grooves in the double helix, a minor groove (the small angle) and a major groove (the big angle). Although the inter-glycosyl angle is about 75°, when one looks at the double helix, the backbones flank the glycosyl bonds, so the ratio of the grooves is about 1.6:1, or 215°:145°, roughly the golden ratio.

It is probably worth discussing briefly the importance of base stacking. Figure 2-2 is a simplified diagram of what is involved in this phenomenon. Base stacking is a special circumstance, partly related to the fact that the bases are flat, and partly related to the highly nitrogenous nature of the bases. The left panel of the drawing shows the two Watson–Crick base pairs, A–T on the left and G–C on the right. The tops of the bases are drawn in orange, to indicate that they are hydrophobic – that is, they don't like to be in aqueous solution very much. However, the edges of the bases, drawn in blue, represent hydrophilic parts of the bases, which very much like interacting with aqueous solution, forming hydrogen bonds with it. The combination of these two traits results in the ability of the DNA bases to stack on each other, like a stack of plates (first pointed out by Astbury[2.4]). This is certainly not a property of all flat molecules, which often tend to organize themselves in a herringbone arrangement.[2.5] The free energies of base stacking and hydrogen bonding are usually readily estimated by using a nearest-neighbor approximation.[2.6]

Non-Watson–Crick DNA pairing. In a fairly simplified nutshell, that's the story of why the double helix is such a successful information-containing system. We haven't considered the quantitative aspects of the distortions, and we haven't worried about conformational variants that might make the presence of, say, two purines, more acceptable. It's useful in this regard to remember that molecules like DNA seek their free energy minima, and that those minima may not conform to an ideal global minimum. Thus, if two short pieces of DNA that can pair perfectly except for a single mismatch are put together in solution, they will pair, and will tolerate the mismatch, because it is the best alternative available.[2.7,2.8] Examples are shown in Figure 2-3. Note that most of these arrangements entail a distortion (relative to standard Watson–Crick base pairing) of the positions of the "glycosyl" bonds that connect the bases to the backbone. An exception might be the lower left image of an alternative A–G pair, where the A has rotated about its glycosyl bond to use other atoms to hydrogen bond with the G.

There are lots of things that can happen when the double helical paradigm is not enforced – not just A–C or G–A mis-pairs within a double helix, but many strange arrangements of nucleotides. To start, there are four different types of A–T pairs: the standard Watson–Crick pair we have discussed, but also a "Hoogsteen" pair, where the A rotates about the glycosyl bond and the acceptor of the hydrogen bond is the imidazole nitrogen. Hoogsteen base pairing is illustrated in the triple helical arrangement[2.9] shown in Figure 2-4; a Watson–Crick base pair is shown in its normal orientation, and a third T is shown hydrogen bonding to the other available site on the A. Two other possibilities are flipping the T about its midline, so that the other carbonyl is pairing with the A, both in the Watson–Crick and Hoogsteen arrangements, so that the glycosyl bonds are roughly *trans* to each other, rather than *cis*, i.e., on the opposite side of the base pair plane, rather than on the same side as in Watson–Crick arrangements. The same kind of triple helical pairing can be seen with Cs and Gs, but one of the Cs must be protonated (implying a low pH), as shown in Figure 2-5.

In the same vein as mentioned above, that DNA molecules will pair with whatever is available, it might be useful to look at a couple of highly favored self-pairing interactions. Arguably the most prominent of these is the G-tetrad structure, shown in Figure 2-6. This is a four-fold symmetric structure that forms readily if G stretches are present, arguably even in the presence of complementary C stretches.[2.10] An even stranger arrangement, called the I-motif, can be seen with oligo-dC stretches. This is shown in

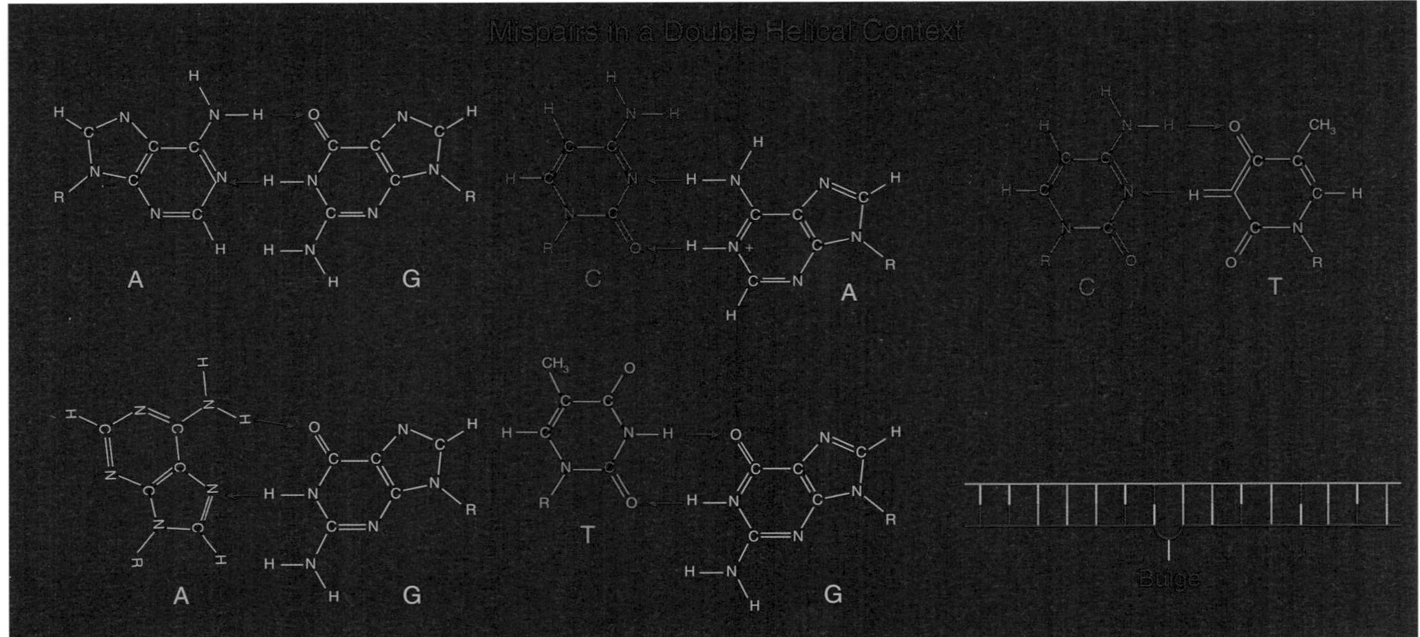

Figure 2-3 *DNA mis-pairs*. The top row shows an A–G mis-pair (with both bases in the "anti" conformation), an A–C mis-pair in which the A has been protonated, and a C–T mis-pair (usually stabilized by a cation in the minor groove). The bottom row shows an A–G mis-pair with the A in the "syn" conformation, a T–G mis-pair, and a base that has been bulged out.

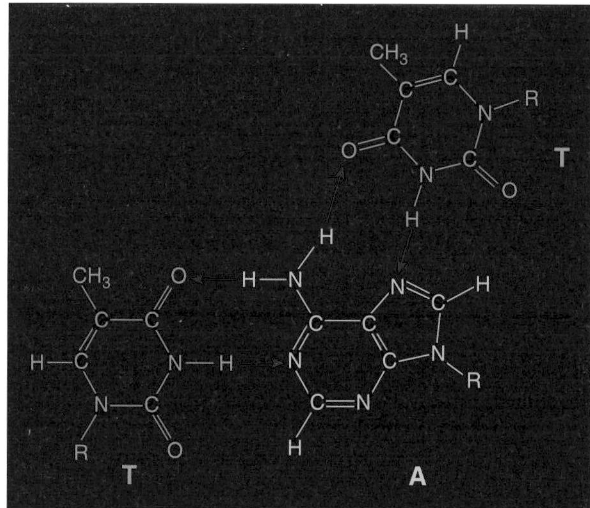

Figure 2-4 *A T–A–T base triple.* In addition to the Watson–Crick T–A pair, there is a second T pairing with the A to form a Hoogsteen pair on the major groove side of the purine.

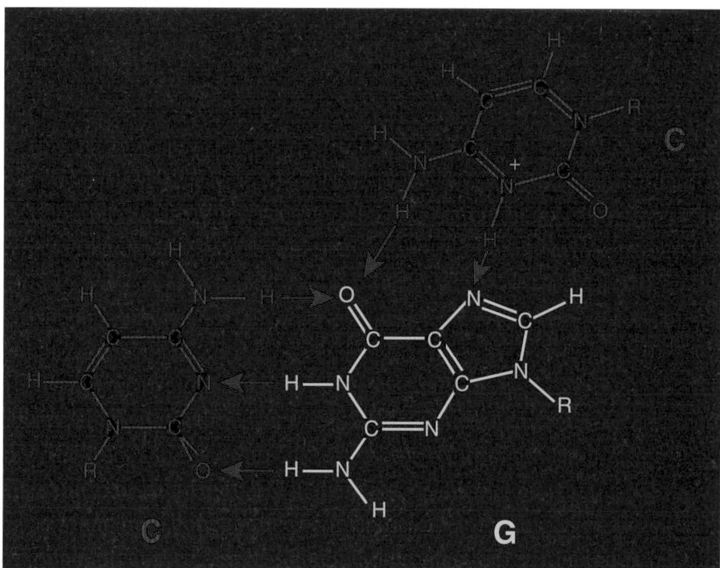

Figure 2-5 *A C–G–C^{+} base triple.* This base triple has the same structure as the T–A–T base triple in Figure 2-4, but the C on the major groove side has been protonated (the pH is lowered) so that there are two hydrogen bonds to stabilize the triple.

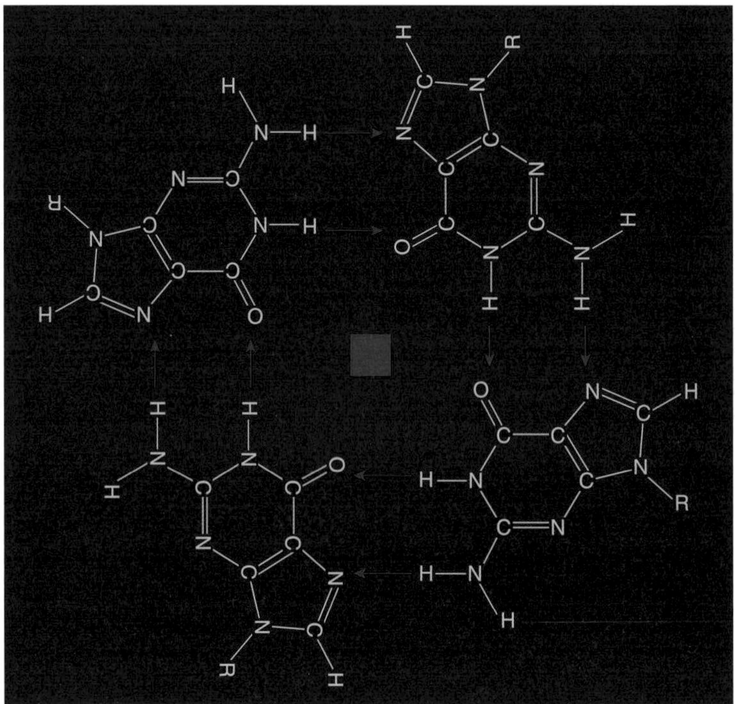

Figure 2-6 *A tetra-G base tetrad.* Four guanine bases form a four-fold symmetric arrangement wherein two donors on the Watson–Crick hydrogen bonding surface donate two hydrogen bonds to the two acceptors on the major groove side of the adjacent base.

Figure 2-7. The image on the left shows C–C pairing as a hemi-protonated species, where one proton is shared between the two N atoms in the middle. On the right we represent this pair of Cs as a rectangle. The diagram indicates that there are alternating pairs of Cs, intercalated between each other and rotated 90° from each other. There are plenty of other examples of strange nucleic acid base pairing, but the point has been made: once the purely double helical context is abandoned, pretty much anything can happen.

Sequence design elements. Given all the things we have seen that can happen when exactly complementary molecules with linear backbones are abandoned (and we've hardly examined a complete list), it seems a good idea to take some precautions when designing molecules whose helix axes are branched. Figure 2-8 shows a simple 4-arm branched molecule consisting of four 16-mers, and designed to contain eight nucleotide pairs in each of its arms. This is a well-characterized DNA branched junction molecule, called J1.[2.11]

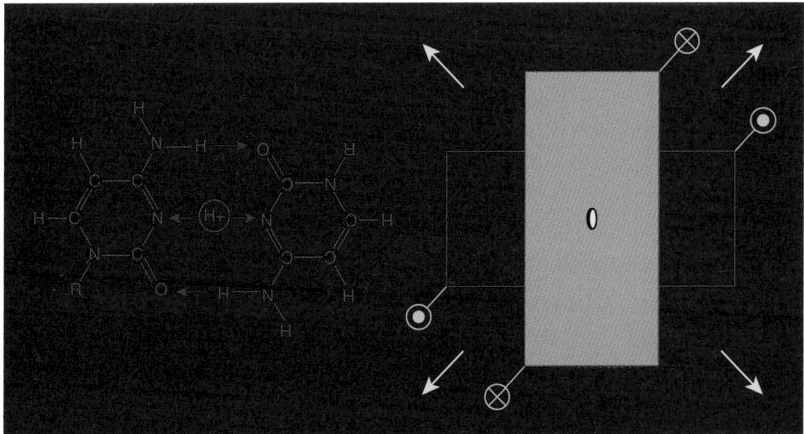

Figure 2-7 *The I-motif.* The left side of the drawing shows a cytosine paired to another cytosine by sharing a proton with it. Note that there is a two-fold axis of symmetry perpendicular to the plane of the page. The right side shows a schematic of the I-motif, with the C–C nucleotide pairs shown as rectangles. Successive C–C pairs are intercalated between other C–C pairs whose backbones are different; one C–C pair is drawn filled and the other one is only outlined. One set of backbones is entering the page and one set is coming out of the page. The base pairs are set at roughly right angles to each other. The inherent symmetry of the structure is shown.

There are a lot of possible sequences that could be used to generate this structure: there are 32 independent nucleotides (eight in each arm) and 32 complementary nucleotides, for a total of 4^{32} possibilities. How was the sequence shown actually selected? Is this the best possible sequence for this architecture? Does it matter if it is?

Ideally, it would be simplest to design this junction if there were 32 base pairs rather than 4. In that case, every position would be unique, and there would be no issues at all about how to select the sequence, except, arguably, how to permute them optimally. Of course, this is not the case. In everything we do in what follows, we assume that forming Watson–Crick base pairs is what DNA strands prefer to do, and that double helices will form if they can. What is the issue then? Why have rules for assigning the sequences? The answer is that we are designing a system that is an excited state of DNA, one that is not a simple double helix. What we want to ensure is that the structure that the DNA assumes is indeed the structure that we have targeted, rather than some other structure that is actually of lower energy.

So what can go wrong when we attempt to form this structure? The answer is simply that other things that are permitted to happen, and that are close enough

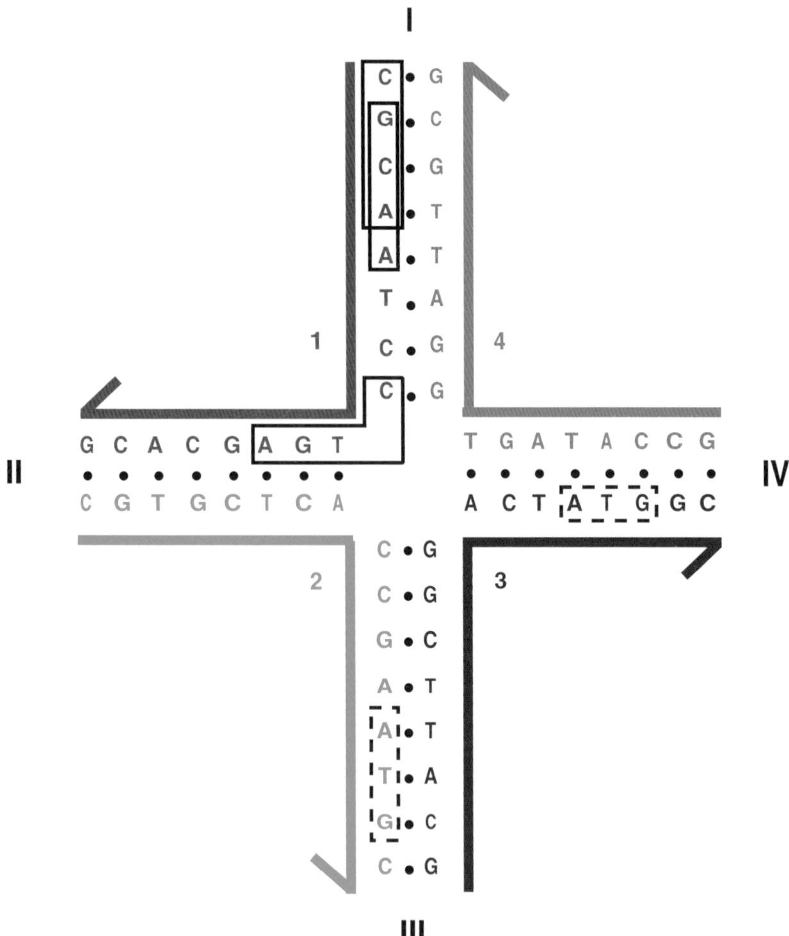

Figure 2-8 *Sequence design of a 4-arm DNA branched junction.* The arms are labeled with Roman numerals, and the strands are labeled with Arabic numerals; arrowheads represent the 3′ ends of strands. All of the 52 tetramers in this sequence are unique. Two on the 5′ end of strand 1 are boxed. Likewise, all of the tetramers flanking the branch point lack contiguous complements; there is no TCAG to complement the CTGA sequence that is boxed. Thus, competition for the octamer arms happens only at the level of trimers, such as the ATG sequences shown in the dashed boxes.

to the free energy of the target structure but are not the target structure, might occur. The easiest way to think about getting alternative structures is in terms of symmetry. *Symmetry in a system is antithetical to control.* Imagine, for example, the extreme case whereby each of the four strands had the same sequence but was able to pair as shown in Figure 2-8. In that case, every strand would be self-complementary (the first eight nucleotides would pair with the last eight), and would likely form hairpins. If hairpins did not form, what would limit the system to four strands? Why not two, three, five, six, or more strands? The inherent problems of sequence symmetry have led to a fairly stringent approach called "sequence symmetry minimization."[2.12]

Sequence symmetry minimization. J1 (Figure 2-8) consists of four 16-nucleotide strands. Since we don't have 32 nucleotide pairs with which to work, one thing we can do is to pretend that short segments of DNA in fact behave as unique units. Thus, we could break up each of the 16-mers into a series of overlapping units, say 14 overlapping trimers or 13 overlapping tetramers. We could insist that each of these short elements be unique (not repeat), and then we could try those sequences to see if we were successful in making our target. It is justified to do this because DNA strands bind to other DNA strands cooperatively, so the binding of any given nucleotide is not exactly independent of the binding of its neighboring nucleotides. There are 64 unique trimers, and 256 unique tetramers. There are really only 240 tetramers available, because 16 of them, such as AGCT, are self-complementary, so we would exclude them from the list. Since we need 52 tetramers, including complements, and since we may want to put other restrictions on the sequence, it is probably easiest to use tetramers as the basic elements of J1, rather than try to find 56 trimers, of which there are only 64. The first two tetramers on strand 1, CGCA and GCAA, have been boxed. These are clearly unique, but, in fact, all of the tetramers are unique.

What other things can go wrong? As pointed out above, the presence of our 4-arm branch point creates a position that is of higher energy than a simple duplex point. It ought to be a good idea to eliminate the chance that any tetrameric element that spans the branch point has a simple Watson–Crick complement someplace in the structure. The boxed CTGA is one of the 12 tetramers that span the branch point, and its complement, TCAG, will not be found in the structure. The whole idea is to make the hard-to-form structure (the branched junction) the most likely thing to happen. The idea is similar to the notion promulgated in Arthur Conan Doyle's Sherlock Holmes mysteries: when the impossible has been eliminated, that which remains, no matter how improbable, is the correct answer.[2.13]

The ATG trimers in the dotted boxes are redundant. In principle, they could wind up opposite the wrong CAT in the structure. However, they are competing with octamers in this case, and such structures are not seen in the case of J1. Since J1 was the first junction to be designed, and the free energetic cost of forming the junction (about 1.05 kcal/mol)[2.14] was unknown, the sequences

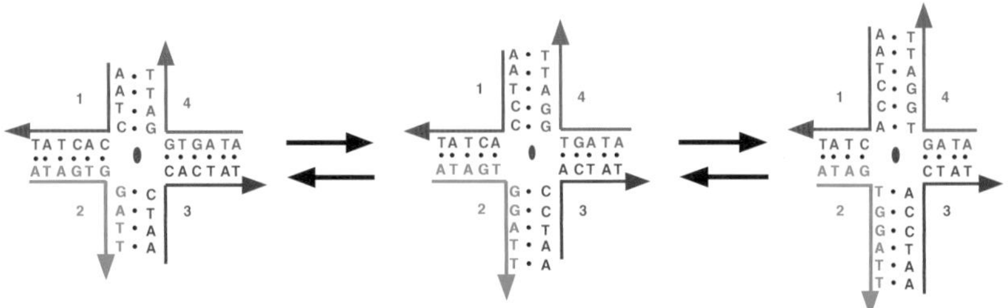

Figure 2-9 *Double-stranded branch migration of a 4-arm junction.* Owing to the dyad sequence symmetry present, the middle junction can isomerize to produce the junction on its left or the one on its right. In moving to the left, the G–C pairs in the vertical arms re-pair to become G–C pairs in the horizontal arms. Likewise, in moving to the right, the A–T pairs in the horizontal arms re-pair to become A–T pairs in the vertical arms. Since these sequences are completely symmetric, the isomerization is likely to continue until the junction resolves as a pair of duplex molecules.

were tested for the contributions of possible sub-sequence (e.g., trimer) pairing;[2.15] these proved to be small, and this computationally expensive step was eliminated in subsequent designs.

A key element in the design of 4-arm junctions is the possibility of branch migration. Branch migration is an isomerization that relocates the site of branching. This phenomenon is shown in Figure 2-9, which shows three junctions. Focusing on the middle junction with five nucleotide pairs in each arm, we see that it has two-fold sequence symmetry about its center. Thus, the sequence of strand 1 is the same as strand 3, and the sequence of strand 2 is the same as strand 4. This situation occurs frequently in biology when recombination intermediates form from two identical, or virtually identical, DNA double helices. If we look at the surroundings of the branch point, we see that it is flanked in both horizontal arms by A–T pairs, and in both vertical arms by G–C pairs. Since the molecules can flop around a bit, it is possible for the pairing arrangements to change. For example, the C on strand 1 could switch pairing partners from the G on strand 4 to the G on strand 2, while the G on strand 4 could also switch to pairing with the C on strand 3. This transition is shown in the change from the central molecule to the one on the left. This isomerization alters the central molecule so that there are now six pairs in the horizontal arms and four in the vertical arms. Just as easily, the A on strand 1 in the central molecule could switch its pairing partner from the T on strand 2 to the T on strand 4, while the A on strand 3 could switch its pairing partner to the T on strand 2. The result of this transition is seen in the molecule on the right, which has four pairs in its horizontal arms and six in its vertical arms. In

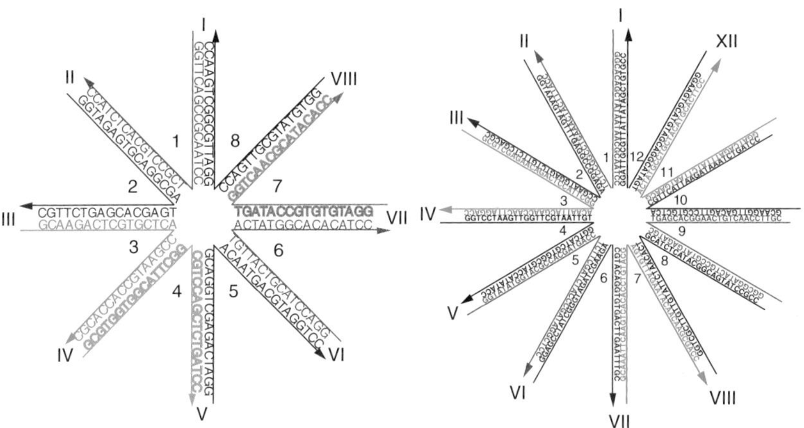

Figure 2-10 *An 8-arm branched junction and a 12-arm branched junction.* The 8-arm junction on the left is stable with 16-mer arms, but the 12-arm junction on the right requires longer arms for stability. Reprinted by permission from the American Chemical Society.[1.9]

principle, this process could go on stochastically until the junction resolves into two linear dimers. The kinetics of this process have been studied extensively.[2.16]

The key to branch migration is the two-fold sequence symmetry that enables it. The J1 junction shown in Figure 2-8 lacks this symmetry. It contains one of each of the four possible base pairs G–C, T–A, C–G, and A–T, going counter-clockwise from arm I to arm IV. What about junctions with more arms? Junctions have been made with up to 12 arms.[2.17,2.18] Examples are shown in Figure 2-10. An 8-arm junction is shown on the left side of Figure 2-10. The junction-flanking pairs there repeat in adjacent positions: arms II and III have T–A pairs, arms IV and V have C–G pairs, arms VI and VII have A–T pairs, and arms VIII and I have G–C pairs. The idea behind this strategy is that the structural constraints of, say, the GC on strand 8 prevent it from pairing with itself, so no branch migration will be seen. Similar strategies were adopted with 5-arm and 6-arm junctions. For the 12-arm junction shown on the right in Figure 2-10, a different approach has been taken: a three-fold symmetric arrangement of junction-flanking bases has been used: GC, TA, CG, AT are used, respectively, in arms I, II, III, and IV. The same sequence is used in arms V, VI, VII, and VIII, as well as in arms IX, X, XI, and XII. Although identical base pairs flank the junction, the loop entropy cost of forming alternate pairing arrangements corresponding to branch migration is prohibitive, so the junctions appear to be stable and well behaved. It is clear that the junctions shown in Figure 2-10 have arms much longer than those of J1 (Figure 2-8). Although octamer arms were found sufficient for 4-arm junctions, they did not offer sufficient stabilization even for 5-arm and 6-arm junctions.[2.17] Those junctions and the first 8-arm junction were built with 16 nucleotide pairs per arm. The 12-arm junction was insufficiently stabilized with 16 nucleotide pairs, so 24 nucleotide pairs were used for it. It is not entirely clear why the extra stability is required. Certainly there is extra charge build-up at the branch point regions of junctions with more arms, but the volume of the branch point region does increase markedly as well. We'll talk more about junctions with many arms in Chapter 3.

There are other things to consider when designing sequences for motifs that may have stability problems. One is to avoid long stretches of Gs. The relative thermodynamics of G4 formation relative to branched structures have not been examined, but, as a rule, it is a good idea not to include long stretches of GGG . . . in the design if it is avoidable. As noted above, it has been reported the G_n stretches can form even in relaxed DNA.[2.10] In a long duplex region, this is probably not a problem, but near junctions it is safe not to use sequences longer than G_2. Short arms are also to be avoided: six nucleotides are about the limit of independent arms.[2.19] In Chapter 3 and beyond, we will be considering other structures with multiple junctions. Empirically, we have found that at least five nucleotides, and preferably more, should be placed between junctions.

We have discussed symmetry at the level of individual bases. What about at the sub-base level? In general, the minimization of sequence symmetry avoids these types of symmetries as well. Thus, long stretches of purines or pyrimidines might lead to trouble; in superhelical DNA, mirror-symmetric sequences of purines lead to H-DNA.[2.20,2.21] A-tract[2.22] and G-tract bending[2.23] are also to be avoided if one is trying to design simple double helical domains. Alternating purine and pyrimidine sequences are another type of symmetry. For example, alternating CG sequences have a tendency to form Z-DNA, a left-handed DNA structure,[2.24] Z-formation is not restricted to those sequences. This has been noted even in relaxed DNA molecules.[2.25] When designing small branched molecules, it is a good idea for the ends to be stable, to avoid fraying, which can lead to instability. In this regard, it is useful to have the ends of sequences end in G–C base pairs (see, e.g., Figure 2-8) to keep the ends intact.

Conditions. One cannot emphasize enough that structure is structure within a given set of conditions. Thus, one must be concerned with temperature, DNA concentration, cation concentrations, and pH when thinking about structure. Standard neutral buffers usually suffice, with a pH around 7.8 or so. However, junctions are most stable in the presence of about 10 mM Mg^{2+}. Low temperatures are usually better for working with small junctions than high temperatures. Many protocols for DNA-modifying enzymes suggest working at 37 °C, but this is often unwise when working with unusual DNA structures, which may unravel in whole or in part at this temperature. We have found that it is best to work at lower temperatures for longer times to keep the DNA stable, rather than to work at the temperatures where the enzymatic activities are nominally optimized.

DNA concentration is often key. We will see in later chapters that DNA concentration must sometimes be optimized in order to get the target structure. Thinking about the J1 molecule (Figure 2-8), it is easy to see that the second half of a molecule of strand 1 will pair with the first half of a molecule of strand 2, and that the second half of a molecule of strand 2 will pair with the first half of a molecule of strand 3, and that the second half of a molecule of strand 3 will pair with the first half of a molecule of strand 4. However, why should the second half of a molecule of strand 4 pair with the first half of the *same* molecule of strand 1, rather a different molecule of strand 1? All of the sequence design that went into designing J1 did nothing to ensure that we would get the 1:1:1:1 monomer of J1, rather than the 2:2:2:2 dimer. That we do not get the dimer[2.11] is only because we are working at a concentration in which the monomer is favored over the dimer. Less well-behaved DNA motifs than the single junction will indeed lead to the dimer, or even the trimer or tetramer at moderate concentrations of DNA.[2.26,2.27] Even a molecule as simple as J1 has a huge number of possible sequences, 4^{32} if we ignore rotational symmetry, or about 1.6×10^{18} (approximating 4^5 as ~1000). This number is far too great to go through and evaluate, either manually or by some algorithm, if we knew enough about the thermodynamics to have such an algorithm. Most of the molecules that we design are much larger than this. We need some way to project the huge sequence space confronting us into a manageable size. As a practical matter, what is done in our laboratory is first to choose those nucleotides we absolutely need in certain places: junction flanking bases, restriction sites (these usually violate the rules), G–C pairs on the ends, etc. Then, we advance with the rest of the sequence a couple of nucleotides at a time, sometimes going back and re-choosing a few. Every time we are choosing a particular pair of nucleotides, we are eliminating a multiplicative factor of 16 from the possibilities. Despite the caveats above, often it does not seem to matter what nucleotides we choose. Of course, as the molecules increase in size, the elements must also increase in length from 4 to 6 or larger to get a sufficient vocabulary to build the molecule. The sub-elements also increase in length, so they are of length 5 for element length 6. These might also run into trouble. Likewise, the process described above is a first-order approximation.

It would not pick out the difference between ACCAATG and ACCGATG if tetramer elements were being used.

Is sequence symmetry minimization necessary? We certainly do it in our laboratory, but Mao[2.28] has shown that there are advantages to *maximizing* sequence symmetry. He has built symmetric DNA motifs whose structures are controlled by the length of a particular cyclic strand. He emphasizes that by maximizing sequence symmetry and controlling structure by solution conditions, one can minimize the labor of purifying multiple strands, and the issues of establishing perfect stoichiometry. The other and highly successful example of ignoring sequence symmetry is DNA origami (Chapter 9).[2.29] In this case, which is advertised as having a pixilation of 60 Å, an arbitrary single-stranded viral DNA is combined with 200–300 "helper" or "staple" strands to make a particular shape. The viral DNA sequence is already set before beginning, and most origami molecules to date have formed the structures they were designed to form. However, for small novel systems, where precision (< 10 Å) is required, it is probably a good idea to perform sequence symmetry minimization. Now that we have seen how to apply sequence design to produce successful motifs, Chapter 3 will teach us how to design a motif itself.

References

2.1 A. Rich, The Era of RNA Awakening: Structural Biology of RNA in the Early Years. *Quart. Revs. Biophys.* **42**, 117–137 (2009).

2.2 M.H. Caruthers, Gene Synthesis Machines: DNA Chemistry and its Uses. *Science* **230**, 281–285 (1985).

2.3 D. Voet, A. Rich, The Crystal Structures of Purines, Pyrimidines and their Intermolecular Complexes. *Prog. Nucl. Acid Res. Mol. Biol.* **10**, 183–265 (1970).

2.4 H.F. Judson, *The Eighth Day of Creation*, New York, Simon & Schuster, p. 114 (1979).

2.5 D.W.J. Cruikshank, A Detailed Refinement of the Crystal and Molecular Structure of Anthracene. *Acta Cryst.* **9**, 915–923 (1956).

2.6 J. SantaLucia, Jr., A Unified View of Polymer, Dumbbell and Oligonucleotide DNA Nearest-Neighbor Thermodynamics. *Proc. Nat. Acad. Sci. (USA)* **95**, 1460–1465 (1998).

2.7 T. Brown, W.N. Hunter, G. Kneale, O. Kennard, Molecular Structure of the G•A Base Pair in DNA and its Implications for the Mechanism of Transversion. *Proc. Nat. Acad. Sci. (USA)* **83**, 2402–2406 (1986).

2.8 D.J. Patel, S.A. Kozlowski, S. Ikuta, K. Itakura, Deoxyguanosine-Deoxyadenine Pairing in the d (C-G-A-G-A-A-T-T-C-G-C-G) Duplex: Conformation and Dynamics at and adjacent to the dG-dA Mismatch Site. *Biochem.* **23**, 3207–3217 (1984).

2.9 G. Felsenfeld, D.R. Davies, A. Rich, Formation of a Three-Stranded Polynucleotide Molecule. *J. Am. Chem. Soc.* **79**, 2023–2024 (1957).

2.10 E.Y.N. Lam, D. Beraldi, D. Tannahill, S. Balasubramanian, G-Quadruplex Structures are Stable and Detectable in Human Genomic DNA. *Nature Comm.* **4**, article 1796 (2013).

2.11 N.R. Kallenbach, R.-I. Ma, N.C. Seeman, An Immobile Nucleic Acid Junction Constructed from Oligonucleotides. *Nature* **305**, 829–831 (1983).

2.12 N.C. Seeman, Nucleic Acid Junctions and Lattices. *J. Theor. Biol.* **99**, 237–247 (1982).

2.13 A.C. Doyle, "The Sign of Four" (first published 1890). In *The Complete Sherlock Holmes*, New York, The Literary Guild, p. 118 (1936).

2.14 M. Lu, Q. Guo, L.A. Marky, N.C. Seeman, N.R. Kallenbach, Thermodynamics of DNA Chain Branching. *J. Mol. Biol.* **223**, 781–789 (1992).

2.15 N.C. Seeman, N.R. Kallenbach, Design of Immobile Nucleic Acid Junctions. *Biophys. J.* **44**, 201–209 (1983).

2.16 P. Hsieh, I.G. Panyutin, DNA Branch Migration. In *Nucleic Acids and Molecular Biology*, ed. F. Eckstein, D.M.J. Lilley, **9**, pp. 42–65, Berlin, Springer-Verlag (1995).

2.17 Y. Wang, J.E. Mueller, B. Kemper, N.C. Seeman, The Assembly and Characterization of 5-Arm and 6-Arm DNA Junctions. *Biochem.* **30**, 5667–5674 (1991).

2.18 X. Wang, N.C. Seeman, The Assembly and Characterization of 8-Arm and 12-Arm DNA Branched Junctions. *J. Am. Chem. Soc.* **129**, 8169–8176 (2007).

2.19 N.R. Kallenbach, R.-I. Ma, A.J. Wand, G.H. Veeneman, J.H. van Boom, N.C. Seeman, Fourth Rank Immobile Nucleic Acid Junctions. *J. Biomol. Struct. Dyn.* **1**, 158–168 (1983).

2.20 S.M. Mirkin, V.L. Lyamichev, K.N. Drushliak, V.N. Dobrynin, S.A. Filippov, M.D. Frank-Kamenetskii, DNA H Form Requires a Homopurine-Homopyrimidine Mirror Repeat. *Nature* **330**, 495–497 (1987).

2.21 H. Htun, J.E. Dahlberg, Topology and Formation of Triple-Stranded H-DNA. *Science* **243**, 1571–1576 (1989).

2.22 T.E. Haran, D.M. Crothers, Cooperativity in A-Tract Structure and Bending Properties of Composite T_nA_n Blocks. *Biochem.* **28**, 2763–2767 (1989).

2.23 N.V. Hud, J. Plavec, A Unified Model for the Origin of DNA Sequence-Directed Curvature. *Biopolymers* **69**, 144–159 (2003).

2.24 A. Rich, A. Nordheim, A.H.-J. Wang, The Chemistry and Biology of Left-Handed Z-DNA. *Ann. Rev. Biochem.* **53**, 791–846 (1984).

2.25 H. Wang, S.M. Du, N.C. Seeman, Tight Single-Stranded DNA Knots. *J. Biomol. Struct. Dyn.* **10**, 853–863 (1993).

2.26 S.M. Du, S. Zhang, N.C. Seeman, DNA Junctions, Antijunctions and Mesojunctions. *Biochem.* **31**, 10955–10963 (1992).

2.27 T.-J. Fu, N.C. Seeman, DNA Double Crossover Structures. *Biochem.* **32**, 3211–3220 (1993).

2.28 Y. He, Y. Tian, Y. Chen, Z. Deng, A.E. Ribbe, C. Mao, Sequence Symmetry as a Tool for Designing DNA Nanostructures. *Angew. Chem. Int. Ed.* **44**, 6694–6696 (2005),

2.29 P.W.K. Rothemund, Scaffolded DNA Origami for Nanoscale Shapes and Patterns. *Nature* **440**, 297–302 (2006).

3
Motif design based on reciprocal exchange

We now know how to design sequences that should be pretty good at leading to a particular motif. We have studied extensively how to make a branched junction with a number of arms, and how to prevent branch migration in one. In this chapter, we are going to discuss several different routes to designing motifs to use as the basis for making objects, lattices, and devices. The key concept in designing motifs is the notion of *reciprocal exchange*.[3.1] This is a process found within biological systems, but we are going to approach it somewhat differently: we are not actually going to perform reciprocal exchange in the laboratory, but merely on paper. We will then use the sequence-selection procedures of the preceding chapter to make the strands that will come together to form the motifs we design.

The notion behind reciprocal exchange is shown in Figure 3-1. The left side of the drawing shows two strands, a red strand and a blue strand. Following reciprocal exchange, we see on the right side a red–blue strand (going from upper left to lower right) and a blue–red strand (going from upper right to lower left). Thus, we still have two strands, but now they are each a mixture of the two strands we had before. We mentioned in Chapter 1 that the strands of DNA have a chemical polarity, called 5′ to 3′ polarity, so that one end of a strand is called the 5′ end and the other end is called the 3′ end. Following the reciprocal exchange shown in Figure 3-1, the red–blue strand contains the 5′ half of the red strand and the 3′ half of the blue strand, while the blue–red strand contains the 5′ half of the blue strand and the 3′ half of the red strand. Resolution in this context means that the crossover is cut and we go back to two juxtaposed bulges or hairpins. This is an idea that derives from recombination chemistry, where crossovers are resolved in one direction or the other to yield different products (e.g., see reference 3.2).

So far, this is just a formalized concept. It would probably help to see what happens in the context of a double helix. This is shown in Figure 3-2, but it

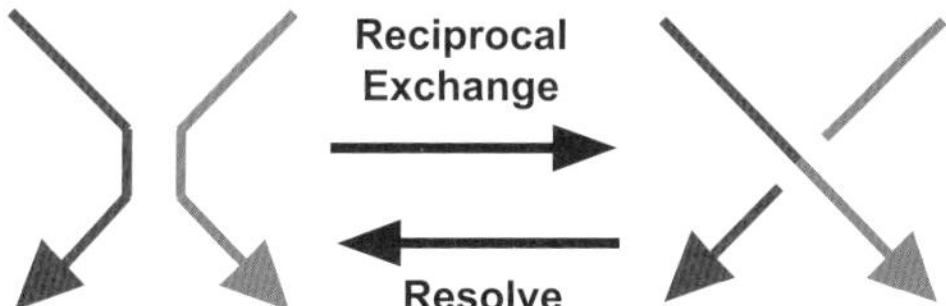

Figure 3-1 *Reciprocal exchange and resolution.* The red strand and the blue strand on the left are effectively cleaved and rejoined in the process of reciprocal exchange. The sign of the node created is arbitrary. Resolution is the reverse operation that restores the original two strands. Reprinted by permission from the American Chemical Society.[3.12]

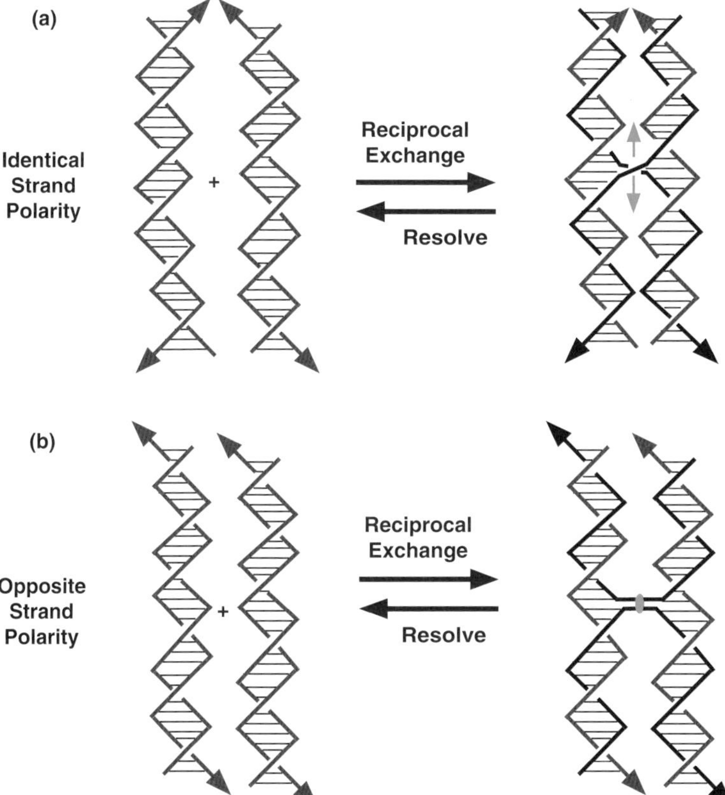

Figure 3-2 *Reciprocal exchange in the context of a double helix.* The arrowheads represent the 3′ ends of the strands. Following reciprocal exchange, the strands are re-colored blue. (a) *Exchange between strands of the same polarity.* The blue strands start at their 5′ ends at the tops of the molecules, reach the middle, cross over to the other side, and then terminate at the bottoms of the molecules. The green dyad symmetry element is vertical in the plane of the page. (b) *Exchange between strands of opposite polarity.* One blue strand starts at the top, goes to the middle, and crosses over, and then reverses direction and ends on the top of the other helical domain. The second blue strand starts on the left bottom domain, crosses over, and ends at the bottom of the other helical domain. Consequently, the green dyad symmetry element is perpendicular to the page. Note that rotating the right helical domain about an axis horizontal in the plane of the page converts this structure to the other one. Reprinted by permission from the American Chemical Society.[3.12]

would be useful to look back at Figure 1-2 first, to recall that the strands of DNA have a polarity, and that it is possible to perform the reciprocal exchange reaction between strands of the same polarity or of opposite polarity. The case of identical polarity is shown in Figure 3-2a, and the case of opposite polarity is shown below it in Figure 3-2b. Two double helices are drawn in red in both cases. Both strands to undergo reciprocal exchange have their 5′ ends at the top of the helices and their 3′ ends (indicated by arrowheads) at the bottom of the helices in panel a. The strands that undergo reciprocal exchange are drawn in blue, following the reaction. If the two strands have the same polarity, one of the new blue strands starts at the top of, say, the right domain, progresses to the middle and then crosses over to the left domain, before going to the bottom; the other blue strand does the opposite, starting at the top of the left domain and crossing over to the right domain. This creates a 4-stranded molecule with a two-fold axis of symmetry (green) vertical in the plane of the page.

In panel b, the strand to undergo reciprocal exchange in the left double helix has its 3′ end at the top and its 5′ end at the bottom, while the strand in the right double helix has its 5′ end at the top and its 3′ end at the bottom. Following reciprocal exchange, one new strand (again blue) starts at the top of the right helix, goes to the middle and crosses over, and then goes back up to the top of the left helix; the other blue product strand starts at the bottom of the left helix, goes up to the middle, crosses over, and proceeds to the bottom of the right double helix. This structure has two-fold symmetry perpendicular to the plane of the page, as indicated by the green lens-like structure at the middle of the 4-stranded molecule. So long as a single crossover event is involved, these two molecules are simply conformers of each other. Flipping the right helix of the product structure in Figure 3-2b end-for-end will yield the product structure seen in Figure 3-2a. We will see shortly that, when more crossover events take place, new topological species are produced as a function of whether the strands are of the same or opposite polarity.

How do these reciprocal exchange products relate to the J1 junction that we discussed extensively above? Figure 3-3 shows the relationships between the different representations and structural isomers of branched junctions. All of the helices here have been unwound, just as in J1 in Figure 2-8. Let's start in the middle at the bottom with a junction that is apparently four-fold symmetric, except for its colors. There are four strands, 1, 2, 3 and 4, and they are labeled with an Arabic numeral the same color as they are. In addition, the arms have been labeled with Roman numerals: the arm formed from strands 4 and 1 is labeled I, the arm formed from strands 1 and 2 is labeled II, the arm formed from strands 2 and 3 is labeled III, and the arm formed from strands 3 and 4 is labeled IV. As usual 3′ ends of strands are indicated by arrowheads.

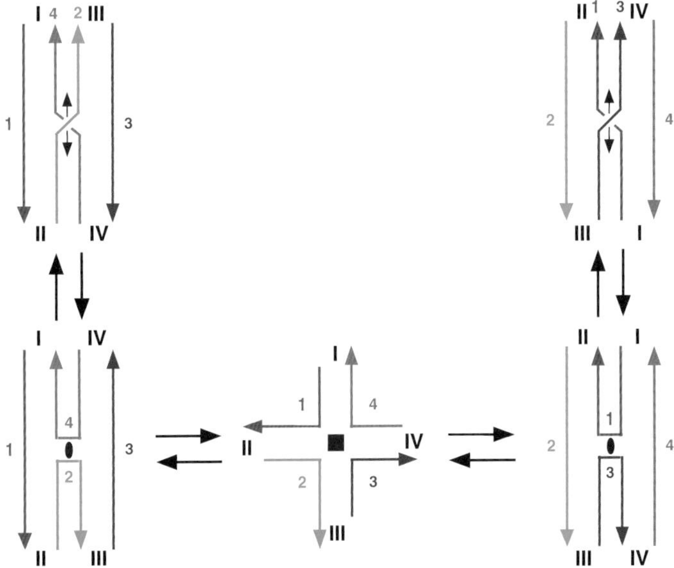

Figure 3-3 *Structural isomerizations of 4-arm branched junctions.* At the center of the drawing is a 4-arm branched junction drawn with four-fold symmetry. The four strands are drawn and numbered (Arabic numerals) in four different colors; the double helical arms are labeled in Roman numerals. The symmetry can be decreased from four-fold to two-fold by two arms stacking on each other in two different ways, as shown moving either right (arm II stacks on arm III, arm I stacks on arm IV) or left (arm I stacks on arm II, arm III stacks on arm IV). In principle (although not seen in the absence of proteins), the one domain can rotate 180° to reorient the two-fold axis (dark-blue symbol) from perpendicular to the page to within the plane of the page.

The arms of 4-arm junctions are capable of stacking on each other, and are seen to do so in the presence of Mg^{2+} cations.[3.3] When they do stack, the four-fold symmetry is destroyed, to be replaced with two-fold symmetry. This can happen in either of two ways, shown to the left and the right of the central four-fold structure. On the left, red strand 1 and purple strand 3 assume a helical conformation, while green strand 2 and cyan strand 4 adopt a bent conformation. On the right, green strand 2 and cyan strand 4 assume helical conformations, while red strand 1 and purple strand 3 adopt a bent conformation. The two transformations that these structures can undergo (going upwards in the figure) relate to Figure 3-2. The right double helical domain of each of the junctions is flipped about a horizontal axis. It is easy to follow this transformation: you see that, on the left, strand 3 flips over and, on the right, strand 4 flips over; so does the strand to which it is paired, changing the orientation of the dyad axis

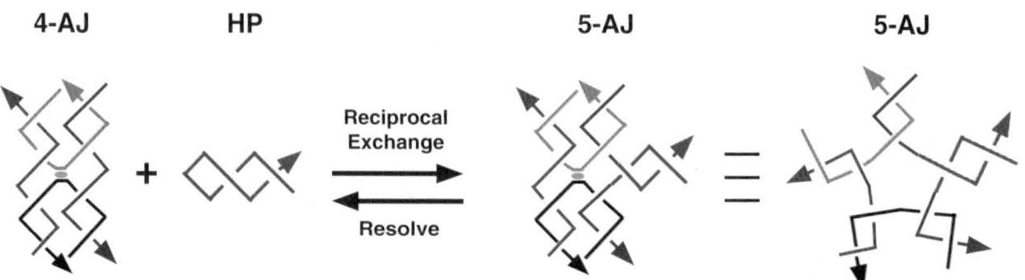

Figure 3-4 *Reciprocal exchange is used to convert a 4-arm junction to a 5-arm junction.* The 4-arm junction (4-AJ) on the left is combined with a hairpin (HP) to produce a 5-arm junction (5-AJ). The 5-arm junction is symmetrized on the right. Reprinted by permission from the American Chemical Society.[3.12]

from perpendicular to the plane of the page to vertical within the plane. This is exactly the difference between the structures in Figures 3-2a and 3-2b, except that we are not showing the helical rotation features. These structures with parallel rather than antiparallel helical strands are not commonly observed, although 4-arm junctions, known as Holliday junctions,[3.4] are often drawn that way in the literature on genetic recombination. The ability to form a stacked structure (with either parallel or antiparallel helical strands) appears to be unique to the 4-arm junction. In principle, other junctions (e.g., the 6-arm junction) could do this, but so far this structure has not been observed.

Application of motif generation to extended networks. So now that we know how to generate motifs by reciprocal exchange, what can we do with this knowledge? One thing we can do is to generate motifs that are of utility. The simplest extension of the protocol shown in Figure 3-2 is to perform reciprocal exchange between a 4-arm junction and a DNA hairpin to yield a 5-arm junction. This is shown in Figure 3-4. There is no obvious limit on the number of times this operation can be done. We saw in Chapter 2 that junctions with as many as 12 arms have been made. What is the purpose of this?

A prominent class of targets for structural DNA nanotechnology consists of DNA stick figures. The edges of these stick figures consist of double helical DNA, and they have a branched junction at their vertices. These are called N-connected objects, and they can be extended to form N-connected lattices.[3.5,3.6] The simplest objects are 3-connected objects. The first non-trivial object made from DNA was a molecule with the connectivity of a cube or a rhombohedron.[3.7] It is a 3-connected object, and its structure is shown in Figure 3-5a. The molecule consists of six different strands, each one corresponding to one face of the cube. Each of the vertices is a 3-arm junction. Figure 3-5b shows a truncated octahedron that has also been constructed from DNA. This is also a 3-connected object, but it has been built from 4-arm

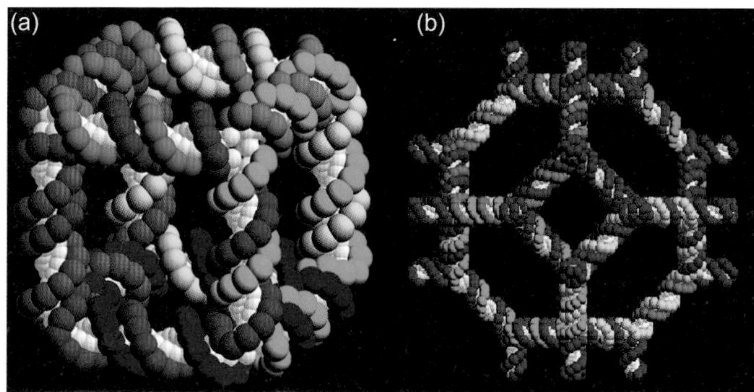

Figure 3-5 *DNA polyhedral catenanes.* (a) *A molecule with the connectivity of a cube.* (b) *A molecule with the connectivity of a truncated octahedron.*

junctions, so every vertex has a double helix sticking out of it so that it can be connected to another such molecule, making a 4-connected lattice.[3.8] The lattice was never built, owing to synthesis problems, but the object was constructed successfully. Both of these objects have been characterized topologically, so we know that their branching is correct, and so is the linking of their strands, but the junctions from which they were built are floppy, so we don't know what their geometries actually look like. We will discuss topological structures and robust motifs in later chapters.

Why build branched junctions with many arms, such as the recently constructed 8-arm and 12-arm junctions?[3.9] The ability to construct 8-arm and 12-arm junctions means that a large number of previously impossible structures can be constructed from branched DNA molecules. Tabulations have been produced[3.6] describing the space-filling networks with 432 symmetry that can be formed from Platonic solids (all faces are the same regular polygon) and Archimedean solids (all faces are regular polygons, but not the same). These structures can, in principle, be formed from the helix axes of DNA molecules; the symmetry is much lower if one considers the entire DNA molecule. Figure 3-6a illustrates an 8-connected network based on the packing of cuboctahedra (red) and octahedra (blue). Each vertex is connected to four edges that form two sides of osculating coplanar squares (say the tops of the cuboctahedra on the lower right of Figure 3-6a), and four others that come from another pair of squares perpendicular to the first pair. The cavities are filled by octahedra.

Figure 3-6b illustrates the basis for the stability of cubic close packing (face-centered cubic packing) which is a 12-connected network. This basis is shown as Buckminster Fuller's "octahedral truss."[3.10] Two edge-sharing octahedra

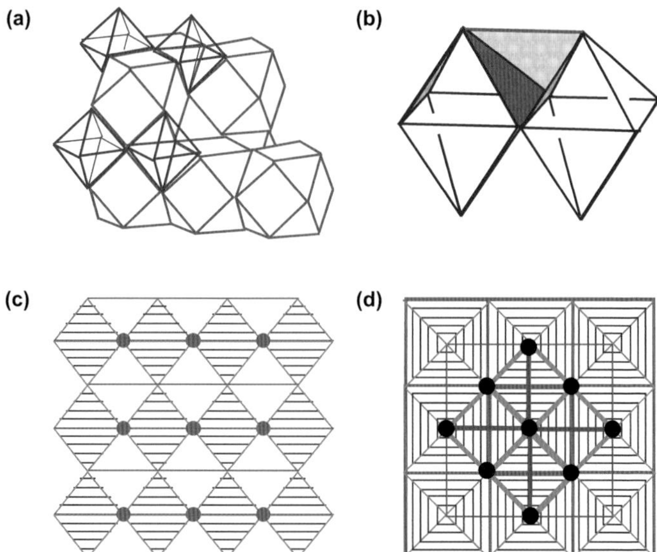

Figure 3-6 *Lattice-like structures.* (a) *An 8-connected lattice built from octahedra and cuboctahedra.* The octahedra are drawn in blue and the cuboctahedra are drawn in red. (b) *Two edge-sharing octahedra joined by an octahedral truss.* The octahedra are drawn in blue, and their tips are connected by a red octahedral truss. The presence of the truss prevents the octahedra from rotating about their common edge. The truss combines with the shared edge and the two faces flanking it to form a tetrahedron, drawn in red. (c) *A 2D bounded projection of a lattice composed of trussed octahedra plus tetrahedra.* The view is down the shared edge, which is drawn as a filled magenta circle. The truss is drawn in red. The faces of the octahedra are ruled, but the faces of the tetrahedra are unshaded. (d) *A top view of the 12-connected lattice consisting of octahedra and tetrahedra.* The visible portions of the octahedra are shown as square contour lines, and their central edges are drawn with a thick magenta line. The octahedral trusses are drawn in red, as in (c). The edges of the central vertex of the central pyramid (top half of the central octahedron) and those edges of the four tetrahedra that are shared with octahedra are drawn in dark blue. The other two edges of those tetrahedra are drawn in thick red (top edges) or magenta (bottom) lines. The nine vertices at the tops or bottoms of the central tetrahedra are shown as black circles. This lattice is often called face-centered cubic packing. One of the face-centered squares is seen in the five black circles on the top plane, at the ends of the four thick red lines. The other four black circles are in a plane behind it. Reprinted by permission from the American Chemical Society.[1.9]

(blue) could wobble back and forth about their common edge, but are kept from wobbling by the truss, drawn in red. This leads to the formation of a tetrahedron, indicated with shaded faces. Figure 3-6c shows a view down the common edge (a magenta dot), illustrating how the truss (top and bottom) could keep a lattice of octahedra from wobbling. Figure 3-6d shows a top view of nine octahedra. Four of the tetrahedra are drawn with thicker edges (red on top, magenta on the bottom, and blue on the sides). The vertices of these tetrahedra are emphasized with black dots to indicate their 12-connected character. Each vertex has four upward-directed edges (forming the bottom of an octahedron above), four downward-directed edges (forming the top of an octahedron below), and four coplanar edges corresponding to the shared edges of four octahedra (as in Figure 3-6b). It is worth pointing out that the octahedral truss structure is just a stick-style representation of cubic close packing.

Having emphasized that the 4-arm junction is special because of its stacking properties opens the door to deriving a variety of different molecules. The relaxed conformation of the 4-arm junctions appears to have an angle of about 60° between the helical domains in the conformation with antiparallel helical strands (on the ends of the bottom row in Figure 3-3).[3.11,3.12] Thus, the helical domain on the right is rotated about a horizontal axis in the plane of the page so that its top is 30° closer to the viewer, and the domain on the left is rotated 30° about the same axis so that its bottom is closer to the viewer. Nevertheless, it is straightforward to design structures with two parallel helix axes by performing two reciprocal exchanges between adjacent helices. This is shown in Figure 3-7. As in Figure 3-2, we have to take two different possibilities into account, whether the strands where the crossover occurs are of the same polarity (Figure 3-7a) or of opposite polarity (Figure 3-7b). The resulting molecules are called double-crossover (DX) molecules.[3.13]

DX molecule species. Figure 3-7 illustrates the case where there is a full turn between the sites of reciprocal exchange. It is clear that a full turn will minimize torsional stress on the molecule. However, it is not necessary for there to be a full turn (or multiples of full turns) between the crossover sites. There can also be a half-turn between the crossover positions. All that is necessary is that two backbones can juxtapose. The meaning of a half-turn can also be tricky for DNA. For those molecules with antiparallel helical strands, it is pretty straightforward, about five inter-nucleotide spacings. However, for those molecules with parallel helical strands, juxtaposition can occur when the separation is either a minor groove spacing or a major groove spacing. These differences lead to five different isomers of DX molecules (Figure 3-8), two antiparallel (DAE and DAO) and three parallel (DPE, DPON, and DPOW), where A and P mean antiparallel and parallel, and E and O mean an even or odd number of double helical half-turns between crossovers. The N and the W further differentiate the odd separations in parallel, odd separation molecules, where the half-turn can be a minor (narrow) groove spacing or a major (wide) groove spacing, respectively.

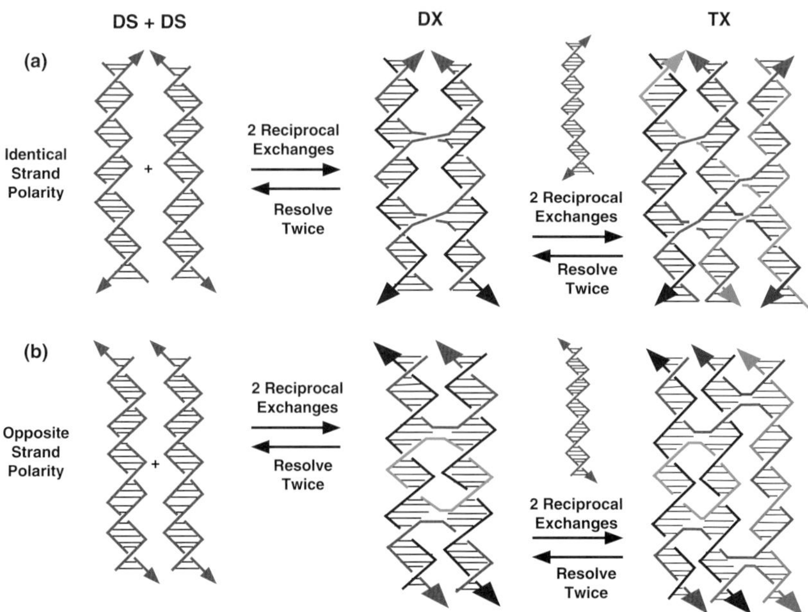

Figure 3-7 *Motif generation involving two crossovers between helices.* This is similar to Figure 3-2, but now two crossover events occur. (a) *Strands of the same polarity undergo reciprocal exchange twice.* A structure known as DX ensues. These DX molecules are not particularly well behaved. Nevertheless, one can imagine continuing the process of two reciprocal exchanges with a third helical domain, leading to a TX motif. (b) *Strands of opposite polarity undergo reciprocal exchange twice.* The DX molecule that results contains five strands (two red, two blue, and one green), in contrast to the 4-stranded molecule in (a). This molecule can also be extended with a third helical domain, leading to a TX molecule that is used frequently. Reprinted by permission from the American Chemical Society.[3.12]

Figure 3-7 shows that one need not be limited to two helices fused together. The right side of the figure shows that a third helix can be added to the first two to produce what is known as a TX molecule, containing three double helical domains (and four crossovers). There is nothing to limit the number of crossovers in DX or TX molecules, given sufficient length of the double helical domains. Two types of TX molecules are shown in which a DX has been augmented to form a TX; in the molecule in Figure 3-7a all crossovers are between strands of the same polarity, whereas in the molecule in Figure 3-7b all crossovers are between strands of opposite polarity.

As a practical matter, structures of the sort shown in Figure 3-7a do not form nearly as readily as those in Figure 3-7b. This may be a consequence of the

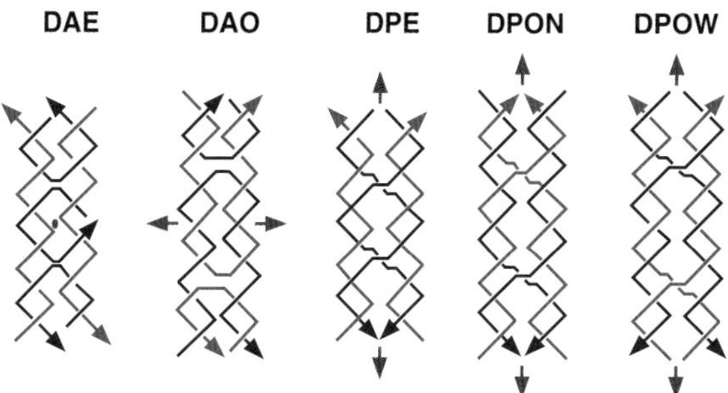

Figure 3-8 *Isomers of DX molecules.* The DAE and DAO molecule are the result of two reciprocal exchanges between strands of opposite polarity. They differ from each other in that there are an even number (DAE) or an odd number (DAO) of double helical half-turns between crossovers. The DAE isomer contains five strands, but the DAO isomer contains four strands. The DPE, DPON, and DPOW isomers result from two reciprocal strand exchanges between strands of the same polarity. The DPE molecule contains an even number of double helical half-turns between crossovers, while DPON and DPOW contain odd numbers of half-turns between crossovers. They differ from each other in that the odd half-turn is a minor (narrow) groove spacing in DPON or a major (wide) groove spacing in DPOW. Reprinted by permission from the American Chemical Society.[3.18]

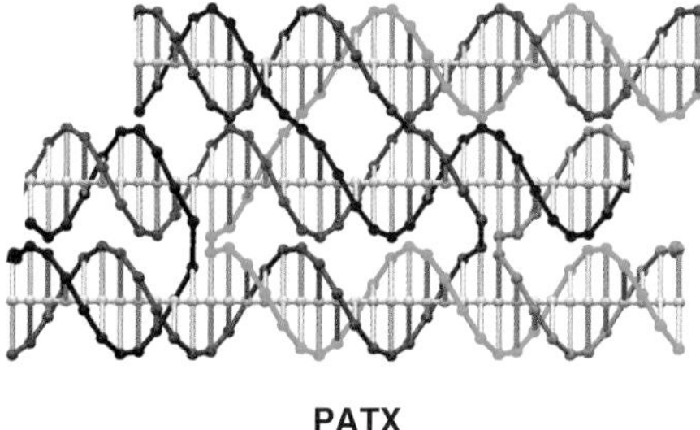

Figure 3-9 *The PATX motif.* This molecule is a TX variant, but mixes reciprocal exchange types. Thus, the upper crossovers (blue–green and brown–magenta) are between strands of the same polarity, but the lower crossovers (black–green and brown–orange) are between strands of opposite polarity. Reprinted by permission from the American Chemical Society.

juxtaposition of phosphates in the same-polarity molecules, but this has not been established definitely.[3.13] Another argument is that those molecules are much more susceptible to kinetic traps involved in their formation.[3.14] It is certainly possible to form molecules with same-polarity linkages, and this can be strongly aided if one makes a TX molecule that mixes the two types of linkages,[3.15] as illustrated in Figure 3-9.

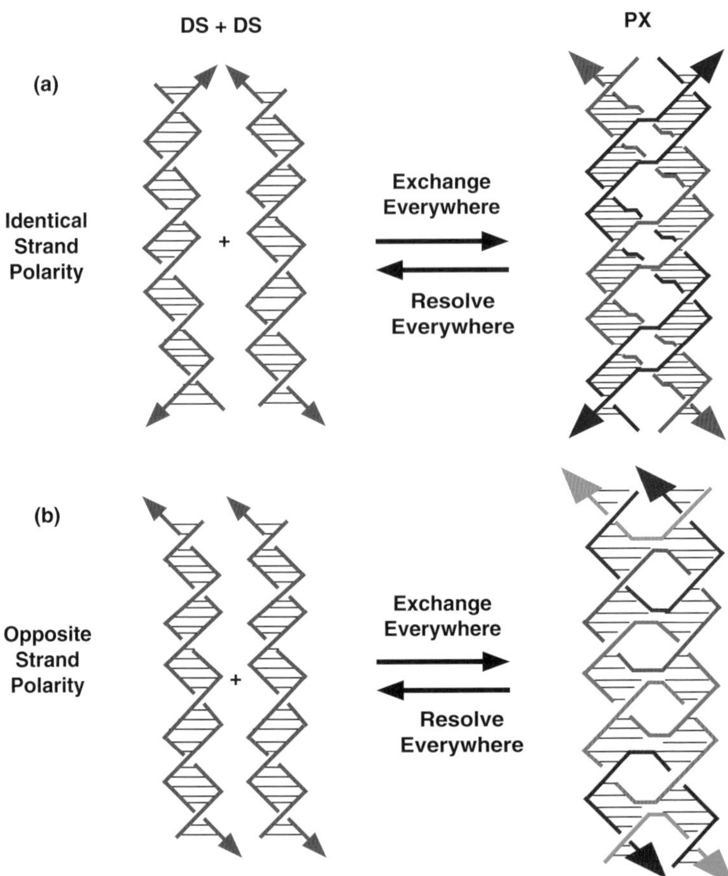

Figure 3-10 *Reciprocal exchange at every possible point.* (a) *Exchange between strands of the same polarity.* The result on the right is the PX (paranemic crossover) motif, which features prominently in structural DNA nanotechnology and in the recognition of homology in supercoiled plasmids. (b) *Exchange between strands of opposite polarity.* This motif has not been reported. Reprinted by permission from the American Chemical Society.

The PX motif. There is at least one other motif that can be derived from reciprocal exchange; this is the PX motif, shown in Figure 3-10. It is derived from the same procedures we have used earlier, but now crossover events occur everywhere. Although we worked primarily with the products of reciprocal exchange between strands of opposite polarity, in this case we will work with motifs that are derived from exchange between strands of the same polarity. There has been no attempt, to my knowledge, to examine the motif derived from Figure 3-10b, but the PX structure in Figure 3-10a (from now on, this is what we will call PX) has been a rich source of utility in structual DNA nanotechnology, and it might have relevance to biological processes as well.[3.16] It is possible to extend the PX molecule to a 3-domain system, as we did with DX and TX, but 3-domain PX molecules are not well behaved.[3.17]

The PX structure has been drawn to look like it consists of a red double helix wrapped around a blue double helix, to yield two different helices fused at all positions. With the color pattern used here, alternating half-turns of DNA have different properties. Let's just look down the left double helical domain from top to bottom. The top half-turn consists of a red strand paired with another red strand. The half-turn below is formed from a red strand (rear) paired with a blue strand (front). The third half-turn contains two blue strands paired to each other, and then the fourth half-turn from the top shows pairing between a blue strand (rear) and a red strand (front). The pattern then repeats with a red–red half-turn, and so on.

Figure 3-11 shows a variety of representations of the PX structure, so that we can think about it in a lot of different ways. All five images show green arrows above and below the motifs; these indicate that the structure in principle has a two-fold axis relating its two helical domains, just like the DPE, DPON, and DPOW motifs discussed above. In addition, small arrowheads above each of the helical domains in the three left-hand motifs indicate the helix axes of the two domains. The three molecules on the left have different coloring schemes. The one labeled PX has a color scheme whereby the two pairs of strands with the same polarity have the same colors: red strands coming up from the bottom and blue strands coming down from the top. This representation emphasizes the two-fold symmetric nature of the molecule. The schematic makes an attempt to differentiate the major (wide) and minor (narrow) grooves. The central (green) dyad axis of the molecule is flanked by alternating minor and major grooves, indicated respectively by N and W.

The schematic to the right of this one, PX-N, is an identical molecule, except that it has been colored differently. The coloring indicates alternating pairs of blue and red strands. These pairs flank the N symbols on the central dyad axis. This is what a PX molecule could look like if it had been produced by a red double helix and a blue double helix forming the PX structure. Note that although this structure looks like it consists of two inter-wrapped double helices, one red and one blue, the structure is more subtle than that. As we noted in the original description of PX molecules, there has actually been an exchange of strands in every other half-turn of each of the two double helices that make up the molecule. The PX-W molecule, in the middle of Figure 3-11, is slightly different from the PX-N molecule, because it has been phased a half-turn away. Thus, there are three Ws and two Ns in the middle, rather than

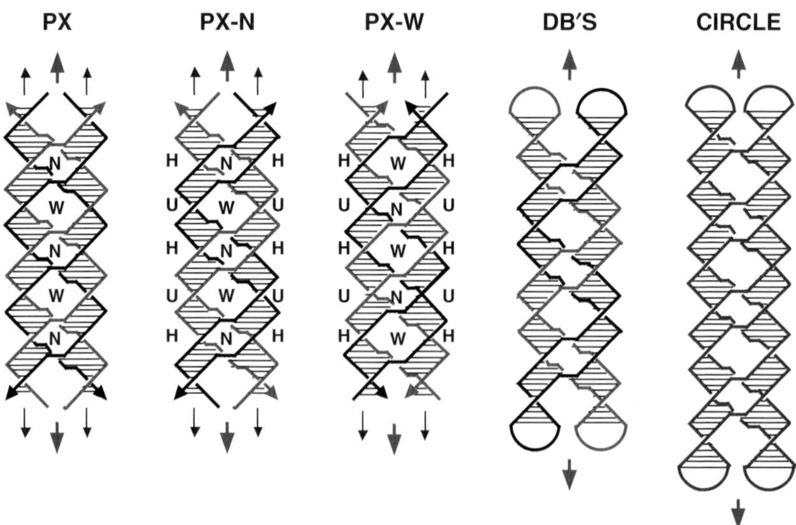

Figure 3-11 *PX variants.* The PX molecule shown at the left is drawn with the strand colorations (dark blue and red) reflecting the central two-fold axis, indicated with thick short arrows. The thin short arrows indicate local helix axes. N and W represent minor and major groove spacings, respectively. The molecule labeled PX-N shows an interaction between two double helices, one dark blue and the other red. The separation of strands of the same color is a minor groove separation. The half-turns labeled H must be homologous if the red strands and the blue strands are to combine to form PX. The sequence of the U half-turns has no impact on the structure. The molecule labeled PX-W is similar, except that the separation of the strands of the same color is a major groove separation. The dumbbell (DB′S) structures emphasize the paranemic character of the PX molecule of this length; the dark-blue and red duplexes can be separated from each other without strand passable events. The CIRCLE structure at the far right results when one extra reciprocal exchange takes place in molecules with loops holding their ends together.

three Ns and two Ws, as there are in PX-N. The coloring reflects this, because the wide grooves are flanked by the same colors, rather than the narrow grooves.

Both the PX-N and PX-W molecules are flanked by a row of H and U labels. These refer to the need for homology or the lack of it if the PX molecules were to be formed from separate pairs of *identical* molecules, one red and one blue. The labels simply indicate that the mixed-color half-turns, consisting of one red strand and one blue strand, need homology (H) to form the molecule; by contrast, the uniformly colored half-turns do not need homology (U) to form the molecule. Homology (the same sequence in two different pieces of DNA) seems to underlie a number of processes in molecular biology. Molecules are said to be "PX-homologous" if they contain the same sequence in the half-turns requiring homology, but not in the half-turns that do not require it.[3.16] It is of course possible to design PX molecules to form from individual strands designed in the same way that J1 was designed in Chapter 2.[3.18] PX-DNA has not been shown to form from relaxed homologous molecules in dilute solution (after all, any solution of DNA oligonucleotides typically will contain about a trillion homologous molecules), but both purely homologous molecules and PX-homologous molecules have been seen to form in superhelical DNA.[3.16] It has been found that the PX molecule forms best from 4-stranded complexes when the minor groove contains five nucleotide pairs and the major groove contains six, seven, or eight nucleotide pairs; at low concentrations, nine nucleotide pairs also work.[3.18] Fewer independent strands (say two cohering hairpins; see below) can also work with four nucleotide pairs in the minor groove.[3.19] Within a single supercoiled complex, 10 and 11 nucleotides can also work in the major groove. The large-major-groove structures that are formed seem to writhe, rather than to maintain their helix axis intact while unwinding.[3.20,3.21]

The molecule second from the right in Figure 3-11 highlights a remarkable fact about PX-DNA. What has been done here is that the PX-N molecule has been capped with hairpins on each of its ends, to produce dumbbells that form the PX structure. If you inspect the structure closely, you will see that the blue double helix and the red double helix are actually unlinked. Such a structure is called "paranemic," meaning that there is no need for strand-passage operations to form the structure. All that is needed is that the red structure and the blue structure with appropriate complementary sequences be in solution at the same time under favorable conditions, and the two molecules will cohere. This notion is illustrated in Figure 3-12, where two DNA triangles are shown cohering via this "PX-cohesion" mechanism. Note that the two triangles are topologically closed structures. PX-cohesion seems to require a lot of Mg^{2+} in solution to work very well, but it certainly has been used successfully.[3.19,3.22] Also note that this is a sequence-specific interaction, so it can be programmed, just like sticky ends.

The drawing on the far right of Figure 3-11 is a simple yet dramatically different variation on the dumbbell structures to its left. What has been done here is that the PX structure has been increased in length by a half-turn over the dumbbell structures. The result is not a pair of dumbbells at all, but a structure that looks at first glance to be a complex knot. However, closer inspection of the molecule shows that although it is a knot, it is the trivial knot, a circle: the molecular backbone can be unwound to yield nothing but a cyclic unknotted single strand.

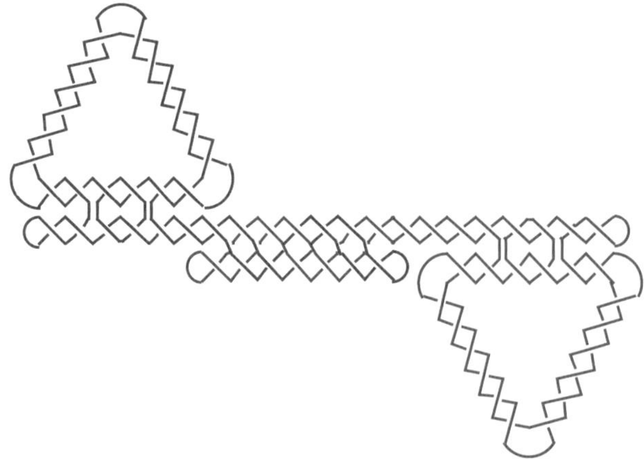

Figure 3-12 *Paranemic cohesion.* Two topologically closed structures are cohesive, because they are complementary in the PX sense. Compare the region of homology to the DB′S structure of Figure 3-11. Reprinted by permission from the American Chemical Society.[3.22]

We are starting to tread in the area of molecular single-stranded topology, and that is the origin of some of the other DNA motifs that have been built. In Chapter 4 we will provide a simple introduction to single-stranded DNA topology, and then discuss some other motifs that can be built using the notions we discuss there.

References

3.1 N.C. Seeman, DNA Nicks and Nodes and Nanotechnology. *Nano Letters* **1**, 22–26 (2001).

3.2 W. Sun, C. Mao, H. Iwasaki, B. Kemper, N.C. Seeman, No Braiding of Holliday Junctions in Positively Supercoiled DNA Molecules. *J. Mol. Biol.* **294**, 683–699 (1999).

3.3 M.E.A. Churchill, T.D. Tullius, N.R. Kallenbach, N.C. Seeman, A Holliday Recombination Intermediate is Twofold Symmetric. *Proc. Nat. Acad. Sci. (USA)* **85**, 4653–4656 (1988).

3.4 A. Rich, D.R. Davies, A New Two-Stranded Helical Structure: Polyadenylic Acid and Polyuridylic Acid. *J. Am. Chem. Soc.* **78**, 3548–3549 (1956).

3.5 A.F. Wells, *Three-Dimensional Networks and Polyhedra*, New York, John Wiley & Sons (1977).

3.6 R. Williams, *The Geometrical Foundation of Natural Structure*, New York, Dover (1979).

3.7 J. Chen, N.C. Seeman, The Synthesis from DNA of a Molecule with the Connectivity of a Cube. *Nature* **350**, 631–633 (1991).

3.8 Y. Zhang, N.C. Seeman, The Construction of a DNA Truncated Octahedron. *J. Am. Chem. Soc.* **116**, 1661–1669 (1994).

3.9 X. Wang, N.C. Seeman, The Assembly and Characterization of 8-Arm and 12-Arm DNA Branched Junctions. *J. Am. Chem. Soc.* **129**, 8169–8176 (2007).

3.10 A. Edmondson, *A Fuller Explanation*, Boston, Birkhauser (1987).

3.11 D.R. Duckett, A.I.H. Murchie, S. Diekmann, E. von Kitzing, B. Kemper, D.M.J. Lilley, The Structure of the Holliday Junction and its Resolution. *Cell* **55**, 79–89 (1988).

3.12 C. Mao, W. Sun, N.C. Seeman, Designed Two-Dimensional DNA Holliday Junction Arrays Visualized by Atomic Force Microscopy. *J. Am. Chem. Soc.* **121**, 5437–5443 (1999).

3.13 T.-J. Fu, N.C. Seeman, DNA Double Crossover Structures. *Biochem.* **32**, 3211–3220 (1993).

3.14 M.T. Kumara, D. Nykypanchuk, W.B. Sherman, Assembly Pathway Analysis of DNA Nanostructures and the Construction of Parallel Motifs. *Nano Letters* **8**, 1971–1977 (2008).

3.15 W. Liu, X. Wang, T. Wang, R. Sha, N.C. Seeman, A PX DNA Triangle Oligomerized Using a Novel Three-Domain Motif. *Nano Letters* **8**, 317–322 (2008).

3.16 X. Wang, X. Zhang, C. Mao, N.C. Seeman, Double-Stranded DNA Homology Produces a Physical Signature. *Proc. Nat. Acad. Sci. (USA)* **107**, 12547–12552 (2010).

3.17 T. Wang, Ph.D. thesis, New York University (2007).

3.18 Z. Shen, H. Yan, T. Wang, N.C. Seeman, Paranemic Crossover DNA: A Generalized Holliday Structure with Applications in Nanotechnology. *J. Am. Chem. Soc.* **126**, 1666–1674 (2004).

3.19 W.M. Shih, J.D. Quispe, G.F. Joyce, A 1.7 kilobase Single-Stranded DNA That Folds into a Nanoscale Octahedron. *Nature* **427**, 618–621 (2004).

3.20 P.K. Maiti, T.A. Pascal, N. Vaidehi, W.A. Goddard, III, The Stability of Seeman JX DNA Topoisomers of Paranemic Crossover (PX) Molecules as a Function of Crossover Number. *Nucl. Acids Res.* **32**, 6047–6056 (2004).

3.21 P.K. Maiti, T.A. Pascal, N. Vaidehi, J. Heo, W.A. Goddard, III, Atomic Level Simulations of Seeman Nanostrutures: The Paranemic Crossover in Salt Solution. *Biophys. J.* **90**, 1463–1479 (2006).

3.22 X. Zhang, H. Yan, Z. Shen, N.C. Seeman, Paranemic Cohesion of Topologically Closed DNA Molecules. *J. Am. Chem. Soc.* **124**, 12940–12941 (2002).

4

Single-stranded DNA topology and motif design

Let's look again at the molecule shown in Figure 3-5a. It looks like a cube built of DNA. However, as we noted earlier, the molecular geometry has really not been characterized, because the 3-arm junctions on its vertices are floppy units.[4.1] Thus, it could look as we have drawn it, or it could look like a rhombohedron (a cube-like structure where one of the body diagonals has been stretched or squashed somewhat). The things we can say for sure about the molecule are that each of the edges is two turns long (the sequence was designed that way) and that each face of the object corresponds to a cyclic single strand of DNA. For example, the front face corresponds to the red strand. Because DNA is a double helix, every turn of the double helix results in the strands being interwound. Since each edge is two turns long, that means that the red strand is linked twice to the four strands of the four faces that flank the front: the green strand on the right, the cyan strand on top, the magenta strand on the left, and the dark-blue strand on the bottom. It is only indirectly linked to the yellow strand at the back. When cyclic molecules are linked together like the links of a chain, they form what is known as a catenane. Of course the hexacatenane corresponding to the cube is a much more complex object than just a simple chain. The molecule in Figure 3-5b, with the connectivity of a truncated octahedron, is a 14-catenane. It is even more complex than the cube-like molecule shown in Figure 3-5a.

Catenanes and knots. It turns out that catenanes are closely related to knots.[4.2] This relationship is indicated in Figure 4-1. The upper left image is of a knot with five nodes and an arbitrary strand polarity. Look at the lower right node in this knot. It is made up of a strand that passes over another strand. You can think of it as four strands: the first half of the strand on top, connected to the second half of the strand on top, the first half of the strand on the bottom, connected to the second half of the strand on the bottom. Now, let's imagine breaking those connections, and reconnecting the strands so that we maintain

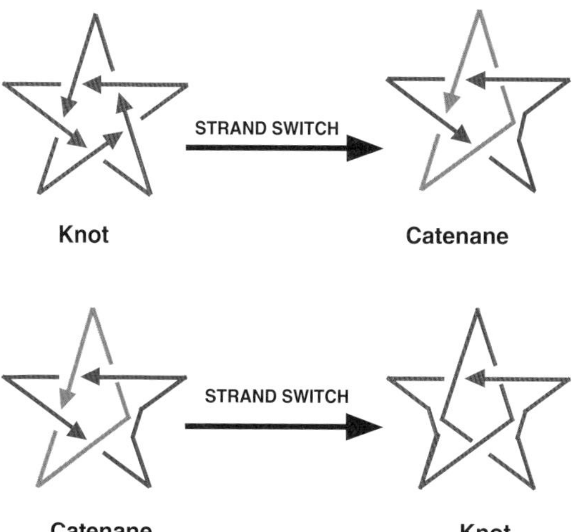

Figure 4-1 *Strand switches that interconvert knots and catenanes.* A 5_1 knot made from a polar strand is shown in the upper left diagram, with the polarity indicated by arrows. Each node consists of two strands, one passing over the other. Each of these strands can be treated as two segments, the one before the crossover and the one after it. The drawing shows that the segments in the lower right node are disconnected and reconnected, top to bottom in each, while preserving polarity. The result is shown in the upper right diagram, a 2-component catenane, rather than a knot. The lower part of the diagram shows this same operation performed on the lower left node of the catenane, resulting in a trefoil knot.

the same polarity. Thus, the first half of the strand on top connects to the second half of the strand on the bottom, and the first half of the strand on the bottom connects to the second half of the strand on top. Performing these two operations produces the blue cyclic molecule and the red cyclic molecule linked together to form a catenane with four nodes, in the upper right of the drawing. Thus, we have converted the knot to a catenane through this strand-switching operation. Now, let's do the same thing to the catenane we have made, which is reproduced on the lower left of Figure 4-1. If we do the same operations on the lower left node, we are left with a knot containing three nodes. Note that a simple ("primitive") knot is a single cyclic species, one that we can draw with a single color. If you think about it a little bit, the operation shown here is just the opposite of reciprocal exchange, i.e., it is resolution. Of course, if we went backwards, we would be doing reciprocal exchange.

Knots and motif design. Given the relationship we have just discussed, we might ask whether knots have anything to contribute to DNA motif design, and whether DNA can contribute to the chemistry of knotted (and linked) species.

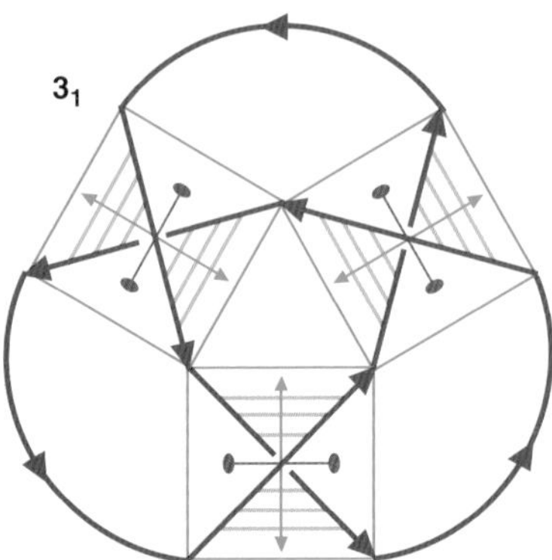

Figure 4-2 *The general relationship between nodes and DNA.* A 3_1 (trefoil) knot is drawn with a polar red strand. It contain three nodes, flanked by green square boxes. The strands of the knot divide each box into four areas, two between parallel strands and two between antiparallel strands. Six blue base pairs are drawn in the regions between antiparallel strands, because the base pairs of DNA are antiparallel. Thus, each box contains a half-turn of DNA, and the helix axis is drawn in green, terminating in two arrowheads. The central two-fold axis of the half-turn is drawn in magenta, and it terminates in two ellipse-shaped dyad symbols.

The answer is a resounding "YES!" Let's first look at the key relationship between single-stranded DNA and topological species. This point is illustrated in Figure 4-2 for the simplest knot, a trefoil knot, sometimes called a 3_1 knot.[4.3] This knot has three nodes, where the strand passes over itself. We have drawn it in red, with polarity, just like the DNA single strand. There is a square box surrounding each of these nodes, with the strands making up the node serving as the diagonals of the box.

Let's look at the bottom node. When we take the polarity into account, the box is divided into four regions, two between parallel stands and two between antiparallel strands. DNA forms its base pairs between antiparallel strands, so we have drawn six light-blue base pairs (about a half-turn of DNA) between the antiparallel strands. The local double helix axis is shown in green, perpendicular to the base pairs. In addition, we have drawn the local dyad axis of this half-turn in magenta, a line terminated in ellipses. Thus, there is a relationship between a half-turn of DNA and a node in a knot or a catenane. The other two nodes of this trefoil knot have been drawn in the same way, just rotated 120° left or right from the bottom node. If we wanted to build this knot out of DNA, we could just design the three half-turns that are shown, and then link them with unstructured DNA (or another polymer) such as oligo-dT to fill in the curved parts of Figure 4-2; indeed, this has been done successfully.[4.4]

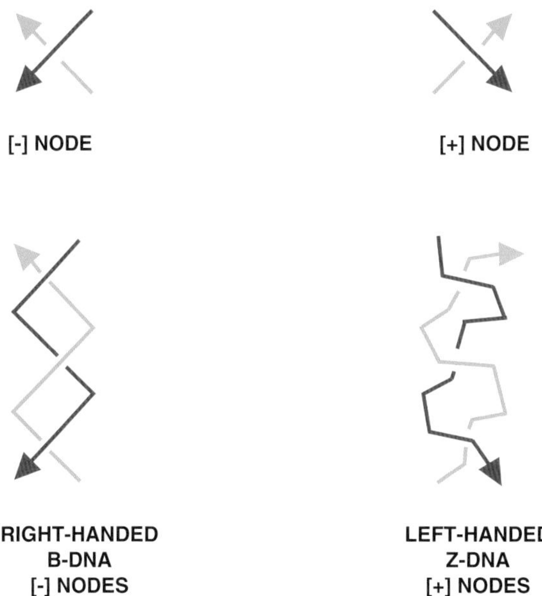

Figure 4-3 *Schematic representations of B-DNA and Z-DNA.* It is clear from the diagram on the left that a half-turn of B-DNA corresponds to a negative node in a knot or a catenane. Similarly, the diagram on the right shows that a half-turn of left-handed Z-DNA corresponds to a positive node.

One of the key features of the DNA double helix is that it is right-handed when made from D-nucleotides, the ones found in nature. That is the handedness of the DNA we have drawn here. The nodes that result from a right-handed double helical structure have been designated by mathematicians[4.2] as having a negative sign. Of course, it is also possible to have nodes with a positive sign, but how do we make them from DNA? Figure 4-3 illustrates this point, and shows us one way to do that. The top part of the figure shows two nodes, a negative node on the left and a positive node on the right. Beneath the negative node is a schematic of B-DNA, the conventional form of DNA with which we have been dealing. Beneath the positive node is a schematic of Z-DNA, a left-handed conformation of DNA discovered by Alexander Rich and his colleagues.[4.5] The left-handed conformation of Z-DNA enables us to make positive nodes. There are two requirements for getting DNA to assume the Z-conformation: an amenable sequence and appropriate solution conditions. Sequences related to . . . CGCGCG . . . form Z-DNA readily when in solutions containing 3–4 M NaCl or 10 mM $Co(NH_3)_6Cl_3$.[4.5]

Left-handed DNA. The use of Z-DNA has enabled the construction of a variety of molecular species that contain positive nodes. The first example of this is the figure-8 knot (also called a 4_1 knot).[4.6] In addition, the trefoil knot containing positive nodes has been built using Z-DNA.[4.7] In fact, by varying solution conditions, it has been possible to build a circle, a 3_1 knot with negative nodes, a 4_1 knot, and a 3_1 knot with positive nodes, all from the same strand of DNA.[4.7] The system used is shown in Figure 4-4. X and its complement X′, and Y and its complement Y′, are represented by bumps on the sides of a square, and they are all about one turn of DNA long; the corners represent long oligo-DT segments. Both X and Y are capable of forming Z-DNA if the conditions are sufficiently favorable, but the Y domain forms Z-DNA more readily than the X domain. If ligated under conditions of sufficiently low ionic strength, the double helix doesn't even form, and all that happens is that a circle forms. Under slightly higher ionic strength conditions, the double helices do form, but they form only B-DNA. Note that this arrangement produces a trefoil knot, although in principle four nodes are present; one of the nodes can be unwound without affecting the topology of the strand. Under conditions of still higher ionic strength, the Y domain converts to the Z-form, but the X-domain remains B-DNA, to yield a knot with two positive and two negative nodes (4_1). If cobalt hexammine is added to the solution, the X domain also becomes Z-DNA, and a trefoil knot with positive nodes results.

The use of Z-DNA also enabled the construction of the first Borromean rings from DNA.[4.8] Figure 4-5a shows an image of Borromean rings, consisting of three linked rings: a red ring, a blue ring, and a green ring. The interesting thing about them is that no pair of them are linked; if any ring is broken, the other two fall apart. They are the symbol of the northern Italian Borromeo family, although they can be found in much older sources.[4.9] The way that they are formed is that the three inner nodes have one sign and the three outer nodes have the opposite sign. Using our rule that a half-turn of DNA corresponds to a node of DNA, we could imagine making this object using three half-turns of B-DNA and three half-turns of Z-DNA. However, it is hard to make three different half-turns of Z-DNA, so 1.5 turns were used instead (Figure 4-5b). What is immediately apparent is that the center of Figure 4-5b is a 3-arm junction made from right-handed DNA. However, since the target here is topological rather than geometrical, it is possible to think of Figure 4-5b as a polar projection, just like the inside of the UN symbol (Figure 4-6a). This image shows the Earth from the North Pole, and every point on the perimeter corresponds to the South Pole. Thus, going back to Figure 4-5b, the center can be thought of as a 3-arm B-DNA branched junction flanking the north pole, and the three left-handed 1.5-turn motifs can be thought of as a 3-arm Z-DNA

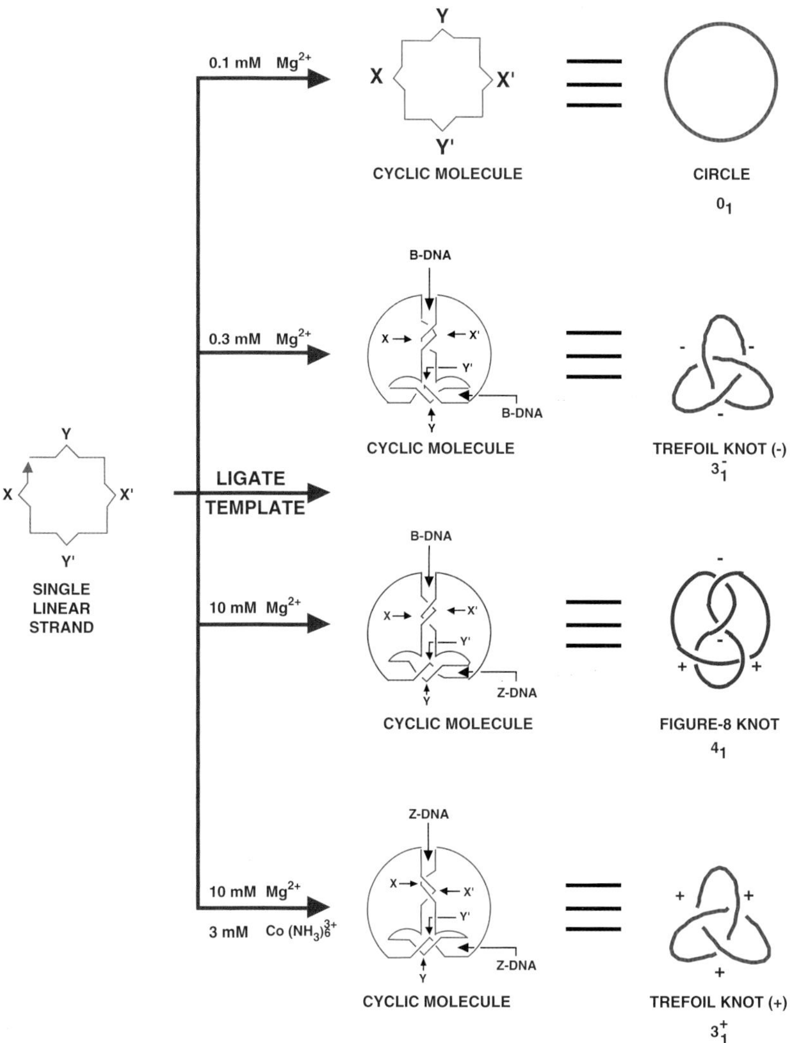

Figure 4-4 *Four different topologies produced from a single 104-mer strand of DNA.* The strand is shown at the left, consisting of two complementary regions, X and X′, and Y and Y′. These two regions have different Z-forming propensities. By ligation at relatively low ionic strength (top), the strands repel each other enough to yield only a circle (orange). At higher ionic strength, both domains form B-DNA, leading to a trefoil (3_1) knot with negative nodes (light blue). At higher ionic strength yet, the Y–Y′ domain forms Z-DNA, yielding a figure-8 (4_1) knot (green). Under more severe Z-forming conditions, the X–X′ domain also forms Z-DNA, yielding a trefoil (3_1) knot with negative nodes (red). Reprinted by permission from the American Chemical Society.[4.4]

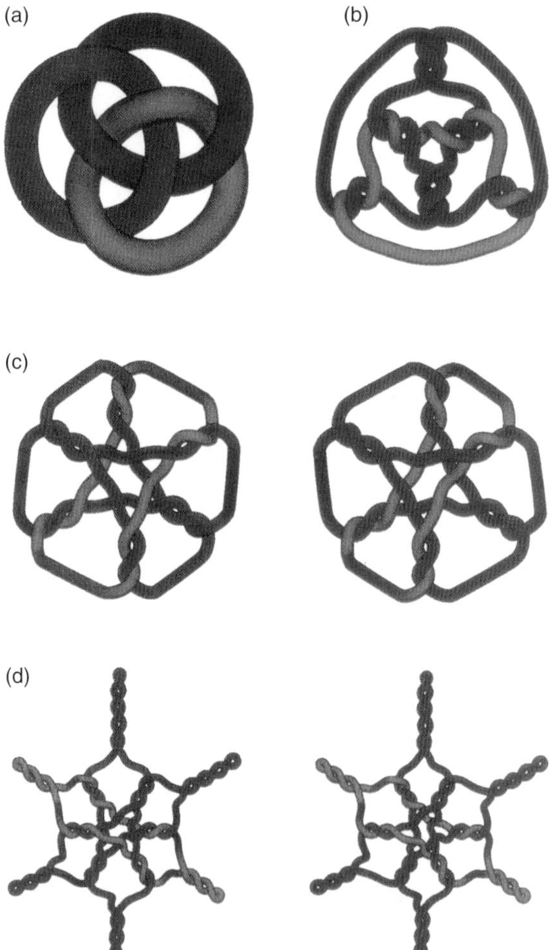

Figure 4-5 *Borromean rings built from DNA.* (a) *Classic Borromean rings.* The three rings – red, blue, and green – are linked as a group, so that if one breaks the other two will fall apart. The three inner nodes are of one sign and the three outer nodes are of the opposite sign. (b) *A variation in which each node or DNA half-turn is replaced with three half-turns.* Cleaving one of these rings will also result in separation of the other two. (c) *A stereographic drawing showing the two sets of 1.5-turn segments as 3-arm branched junctions.* The drawing in (b) has been redrawn here so that the junction made from B-DNA (negative nodes) flanks the north pole of an imaginary sphere, while the junction made from Z-DNA (positive nodes) flanks the south pole. (d) *A stereographic drawing in which the equator has been decorated with hairpin loops.* The hairpins do not affect the topology, but they are used as ligation sites in the construction and restriction sites in the analysis. The three cyclic strands are of different lengths.[4.8]

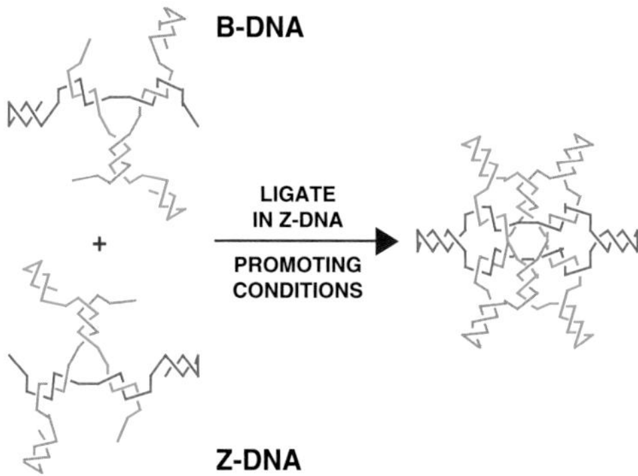

Figure 4-6 *The construction of Borromean rings.* (a) *The United Nations symbol.* The central portion of this symbol is a polar projection of Earth. It is the same sort of representation of our planet as Figure 4-6b is of the Borromean rings. (b) *The components that were ligated to make Borromean rings.* One junction is made of B-DNA and the other of Z-DNA, and they form the structure when ligated in Z-promoting conditions. The hairpins at the equator are all the same length in this representation.

branched junction flanking the south pole. This relationship is shown in stereographic projection in Figure 4-5c. For ease of synthesis, and for proof of synthesis, a lot of turns of DNA were added to the "equator," as shown in the stereographic projection of Figure 4-5d; this addition does not affect the overall topology of the structure. The synthesis of the molecule is shown in Figure 4-6b. Two 3-arm branched junctions are ligated together in the helices at the equator of the molecule. This drawing does not show that the rings are of different lengths to aid in the proof of construction.

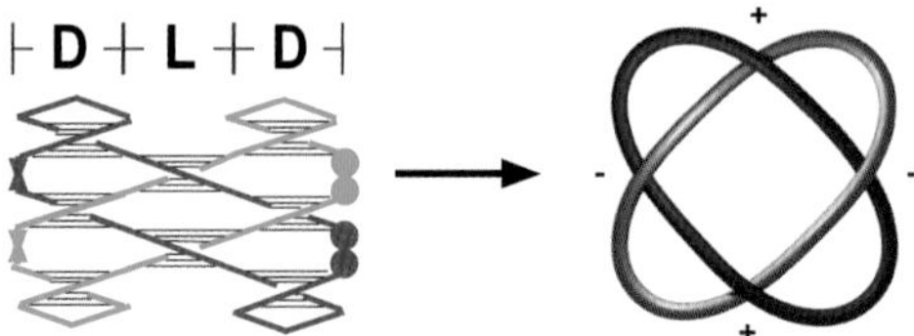

Figure 4-7 *Construction of a Solomon's knot from DNA.* The two interlocking strands at the left contain D-nucleotides in the helices indicated by the D symbol, and L-nucleotides in the central portion under the L symbol. The fused filled circles represent 5′, 5′ linkages, and the vertical bowties at the far left represent 3′, 3′ linkages. A smoothed version of the knot is shown at the right, with the signs of the nodes indicated. Reprinted by permission from John Wiley & Sons.[4.10]

Recently, L-nucleotide phosphoramidites have become commercially available (ChemGenes, Wilmington, MA). They, too, can be used to produce positive nodes, with no sequence requirements, thus enabling much greater sequence diversity and structures that would otherwise be difficult to achieve.[4.10] For example, the toroidal link shown at the right of Figure 4-7 was built using two half-turns of B-DNA made from D-nucleotides and two half-turns of B-DNA made from L-nucleotides. This is seen in the strand diagram on the left of Figure 4-7. The cyclic nature of the two strands was needed to demonstrate the success of the woven braiding that is key to the structure. The need to maintain antiparallelism between the paired strands led to the presence of 5′, 5′ linkages (fused circles in the two strands on the left) and 3′, 3′ linkages (bowtie structures on the left). DNA is arguably the best topological synthon available, but proof of synthesis is often very difficult because direct structural observation of the products is not readily available yet.

Knots and half-turns of DNA. It is an interesting and useful exercise to see how to build a variety of knots using the relationship between a node and a half-turn of DNA. Figure 4-8a shows a collection of half-turns that would lead to the construction of the knot that has been designated 6_3. Three of the half-turns (#1, #2, #4) have negative signs and their helix axes (perpendicular to their base pairs) converge on an area containing a filled triangle. This is a conventional 3-arm branched junction, made with right-handed DNA. The helix axes of the other three half-turns (#3, #5, #6) have positive signs, and their helix axes converge on another filled triangle. This is also a conventional branched junction, but it would be made with left-handed DNA – either Z-DNA, or B-DNA with L-nucleotides. Thus, connecting larger constructs, 3-arm junctions in this case, can lead to a knotted topology.

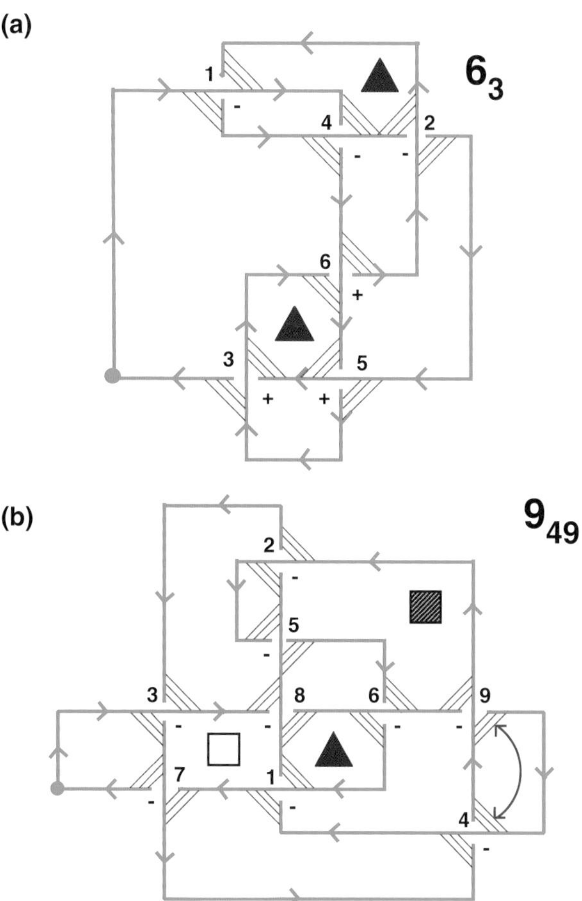

Figure 4-8 *The construction of single-stranded DNA knots.* The strands in both are indicated by aqua-colored strands, with polarities indicated by arrowheads along them. The nodes are numbered in an arbitrary fashion. The signs of the nodes are shown. Each node is flanked by a half-turn of base pairs, showing how it could be constructed from DNA. (a) *The 6_3 knot.* This knot consists of two 3-arm branched junctions, one junction with negative nodes formed from nodes 1, 2, and 4, and the other with negative nodes formed from nodes 3, 5, and 6. Because the nodes can be grouped in this fashion, a solid triangle has been placed at the branch point of each of these junctions. The helix axes of each node point towards the branch point. (b) *The 9_{49} knot.* The nine nodes in this knot are all negative. The groupings chosen are as follows: a double-headed curved arrow represents stacking nodes 4 and 9 of DNA to form a full turn; a solid triangle represents the center of a 3-arm junction formed by nodes 1, 8, and 6; a shaded square is at the center of a mesojunction formed by nodes 2, 5, 6, and 9; nodes 1, 7, 3, and 8 flank an antijunction that is represented by an empty square. Other groupings have not been indicated. Reprinted by permission from Springer.[4.3]

Knots lead to antijunctions and mesojunctions. Figure 4-8b shows another case, a knot called 9_{49}; all of its nodes are negative. Three half-turns (#1, #6, #8) correspond to another 3-arm branched junction, just as in Figure 4-8a. The two half-turns corresponding to nodes #4 and #9 have been combined into a full turn of DNA duplex. But what is that open-square symbol in the middle of nodes #1, #7, #3, and #8? The helix axes that correspond to those nodes are not pointing towards the center of the rectangle; rather, they are directed in a circumferential direction. It is the dyad axes of each of the half-turns that are directed towards the center. This is a new type of structural motif, and it has been called an antijunction.[4.11] It is possible for both types of half-turns to flank a cavity: look at the cavity flanked by the half-turns representing nodes #2, #5, #8, and #9. Nodes #8 and #9 have their helix axes directed towards the center of the cavity, but nodes #2 and #5 have their helix axes directed circumferentially. This type of motif is called a mesojunction,[4.11] and is represented here by a shaded square.

Antijunctions and mesojunctions can only be made within certain limits. If one insists that all nucleotides be pairable, and that long non-pairing loops are forbidden, then one must change the orientations of two helices simultaneously. Thus, no 3-arm antijunction can be made from three strands of DNA that flank a triangle, but a 3-strand mesojunction is feasible. Likewise, no arrangement of three helices and one helix can be made, but two different 4-strand mesojunctions can be made containing two radial and two tangential helices.

Figure 4-9 shows the junctions, antijunctions, and mesojunctions that flank triangles or squares. This diagram shows each of the double helical arrangements to contain six base pairs (thin black lines), just about a single half-turn of DNA, at the vertices of the triangles or squares. The species are named by the number of helix axes directed at the centers of the polygons. Thus, the 3-arm and 4-arm junctions, with all 3 or 4 of their arms directed at the centers of their polygons, are termed 3_3 and 4_4, respectively. The strands of DNA are shown by long lines with an arrowhead, indicating the $5' \rightarrow 3'$ polarities of the strands. The helix axes are drawn as blue double-headed arrows perpendicular to the base pairs. Perpendicular to the helix axes are the primary two-fold axes of each of the half-turns. They are drawn in a maroon color, and consist of a line ending in ellipses on either end. The antijunction is labeled 4_0, because there are no helix axes pointing towards the center of the square. The mesojunctions are on the second line of the diagram. These contain a mixture of double helical domains pointing towards the center of the polygon or pointing circumferentially. The 3_1 mesojunction contains two domains with their two-fold axes pointing towards the center of the triangle, and one domain with its helix axis

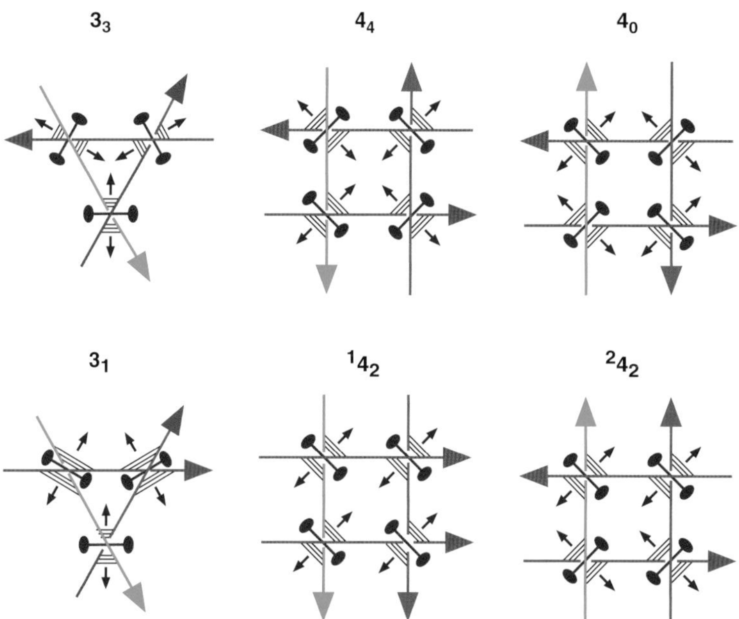

Figure 4-9 *Single-half-turn generalized junction constructs.* The 3-strand junction (3_3), the 4-strand junction (4_4), and the 4-strand antijunction (4_0) are shown in the top row, while the 3-strand mesojunction (3_1) and the first ($^14_2$) and second ($^24_2$) 4-strand mesojunctions are shown in the bottom row. The strands are drawn as arrows of different colors, the helix axes of the nodes are shown as blue double headed arrows, and the two-fold axes of the nodes are represented as ellipse-flanked deep-red lines. The subscripts represent the number of helix axes that point towards the center of the triangle or the square. The superscript on the 4-strand mesojunctions is just a counting index. Reprinted by permission from the American Chemical Society.[4.11]

pointing there. There are two different kinds of 4_2 mesojunctions; they are named with a pre-superscript as a serial number.

The sequences of the component strands of each of the structures shown can be derived by switching the 5′ and 3′ sections of the strands that form the branched junctions. Higher-order antijunctions and mesojunctions that combine radial and tangential components to different extents are possible in principle. Table 4–1 lists the antijunctions and mesojunctions for complexes containing as many as 12 strands. The entries in the table are derived from the formula for determining the number of different necklaces that can be made with two differently colored beads.[4.12] The bulk of the entries in the table are in fact mesojunctions. Conventional branched junctions correspond to the unitary entries down the first column of the table. Antijunctions correspond to the unit

Table 4–1 *Populations of junctions, antijunctions, and mesojunctions.*[a]

Number of arms	Number of circumferential arms 0	2	4	6	8	10	12	Total
3	**1**	1						2
4	**1**	2	*1*					4
5	**1**	2	1					4
6	**1**	3	3	*1*				8
7	**1**	3	4	1				9
8	**1**	4	8	4	*1*			18
9	**1**	4	10	7	1			23
10	**1**	5	16	16	5	*1*		44
11	**1**	5	20	26	10	1		63
12	**1**	6	29	50	29	5	*1*	122

[a] Conventional junctions are in **bold**. Antijunctions are in *italics*. All other structures indicated are mesojunctions. The total listed on the right refers to the total number of structures, with the number of arms indicated on the left. The number of arms is equivalent to the number of strands.

entries on the right edge of each row for even-stranded complexes; according to our rule for switching helix directions pairwise, there are no antijunctions for odd-stranded complexes. One of the interesting things about 4-arm antijunctions is that they can be alternated with 4-arm junctions if they are all drawn in a tetragonal lattice. This is evident from Figure 4-10, where the notation used for junctions and antijunctions is the same as that in Figure 4-8.

Figure 4-11 shows what these molecules look like with three half-turns in each domain. It is clear from this diagram that the length of the domains in half-turns can affect the structures that the arrangement can form. It is possible to have even numbers of half-turns as well, although we won't discuss them here.[4.13] Even though they are formally very interesting, antijunctions and mesojunctions have not proved to be particularly important motifs in structural DNA nanotechnology. This is because they are inherently less well stacked, or perhaps we should say stackable, than conventional branched junctions. Consider a 4_0 antijunction in Figure 4-11. It's nice to think of it the way we have drawn it. However, only at low DNA concentrations is this structure likely, because a lot of stacking is lost at the ends of each double helical domain. When in doubt, one should always assume that nucleic acid structures will maximize their stacking interactions. This maximization can be achieved by the set of four domains shown if they just form an unending open series of nicked duplex molecules, not limited at all to the closed four strand complex

Figure 4-10 *A 2D array of alternating DNA junctions and antijunctions.* Each junction has a solid square at its center and each antijunction has an empty square at its center. Reprinted with permission from the American Chemical Society.[4.11]

shown in Figure 4-10. The average length of the series will be a function of concentration, so it is possible to make a 4-strand antijunction at very low concentrations.[4.11] The same is true of the various three-half-turn mesojunctions shown in Figure 4-10 and the two-half-turn mesojunctions also characterized.

This is an important object lesson in structural DNA nanotechnology: no matter how clever or elegant we may think a particular motif may be, we should remember that molecular strain and unmaximized stacking can result in a different structure. Central to this point is that one is *not* putting a single copy of each strand in the pot containing the reaction components; at a minimum, there are a billion or perhaps a trillion or more copies of each strand that are free to find their own free-energy minimum. Although entropic considerations suggest that a single copy of each strand will join together into a closed complex, there are many examples where multiple copies of strands can form closed or open complexes that are not what the designer ordered.

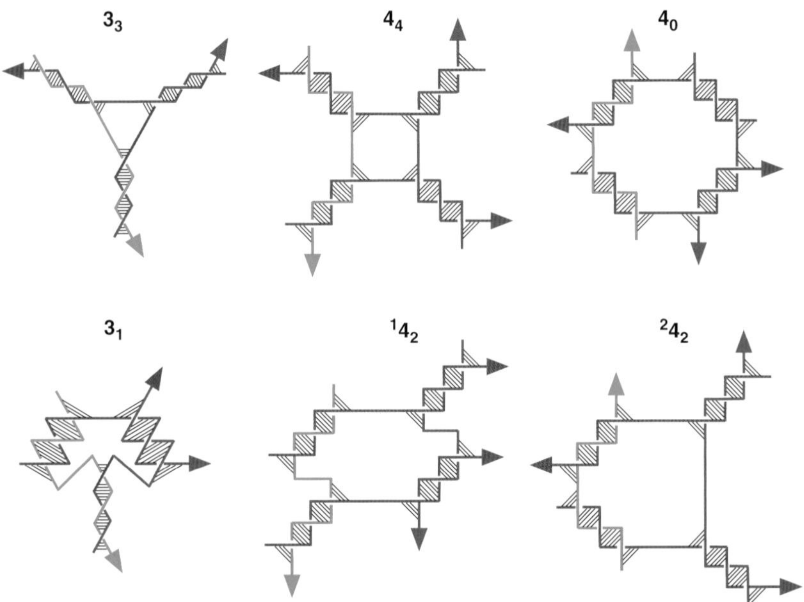

Figure 4-11 *Triple-half-turn generalized junction constructs.* These are the same species as shown in Figure 4-9, except that the half-turns have been replaced by triple-half-turn units. Reprinted by permission from the American Chemical Society.[4.11]

Switchback DNA. An intriguing example of this phenomenon is switchback DNA, which in fact led to the discovery of the exciting PX motif. The DNA polyhedra shown in Figure 3-5 are in fact catenanes. If we consider the intimate relationship between catenanes and knots shown in Figure 4-1, it is reasonable to conjecture whether one could make polygonal and polyhedral knots. The structure corresponding to a knotted triangle is shown in Figure 4-12. Figure 4-12a shows a triangle made from a knot. The blue outline contains within it the right-handed DNA strand drawn in red. Each of three nodes per edge is indicated by a half-turn of DNA, and the helix axis of the DNA is clearly parallel to the edges of the triangle. Unfortunately, if we follow the polarity of the red strand, we see that the strands are parallel. This is not what conventional DNA usually likes to do. However, we are not necessarily restricted to putting the helix axes of the DNA parallel to the edges of the triangle. Figure 4-12b shows that we can reorient the helix axes perpendicular to the edges of the triangle. Thus, there are three helix axes per edge. A minor problem arises in that, by reorienting the helix axes, we have switched from right-handed to left-handed DNA. We can deal with this problem readily by making the long helix

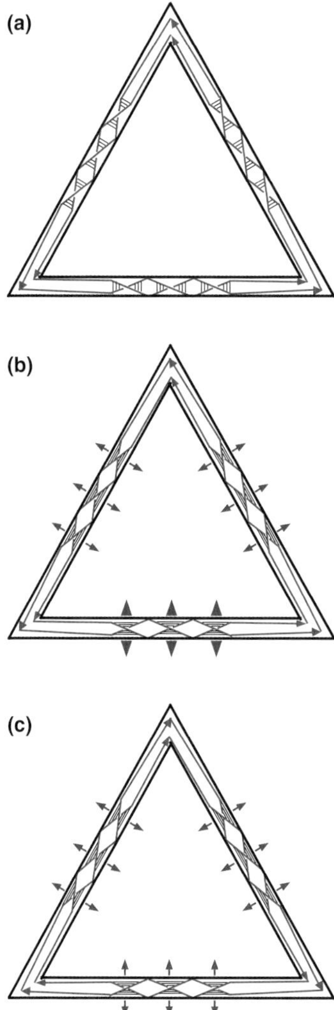

Figure 4-12 *A triangular knot.* (a) *The initial construction.* Note that the strands wind up parallel rather than antiparallel, with the base pairing indicated. (b) *A solution to the parallel-strand problem.* The pairing is indicated in the perpendicular direction. This makes the local helices left-handed, rather than right-handed. (c) *Switching the directions of the strands solves the left-handed problem.*

axis parallel to the triangle edges a left-handed helix, thus making the local helix axes right-handed, as shown in Figure 4-12c.

Thus, in principle, we can make knotted geometrical figures. The motif for each of the edges is an arrangement of DNA known as "switchback DNA,"[4.14]

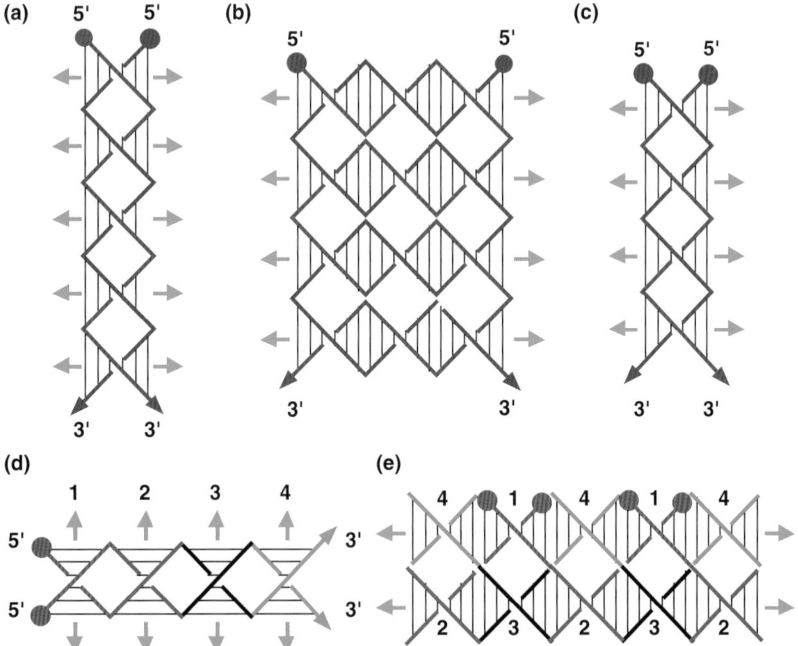

Figure 4-13 *Switchback DNA.* (a) *Two strands forming a 5-domain switchback motif.* This molecule is similar to the triangle edges in Figure 4-12c. The red and blue strands are complementary, but not self-complementary. (b) *A three-half-turn version of the molecule in (a).* (c) *A 4-domain switchback molecule.* (d) *A self-complementary version of the molecule in (c).* Note that this is complementarity in the switchback sense, with dyad axes in each 6-nucleotide unit. (e) *A different pairing of the molecule in (d).* This molecule maximizes the base stacking, and is similar to the PX structure.[3.16]

and it is not limited to a single half-turn of DNA. Thus in Figures 4-13a and 4-13c we see switchback structures with a single half-turn per double helical domain, but Figure 4-13b shows a switchback structure with three half-turns per double helical domain. It is possible to make a molecule that is self-complementary, in the switchback sense, if each of its domains is self-complementary; thus the structures of Figures 4-13a and 4-13c can be made out of a series of 6-nucleotide palindromes, such as 6-base restriction sites. However, there the switchback motif seems to violate the same stacking rules that are violated by antijunctions and mesojunctions. What happens in the laboratory? The answer is a mixed set of results. If an odd number of half-turn domains are used (e.g., Figure 4-13a), the material seems to behave similarly to a duplex containing the same number of nucleotide pairs. However, given the opportunity to do something else, it does. If a switchback molecule containing an even number of half-turns (such as the 4-domain molecule shown in Figure 4-13c) is assembled, the construct behaves like a much longer species.[4.15] Figures 4-13d and 4-13e show what happens. Figure 4-13d redraws the molecule of Figure 4-13c as a switchback self-complementary molecule, showing individual color coding for each half-turn; thus, from the 5′ end to the 3′ end, each strand has 6-nucleotide segments that are magenta (segment 1), light blue (segment 2), dark blue (segment 3), and green (segment 4), respectively. Figure 4-13e shows the sort of rearrangement that actually occurs to maximize the stacking. Two double helical domains are formed, and there is far more base stacking. Follow one strand from beginning to end, from segment 1 to 4, sequentially. Each segment is paired with its complement. The drawing has been simplified somewhat, to show this point more clearly; in fact, the start and end points are actually at the middle (where the crossovers are drawn), rather than on the edges. In the end, the structure closely resembles the PX structure discussed in Chapter 3, and in fact, this is the way in which the PX structure was originally discovered, although the 6:6 motif is not really optimal.[4.15]

The combination of PX-DNA and the availability of L-nucleotides opens a new possibility. PX-DNA interwinds double helices, in the same way that conventional DNA interwinds the two strands of the conventional Watson–Crick double helices. Consequently, one could imagine that double-stranded knots and catenanes could be built, using the same rules described above for the construction of single-stranded knots and catenanes. Negative nodes could be generated from conventional PX-DNA. Positive nodes could of course be generated from PX-DNA built using L-nucleotides.[4.16]

The generalization of complementarity. We are all comfortable with the notion of base pairing complementarity introduced for DNA by Watson and Crick. However, this is a specialized case of complementarity, where there are only two strands and no branch points involved. The essence of everything that we do in structural DNA nanotechnology is that branch points are involved, and that unusual motifs are exploited. This requires us to generalize our notions of complementarity. Figure 4-14 shows what we mean here, when branch points are involved. The image at the far left is a representation of a conventional linear duplex molecule: strand 2 complements strand 1. To its right, strand 2 has been nicked, so that now there are two strands, 2 and 3, that together complement strand 1. Proceeding to the right, we see a 3-arm junction, where again strand 2 and strand 3 together constitute the complement to strand 1. In the next panel, strand 2 and strand 3 of the 3-arm junction have been fused into a single strand via a hairpin loop. This single strand is also complementary to strand 1. At the far right, we see a 4-arm junction. The three strands 2, 3, and 4 together

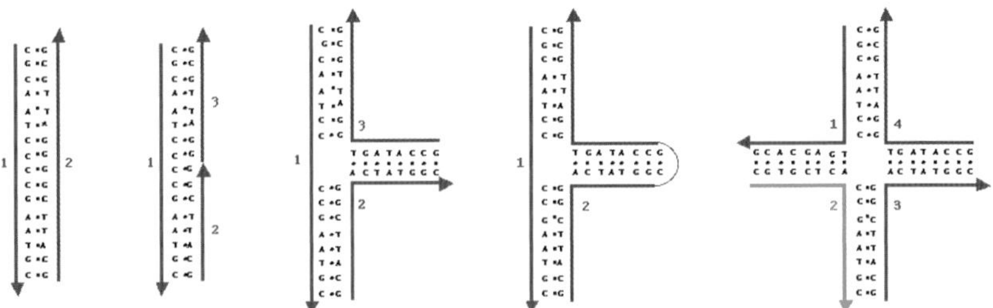

Figure 4-14 *Generalized complementarity.* Five different complements to strand 1 are shown. A traditional Watson–Crick complementarity is shown between strands 1 and 2 at left. To its right, strand 2 has been nicked, so the complement is the combination of strands 2 and 3. In the middle, another helical domain has been added to the nick, so the complement is still strands 2 and 3, but now they have mutual complementarity as well. Moving right, a hairpin loop has joined strands 2 and 3, so the complement is a single strand again, but one with an internal double helical domain. At the far right, strand 1 is complemented by the three other strands of a 4-arm branched junction. Copyright Georg Thieme Verlag KG.[4.17]

form the complement to strand 1, although strand 3 does not interact with it at all. Strands 2, 3, and 4 could, through the addition of two more hairpin loops, produce another single-stranded complement to strand 1. The whole point here is that, once one leaves the constraints of linear duplex DNA, the complement is not well defined. There is no limit on how long the hairpin loop-containing strand in the second-to-right drawing could be. One could speculate that linear duplex DNA has been selected by evolution because it is the only system with a well-defined complement.[4.17]

All the examples in Figure 4-14 rely on traditional double helices. Another feature of a traditional double helix is potential self-complementarity. For example, the sequence AGGCTT-GACTAG-CTAGTC-AAGCCT (dashes only for clarity) is complementary to itself; such sequences are sometimes called palindromes, although "palindrome" really means mirror symmetry (e.g., the quote apocryphally attributed to Napoleon, "Able was I ere I saw Elba"). In our new non-linear-duplex mindset, other rules may apply. For example, if we examine switchback DNA, self-complementarity is defined differently. In the case of a four-half-turn switchback DNA (Figure 4-13c), the sequence TAATTA-GAATTC-GTTAAC-CCGCGG is self-complementary in the switchback sense, even though it is not a traditional palindrome. It goes without saying that even regular complementarity is different in switchback DNA. The traditional Watson–Crick complement of AAAAAA-GGGGGG-TTTTTT-CCCCCC is of course GGGGGG-AAAAAA-CCCCCC-TTTTTT. However, in the switchback sense, its complement is TTTTTT-CCCCCC-AAAAAA-GGGGGG. In this chapter, we have seen how topological considerations have led to new motifs and to new principles. Now it is time to abandon the theoretical approach and to see how we can connect the motifs we generate with the Procrustean bed of experiment.

References

4.1 R.-I. Ma, N.R. Kallenbach, R.D. Sheardy, M.L. Petrillo, N.C. Seeman, 3-Arm Nucleic Acid Junctions Are Flexible. *Nucl. Acids Res.* **14**, 9745–9753 (1986).

4.2 J.H. White, K.C. Millett, N.R. Cozzarelli, Description of the Topological Entanglement of DNA Catenanes and Knots by a Powerful Method Involving Strand Passage and Recombination. *J. Mol. Biol.* **197**, 585–603, (1987).

4.3 N.C. Seeman, The Design of Single-Stranded Nucleic Acid Knots. *Mol. Eng.* **2**, 297–307 (1992).

4.4 S.M. Du, N.C. Seeman, The Construction of a Trefoil Knot from a DNA Branched Junction Motif. *Biopolymers* **34**, 31–37 (1994).

4.5 A. Rich, A. Nordheim, A.H.-J. Wang, The Chemistry and Biology of Left-Handed Z-DNA. *Ann. Rev. Biochem.* **53**, 791–846 (1984).

4.6 S.M. Du, N.C. Seeman, The Synthesis of a DNA Knot Containing Both Positive and Negative Nodes. *J. Am. Chem. Soc.* **114**, 9652–9655 (1992).

4.7 S.M. Du, B.D. Stollar, N.C. Seeman, A Synthetic DNA Molecule in Three Knotted Topologies. *J. Am. Chem. Soc.* **117**, 1194–1200 (1995).

4.8 C. Mao, W. Sun, N.C. Seeman, Assembly of Borromean Rings from DNA. *Nature* **386**, 137–138 (1997).

4.9 A. Lakshminarayan, Borromean Rings and Prime Knots in an Ancient Temple. *Resonance (Indian Acad. Sci.)* **12**(5), 41–47 (2007).

4.10 T. Ciengshin, R. Sha, N.C. Seeman, Automatic Molecular Weaving Prototyped Using Single-Stranded DNA. *Angew. Chemie Int. Ed.* **50**, 4419–4422 (2011).

4.11 S.M. Du, S. Zhang, N.C. Seeman, DNA Junctions, Antijunctions and Mesojunctions. *Biochem.* **31**, 10955–10963 (1992).

4.12 G.M. Constantine, *Combinatorial Theory and Statistical Design*, New York, John Wiley & Sons, p. 235 (1987).

4.13 H. Wang, N.C. Seeman, Structural Domains of DNA Mesojunctions. *Biochem.* **34**, 920–929 (1995).

4.14 T. Wang, Ph.D. thesis, New York University (2007).

4.15 Z. Shen, Ph.D. thesis, New York University (1999).

4.16 N. Baas, N.C. Seeman, A. Stacey, Synthesizing Topological Links. *J. Math. Chem.* **53**, 183–199 (2015).

4.17 N.C. Seeman, In the Nick of Space: Generalized Nucleic Acid Complementarity and the Development of DNA Nanotechnology. *Synlett* **2000**, 1536–1548 (2000).

5
Experimental techniques

The preceding chapters have emphasized theoretical methods for designing sequences and motifs in structural DNA nanotechnology. We have implicitly assumed that there was some way to put DNA strands together to produce target products. In this chapter, we will discuss how this is done. Interspersed with the theoretical methods have been obscure comments that such-and-such motif forms well or does not form well, or perhaps does not form at all. It is now time to discuss the experimental techniques that give us the right to make such claims. Many of the construction and analysis experiments done so far in structural DNA nanotechnology have been done on a scale ranging from femtomoles to nanomoles, which is the scale on which the techniques of molecular biology are employed. Structural DNA nanotechnology is indeed somewhat parasitic on molecular biology, both in terms of its methods and in terms of the materials that are commercially available because of the molecular biology and biotechnology enterprises. These range from enzymes, such as restriction enzymes, ligases, and topoisomerases, to specialized components for synthesizing and modifying DNA strands.

5.1 Construction methods

5.1.1 Synthesis and purification

Solid-support methodology for the synthesis of DNA-containing designated sequences[5.1] is the central enabling methodology for the pursuit of structural DNA nanotechnology. This field is still at the stage where the key principles are being elucidated. Consequently, it is both economical and convenient to perform syntheses on relatively small scales, e.g., 10–200 nm. On a well-tuned synthesizer, one can achieve apparent step yields of 99.2% or so based on trityl group deprotection; these step yields lead to about 45% yield of crude target product in the synthesis of a 100-mer. Despite numerous attempts at alternative approaches, purification by denaturing gel remains the most reliable (and tedious) method of separating target molecules from failure products. It is worth pointing out that DNA of this length frequently has suffered damage during the synthesis, so reproducing (and amplifying) it with polymerase chain reaction (PCR) may be advisable in cases where high chemical purity is an issue.[5.2,5.3] These cases include the accurate measurement of physical properties, and crystal formation. PCR is a very good method to amplify small numbers of molecules up to the picomole scale; however, if larger quantities are needed, say for crystallization, synthesis followed by an orthogonal dimension of HPLC is likely to be the most effective way of getting larger quantities. In recent years, large parallel syntheses of very small quantities (< 20 nmol) have been available commercially, thereby enabling DNA origami and DNA bricks (see Chapter 9). Future issues involving synthesis will likely entail scale-up to larger quantities for specific applications.

5.1.2 Hybridization

The sequence-dependent hybridization[5.4] of complementary nucleic acid molecules, or parts of molecules, is implicit in virtually all forms of DNA nanotechnology. In our experience, the design algorithm discussed in Chapter 2 has proved very reliable in producing unusual DNA motifs. However, it is predicated on estimates of equilibrium structures, and it is key that kinetic traps be avoided.[5.4] Consequently, all samples are heated to 90 °C for 5 minutes, and then cooled slowly. Relatively quick protocols entail a number of stages in heating blocks, say 20 minutes at each of 65 °C, 45 °C, 37 °C, room temperature, and perhaps a lower temperature. To form simple arrays, the 90 °C solution can be put in a styrofoam box and cooled to room temperature over about 40 hours. In a sense, it is amazing that 4-tile arrays, with 20 unique strands can form a 2D array just by heating and cooling. For more complex arrays, a thermocycling protocol is sometimes used with pre-formed tiles. The presence of Mg^{2+} or other multi-valent cations in solution appears to be required for the stability of small branched molecules on gels.[5.5]

5.1.3 Phosphorylation and ligation

Molecules that are to be ligated must contain phosphates on their 5′ ends. Phosphates may be added chemically as a final step in the synthesis, or they may be added enzymatically using polynucleotide kinase. We have found that phosphorylation of unpurified DNA with ^{32}P-labeled phosphate may occasionally result in the labeling of a failure sequence, one whose secondary structure is more accessible to the enzyme than the target strand. Neither chemical nor enzymatic phosphorylation (nor even both) produces completely phosphorylated material. If this is critical to one's experiment, one can produce the most highly phosphorylated strands by restricting a molecule containing the sequence of interest.[5.6]

Ligation of DNA is central to many aspects of DNA nanotechnology. We have found that enzymatic ligation is superior to chemical ligation,[5.7] but it is still not very effective.[5.8] Virtually stoichiometric quantities of ligase are often needed to produce relatively limited yields of product. The origin of this problem is not well understood, but it may arise from the presence of unphosphorylated ends or branch points titrating out the ligase. Branched molecules with sticky ends often ligate more poorly than blunt-ended linear molecules. Minimizing the number of ligations of branched molecules in a preparative protocol is strongly advised; we have found that it is usually best to do more synthesis and less ligation.

The topological state of a ligation product is often a key feature of a target molecule. If the molecule contains unpaired segments, it can be useful to employ topological protection.[5.9,5.10] This is a technique whereby the single-stranded regions are paired temporarily with complements during ligation, thus eliminating any braiding they may undergo. The notion is illustrated in Figure 5-1. This procedure has worked successfully in the synthesis of specific catenanes, but has not been as useful in the synthesis of knots. This may be owing to the single-stranded segments in knots usually consisting of oligo-dT; the addition of oligo-dA could lead to the formation of DNA triplexes, whose rotation defeats the purpose of topological protection.

5.1.4 Solid-support methodology

Control over the products of DNA ligation is critical for the construction of complex motifs. The initial complex object, a DNA cube (Figure 3-5a), was synthesized entirely in solution.[5.8] This approach afforded only limited control over the edges that were designed to ligate at any particular step: they could be phosphorylated, or not. After the first partial products were obtained, even this level of control was lacking, because there was no control over which 5′ ends were phosphorylated on the second step. A solution to the lack of control was the development of a solid-support methodology, allowing successive double helical edges of the target molecule to be ligated individually.[5.11] The starting unit was added to the solid support, with all of its sticky ends protected as DNA hairpins. These could be deprotected individually by restriction enzymes,

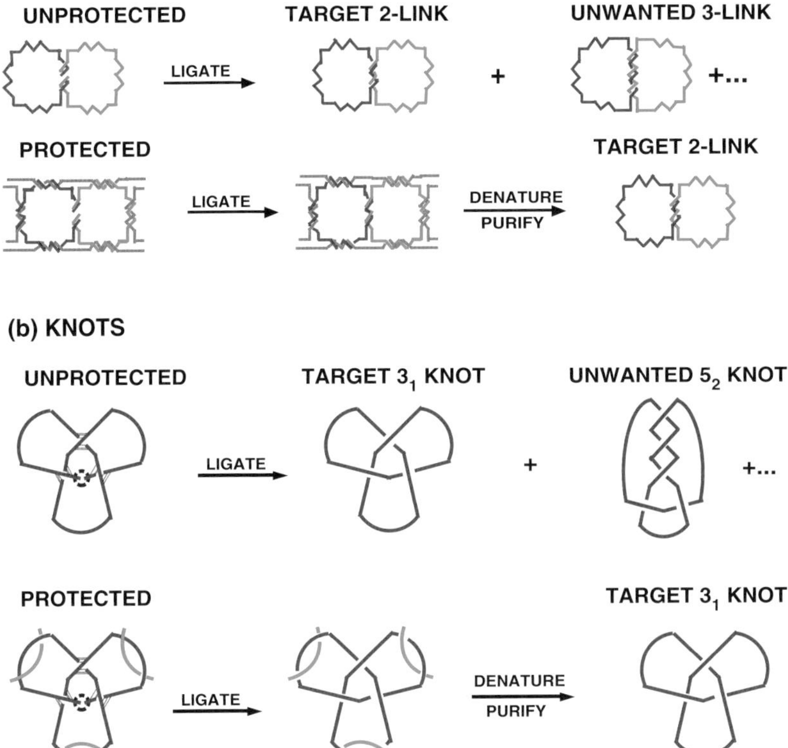

Figure 5-1 *Topological protection.* (a) *Topological protection of catenanes.* The drawing indicates how uncontrolled ligation of single-stranded substrates can lead to multiple products, but that this result is lessened when topological protection is used. (b) *Topological protection of knots.* The same principle as (a), of converting single-stranded DNA to double-stranded DNA to prevent accidental linking, is shown for a trefoil knot.

and then ligated to another molecule whose only unclosed helix contained the complementary sticky end. Thus, every target intermediate is topologically closed, permitting intermediate purification by exonuclease treatment.

The solid-support based construction of a quadrilateral is shown in Figure 5-2. The solid support is shown as a filled cross-hatched circle. Two strands have been crosslinked below it, to render the structure immune to denaturation. A 3-arm junction is attached to the crosslinked region, with one arm having a restriction site named 1 and the other arm having a restriction site named 2. In the first step, arm 1 is restricted and a second junction is added on.

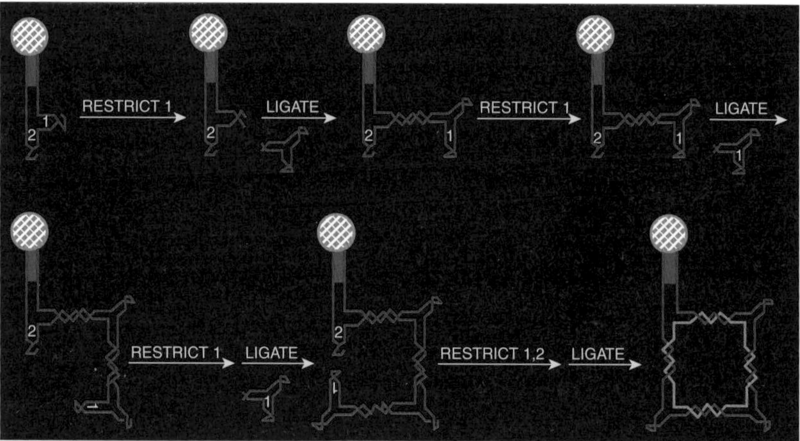

Figure 5-2 *Solid-support-based synthesis of a quadrilateral.* The solid support is shown as a cross-hatched circle, and the crosslinking of two strands is indicated by a filled blue area between. Restriction enzyme 1 is applied to a 3-arm junction, liberating a sticky end, and a new 3-arm junction, also bearing restriction site 1, is added to it. This process is repeated twice more. At that point, both restriction enzyme 1 and restriction enzyme 2 are applied, creating an intramolecular sticky end that can be ligated to convert the single-stranded structure into a quadrilateral-like catenane. This is the opposite of the transition shown in the second step of Figure 4-1. Reprinted by permission from the American Chemical Society.[5.11]

This process is repeated twice, to put a total of four junctions in the quadrilateral. In the final step, both arms 1 and 2 are restricted, and they are fused by ligation. Note that the single strand has now become a catenane. This process is the reverse of the strand-switch removal of nodes shown in Figure 4-1.

One major advantage of the solid-support approach is that intermolecular reactions can be conducted at high concentrations but crosslinking between growing objects during intramolecular reactions can be minimized. It is worth noting that sticky ends should *not* be self-complementary, as this decreases control unacceptably. It is important that the growing objects be sufficiently far apart that reactions that are designated as intramolecular do not suffer from crosstalk between growing objects; sometimes it is necessary to poison the support so as to make sure that the growing objects will not interact. Figure 5-3 illustrates the synthesis of the truncated octahedron shown in Figure 3-3b.[5.12] The box in the upper left corner shows the initial connectivity of the six squares of the object. Each of these squares is built from a 4-arm junction, rather than 3-arm junctions. Once the solid support has been prepared with sufficiently light loading, square 1 is attached to it, as shown below the box. At that point, an intermolecular reaction is performed, attaching a 4-square complex to the

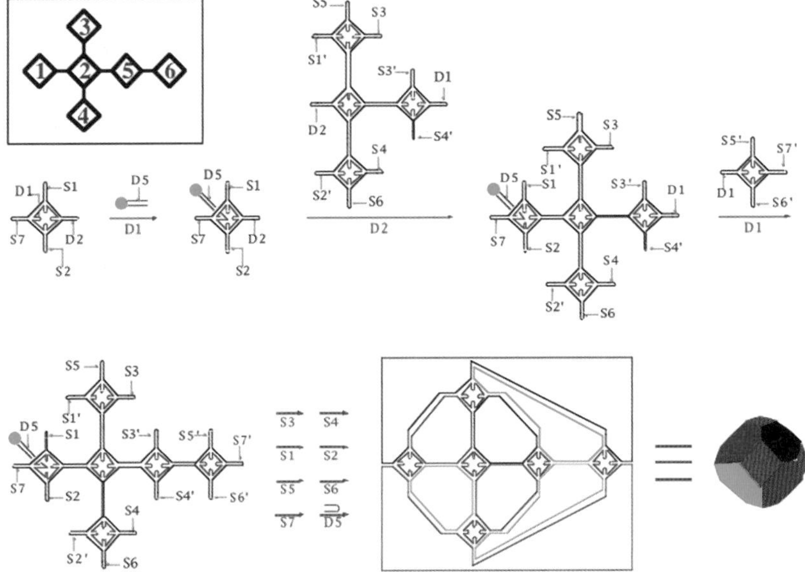

Figure 5-3 *Solid-support-based construction of a truncated octahedron.* The overall scheme product of intermolecular additions is shown in the upper left, along with the numbering scheme. Intermolecular restriction sites are indicated by D followed by a number, and intramolecular pairs of sites by S or S′ followed by a number. In the first step, the first square is attached to a solid support (green circle). A 4-square complex is added using D5, and then a single complex is added using D1. Intramolecular reactions are then performed to produce the final object. Reprinted by permission from the American Chemical Society.[3.8]

original square through restriction and ligation at the site labeled D2. The final square is then attached at the site labeled D1 by the same process. Seven intramolecular reactions then are performed at the sites named S1, . . ., S7, combining them with S1′, . . ., S7′, respectively, again with restriction and ligation. Virtually every one of these reactions went with quantitative yield, except for the S6–S6′ connection, which went at 1% yield, possibly because this was the point where the 2D nature of the square array became a 3D object, and there was a lot of strain involved.

5.2 Characterization in structural DNA nanotechnology

5.2.1 Characterization of motifs

DNA nanotechnology has produced a large variety of complex motifs derived by merging the notions of reciprocal exchange between strands (Chapter 3) and the generalization of complementarity (Chapter 4). If you produce a set of strands designed to form such a motif, you must demonstrate that the target motif actually has been obtained. These species are quite large, and are not readily tractable using techniques that produce unambiguous structural results in synthetic organic or inorganic chemistry, such as X-ray crystallography or NMR. Consequently a battery of analytical procedures has been developed or adapted to establish the features of these unusual DNA motifs.

Partial product gels

The very first thing to do with a new motif is to demonstrate that it forms cleanly. A routine way to do this is to prepare a non-denaturing gel containing the stoichiometric complex and its sub-complexes; for example, for a 4-arm junction, such a gel would have lanes containing each strand, each of the six strand pairs, the four 3-strand complexes, and the 4-strand target complex.[5.13] It is important to work out the stoichiometry of the strands first in a pairwise fashion. Titrating each strand against one of its partial complements on gels is straightforward; the end point is the ratio that shows no excess of either strand. An example of such a gel is shown in Figure 5-4. All of the combinations of strands are shown, and the target 4-strand complex is shown clearly in lane k. There are neither multimers nor breakdown products seen in lane k, indicating

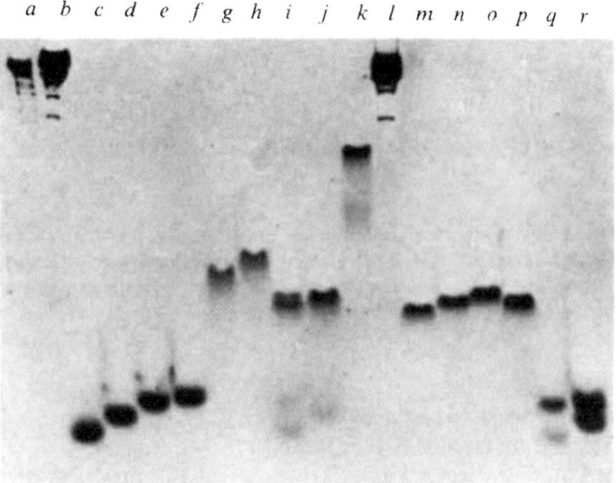

Figure 5-4 *A non-denaturing polyacrylamide gel of the stoichiometric mixtures of the components of the J1 4-arm branched junction.* Lanes a, b, and l contain markers. The individual strands are in lanes c, d, e, and f. The four strand-triples are in lanes g, h, i, and j. Lane k contains the mixture of all four strands. Lanes m–p contain the mixtures of adjacent pairs of strands, while lanes q and r contain the diagonally opposite pairs.[2.11]

that the structure forms cleanly. Note that the 3-strand complexes do not behave this way, with a certain amount of dissociation clearly visible. The single strands do not migrate identically, because this is a non-denaturing gel, and the transient secondary structures available to all of them are not quite the same, affecting their ultimate mobilities under these conditions. It is not uncommon for individual strands to migrate as multimers, and this occurrence should not be regarded as a problem; the SEQUIN algorithm is designed for the final complex, not the partial products.

Sometimes there are simply too many strands to do a thorough version of the analysis shown in Figure 5-4. Another solution to this problem, which also allows for troubleshooting, is the type of gel shown in Figure 5-5. Figure 5-5a shows a 9-strand complex that acts as a cassette[5.14] containing one state of a DNA-based nanomechanical device, known as a PX-JX_2 device (Chapter 10). Getting this arrangement to come together properly was not simple, and several modifications had to be made to the strands. The gel in Figure 5-5b demonstrates the final successful assembly. Each of the strands in Figure 5-5a has been labeled with a trace of radioactive ^{32}P in each of the lanes of Figure 5-5b. A single band of the same mobility in each of the lanes indicates that there is a single product of the same molecular weight in each lane. Before the design problems were solved, both breakdown products and multimerization products were visible in some of the lanes. The successful assembly of a motif is characterized by a single band of roughly the expected molecular weight. Migration under non-denaturing conditions is a function of many factors, including surface area, molecular weight, and shape. Nevertheless, target motifs tend to migrate roughly in the vicinity of linear duplex markers of similar molecular weight. Ferguson analysis (see below) should always be performed to be sure that one's results are not affected by choice of gel percentage. There are several ways that a motif can fail this analysis. (1) It can be unstable, producing bands that migrate more rapidly than the target band. (2) Another form of failure is characterized by bands corresponding to multimers of the target complex. These can arise from a system that is stressed by torsional or electrostatic features of the molecular design; e.g., parallel DX molecules generate 8-strand dimers, 12-strand trimers, and 16-strand tetramers at high concentrations.[5.15] This is shown in Figure 5-6, where parallel DX molecules are seen to multimerize. If compatible with one's goals, multimers can be avoided at high concentrations if the target complex is converted to a topologically closed form.[5.16] Sticky ends can be an artificial source of multimer bands;[5.17] a new motif should be assayed first either with blunt ends, or analyzed on gels run at temperatures that preclude inter-motif cohesion. (3) A third form of failure is a smear, resulting from the formation of an

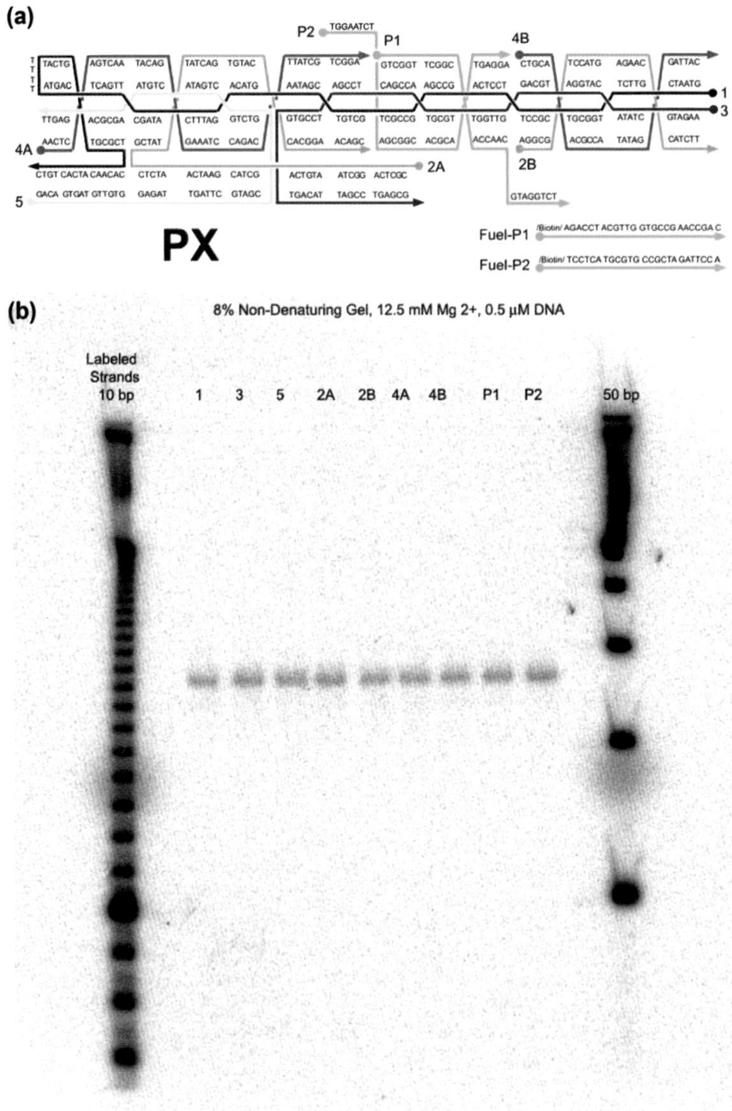

Figure 5-5 *A gel to demonstrate the presence of all strands in a successful DNA assembly.* (a) *A schematic drawing of a complex DNA machine in the PX state.* The strand numbers are indicated, and the strands are drawn with different colors, with 5′ ends indicated by filled circles and 3′ ends by arrowheads. (b) *A non-denaturing polyacrylamide gel autoradiogram demonstrating its assembly.* The left-most lane contains a ladder of 10 nucleotide pair markers and the right-most contains a 50-nucleotide ladder. In the intervening lanes, the label indicates the strand that has been labeled with ^{32}P. Note that each lane contains a single species of identical mobility. Reprinted by permission from the AAAS.[5.14]

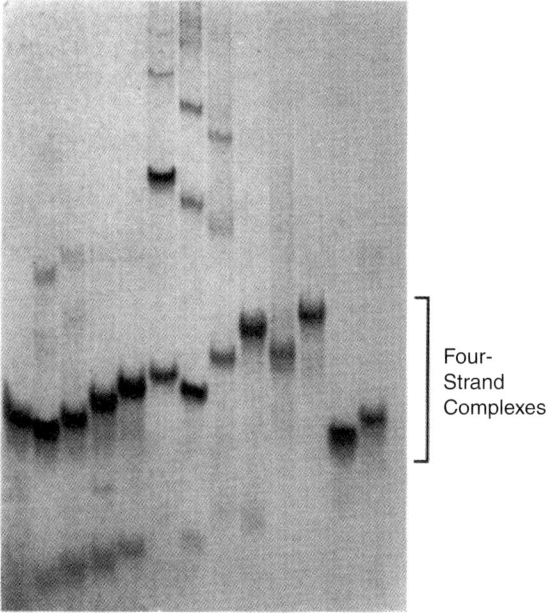

Figure 5-6 *Non-denaturing gels of double-crossover molecules and related junctions.* This is an 8% native polyacrylamide gel containing stoichiometric mixtures of the complexes. Lane 2 contains a DPE molecule with one turn in the central region and eight nucleotide pairs in its external arms. Lane 1 contains a control junction, which contains the domains with the same sequence but lacking the second crossover. Lanes 3, 4, and 5 contain the same DPE molecule, with external arms containing 9, 10, and 11 nucleotide pairs, respectively. Bands corresponding to less than an entire complex can be noted in the more rapidly moving bands near the bottom of the gel. Lane 6 contains a DPOW molecule, lane 7 contains a DPON molecule, and lane 8 contains a DPE-type molecule, DPE-1, with two helical turns in its central region. Each of these parallel molecules is characterized by the presence of multimers in equilibrium with the 4-strand monomer. Lane 9 contains a control junction, whose domains are designed to contain the sequences as DPE-1. Lane 10 contains a DAO molecule with 1.5 turns in its central portion, and lane 11 contains a control junction for the DAO molecule. Lane 12 contains a DAE molecule, composed of five strands of DNA, and lane 13 contains a control junction for it. The notation "Four-Strand Complexes" on the right of the drawing is meant to include the DAE 5-strand complex. Note the absence of multimers or breakdown in the lanes containing the antiparallel double-crossover molecules (10 and 12), in contrast to the lanes containing the parallel double-crossover molecules. Reprinted by permission from the American Chemical Society.[2.27]

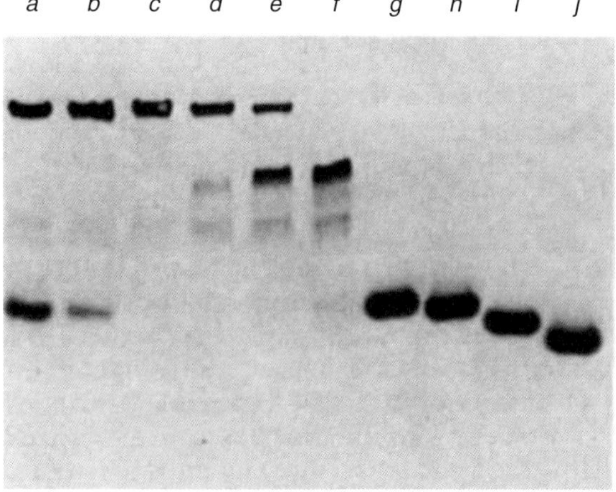

Figure 5-7 *Stoichiometric titration of three strands of a 4-arm DNA branched junction with a fourth.* In this experiment, electrophoresis of mixtures containing different ratios of two components was performed: component 1 consists of an equimolar mixture of J1 strands 1, 2, and 4, while component 2 consists of strand 3 alone. Lanes g–j contain 8 μg of free strands 4, 3, 2, and 1, respectively. Lane f contains 6 μg of component 1 alone. Lanes a–e contain 6 μg of component 1 and the following amounts (μg) of component 2: a, 4; b, 3; c, 2; d, 1; e, 0.5.[2.11]

open complex (e.g., strand 1:2:3:4:1:2:3: . . .), rather than a closed complex (1:2:3:4); as noted in Chapter 4; this behavior was seen with the 4-strand antijunction complex at high concentrations.[5.18]

In addition to knowing the constituents of a new motif, it is also important to establish the stoichiometry of the complex. This is easily done for an N-strand complex by titrating an (N−1)-strand stoichiometric complex with the missing strand. For example, in a 4-strand motif, a 1:1:1 3-strand complex can be titrated with the fourth strand.[5.13] The titration process is monitored readily on a non-denaturing gel. Assuming 1:1:1:1 stoichiometry, a 1:1:1:0.5 mixture will show bands for both the partial complex and the target motif. A 1:1:1:1 mixture should show a band containing only the target motif, and a 1:1:1:2 mixture should show bands corresponding both to the target motif and to the excess single strand. An example of this experiment is shown in Figure 5-7.

Shape analysis

The qualitative shape of a DNA motif can be compared usefully and conveniently with standards by means of a Ferguson plot.[5.19] This is a plot of log(mobility) vs. acrylamide concentration, and its slope is proportional to the friction constant of the molecule. Ferguson analysis has been used to characterize

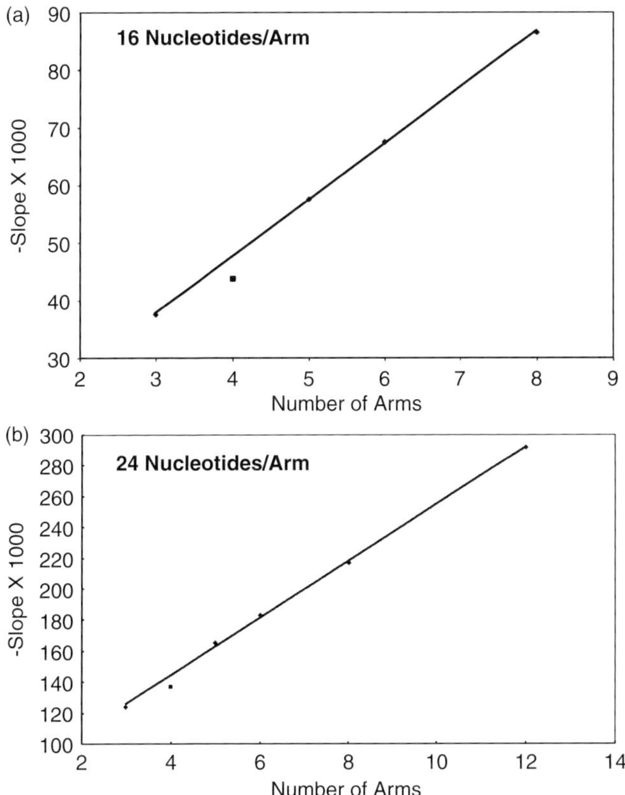

Figure 5-8 *Ferguson slopes of DNA branched junctions.* (a) *Ferguson slopes of 3-arm, 4-arm, 5-arm, 6-arm, and 8-arm junctions (each arm contains 16 nucleotide pairs).* Gels were run at 4 °C. The Ferguson slopes of the junctions are as follows: 3-arm (0.038), 4-arm (0.044), 5-arm (0.058), 6-arm (0.068), 8-arm (0.087). The line was fitted excluding the 4-arm junction, whose position was then plotted. The increment for the 4-arm junction is about 38% lower than expected at 4 °C, on the basis of the relationship between the other four junctions. (b) *Ferguson slopes of 3-arm, 4-arm, 5-arm, 6-arm, 8-arm, and 12-arm junctions.* Gels were run at 4 °C. The Ferguson slopes of the junctions, given as –slope, are as follows: 3-arm (0.124), 4-arm (0.137), 5-arm (0.165), 6-arm (0.183), 8-arm (0.217), 12-arm (0.292). The line was fitted excluding the 4-arm junction, whose position was then plotted. The increment for the 4-arm junction is about 35% lower than expected at 4 °C, on the basis of the relationship between the other five junctions. Reprinted by permission from the American Chemical Society.[1.9]

the stacked character of 4-arm junctions in comparison to 3-, 5-, 6-, 8-, and 12-arm junctions.[5.20,5.21] The slopes of the Ferguson plots of these junctions are shown in Figure 5-8; it is clear that there is something unusual about the 4-arm junction, namely that its arms are stacked in pairs to form two domains, which is unique to this junction.[5.21] The Ferguson plot also has been the means to demonstrate that the addition of helices to linear duplex and to DX molecules results in similar changes in friction constant.[5.22] Likewise, the qualitative structure of PX-DNA has been compared usefully to that of double-crossover molecules to establish their similar shapes.[5.23] Both denaturing and non-denaturing Ferguson analyses have been applied to a DNA cube-like molecule and its subcatenanes.[5.24]

Cooper and Hagerman[5.25] developed a method for the analysis of the shapes of DNA branched junctions. By adding long reporter duplexes pairwise to the arms of 4-arm junctions, they established the basic structural features of DNA branched junctions. Their results suggest that gel mobility is a monotonically increasing function of the angle between the two extended arms. This approach was modified by Lilley and his colleagues[5.26] for the conventional Holliday junction, and was utilized effectively as well for bowtie junctions, which contain 3′, 3′ and 5′, 5′ linkages in the crossover strands of the branched species.[5.27] The same pairwise approach has been applied to junctions using transient electric birefringence[5.28] and fluorescence resonance energy transfer (FRET) to establish details of the branched junction structure and dynamics.[5.29,5.30] FRET is a very powerful technique for obtaining distance information, and it has been used to demonstrate the operation of several DNA-based nanomechanical devices (Chapter 10).[5.31, 5.32, 5.33]

Thermal analysis

The relative stabilities of new motifs are often usefully analyzed by melting them while monitoring a feature such as OD_{260}[5.34] or circular dichroism.[5.5] Melting a complex DNA motif often reveals features that are more characteristic of the compositions of individual helical domains than of the molecular complex itself. It is often useful to pay particular attention to pre-melting transitions, because they may be more diagnostic of features of the complex that differentiate it from linear duplex DNA or other topologies of the same composition. Another technique of great value is analysis of the motif on a perpendicular denaturing gradient gel.[5.35,5.36] This method enables one to see parts of the complex fall apart, rather than monitoring the details of the intradomain de-stacking transition; it is also applicable to the small quantities available at the early analytical stage, not requiring the microgram+ samples for optical melts. Calorimetry has also been applied usefully to DNA motifs.[5.36, 5.37] If you use techniques like calorimetry that typically require large concentrations of material, it is critical to run a gel at the same concentration as the analysis is conducted (or as near to it as you can get), to make sure that no multimerization has occurred in your sample.

Hydroxyl radical autofootprinting

The highest-resolution technique that is used conveniently to characterize the structures of unusual DNA motifs is hydroxyl radical autofootprinting.[5.38] This is a variation on the technique of hydroxyl radical footprinting, used to establish the binding sites of proteins on DNA. Branched junctions,[5.20,5.21,5.38] tethered junctions,[5.39] antijunctions, and mesojunctions,[5.40] as well as DX,[5.15] TX,[5.22] and PX[5.23] molecules, have all been analyzed by this method. The analysis is performed by labeling a component strand of the complex and exposing it to hydroxyl radicals. The pattern of the strand as part of the complex is compared with the pattern of the strand paired to form linear duplex DNA. The key feature noted at crossover sites in these analyses is decreased susceptibility to attack. Decreased susceptibility is interpreted to suggest that access of the hydroxyl radical may be limited by steric factors at the sites where it is detected.[5.41] Likewise, similarity to the duplex pattern at points of potential flexure is assumed to indicate that the strand has adopted a conventional helical structure in the complex, whether or not it is required by the secondary structure.

In studies of junctions, DX molecules, and mesojunctions, protection has been seen particularly at the crossover sites, but also at non-crossover sites where strands from two adjacent parallel or antiparallel domains appear to occlude each other's surfaces, thereby preventing access by hydroxyl radicals.[5.15,5.22,5.39] Thus, crossover sites can be located reliably by hydroxyl radical autofootprinting analysis, but it is not always possible to distinguish them unambiguously from juxtapositions of backbone strands. The technique is particularly powerful as a method to establish whether crossovers occur where they are expected to occur. Those who wish to construct new motifs are strongly advised to ascertain by hydroxyl radical autofootprinting that they have formed the molecular topologies they have designed. An example of the hydroxyl radical protection pattern of a DNA triple crossover molecule[5.22] is shown in Figure 5-9. Protection sites both at the crossover points and at duplex juxtapositions are visible.

5.2.2 Characterization of ligation targets

Denaturing gels

Ligated species are extremely difficult to analyze. The simplest ligated species are linear multimers of a given motif. Junctions,[5.34,5.42] DX molecules,[5.17,5.43] and DNA triangles[5.44] are all examples of unusual DNA motifs that have been oligomerized in one dimension. The trick to analyzing these systems is for them to contain a "reporter" strand. This is a strand whose fate reports the fate of the complex. Examples of reporter strands are shown in Figure 5-10, where the ligation products of DAE, DAE+J, and DAO molecules are shown. The red

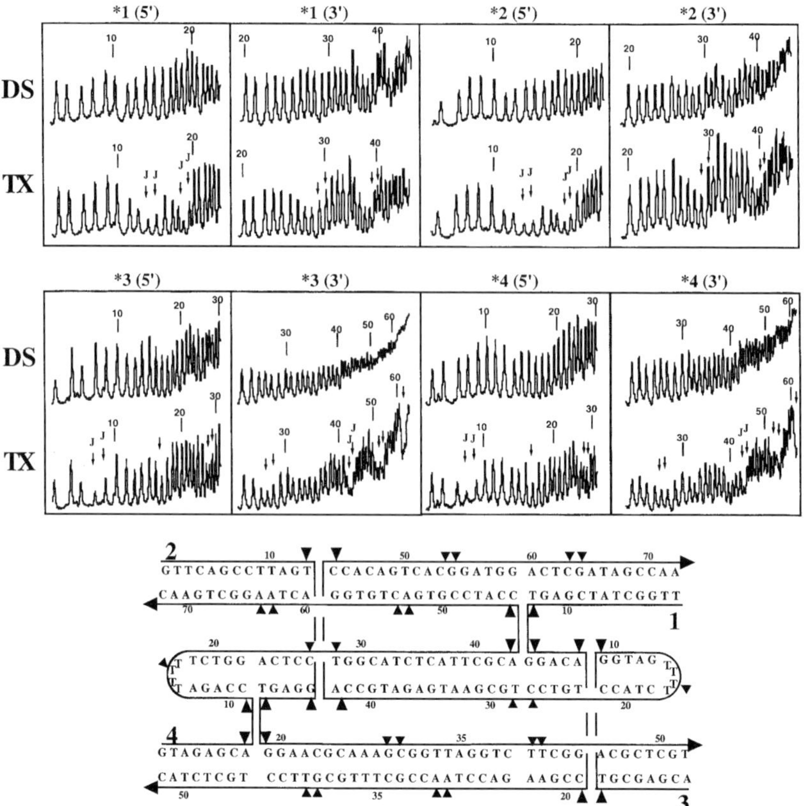

Figure 5-9 *Hydroxyl radical autofootprinting of a TX molecule.* The top portion of the figure contains densitometer scans of autoradiograms for each strand of a TX molecule. The data for each strand are shown twice, once for its 5′ end and once for its 3′ end, as indicated above the appropriate panel. Susceptibility to hydroxyl radical attack is compared for each strand when incorporated into the TX molecule (TX) and when paired with its traditional Watson–Crick complement (DS). Nucleotide numbers are indicated above every tenth nucleotide. The two nucleotides flanking expected crossover positions are indicated by two Js. Note the correlation between the Js and protection in all cases. Additional protection is seen at further locations, indicating occlusion a turn away from the crossover points on the crossover strands, and about 4 nucleotides 3′ to the crossovers on the helical strands. The data are summarized on a molecular drawing below the scans. Sites of protection are indicated by triangles pointing towards the protected nucleotide; the extent of protection is indicated qualitatively by the sizes of the triangles. Reprinted by permission from the American Chemical Society.[5.22]

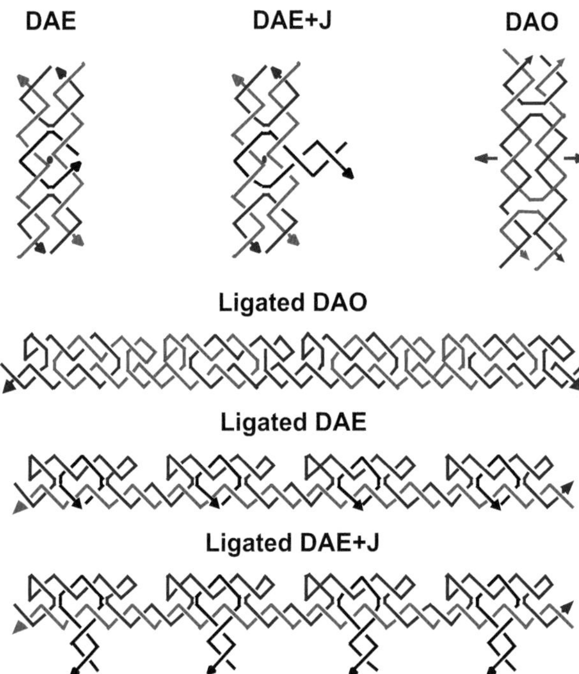

Figure 5-10 *Reporter strands.* Three DX molecules are shown at the top: a DAE molecule, a DAE+J molecule, and a DAO molecule. Below them, versions of each molecule ligated in their bottom domain are illustrated. Reporter strands are shown in red in both the DAE molecule and the DAE+J molecule. By contrast, the ligation of the DAO molecule results in a catenane that does not have a reporter strand. Reprinted by permission from the American Chemical Society.[5.16]

strand in the DAE and DAE+J ligations is a reporter strand, but there is no reporter strand in the DAO ligation. It is important to extract the reporter strands from other strands in the ligation product to which they might be catenated; this is usually done by restriction. The reporter strand can be sized in comparison with linear and cyclic markers.[5.17] Further examples of the use of reporter strands will be discussed in Chapter 6. In addition to providing an estimate of the products, ligation experiments can be used to estimate the stiffness of a given motif, either qualitatively[5.17,5.34,5.42] or quantitatively.[5.6]

More complex target ligation products are likely to be catenanes, or perhaps knots. This is a consequence of the plectonemic nature of the DNA strands; indeed, as noted in Chapter 4, DNA is an almost ideal synthon for topological construction. Catenanes are usually separable on denaturing gels of one[5.8] or two[5.2] dimensions. Catenanes of greater linking number migrate more rapidly on denaturing gels than similar catenanes of the same molecular composition but lower linking number.[5.10] This seems to be true of knots as well, although it is key to ensure that you have not selected a fortuitous acrylamide percentage for your analysis. A reliable way to ensure independence from gel artifacts is to perform a denaturing sedimentation analysis in comparison with markers. If your material appears to be behaving anomalously on denaturing gels, it is useful to check that the sample is completely denatured. The Fischer–Lerman[5.35] 100% denaturing conditions (7 M urea + 40% formamide) often denature samples that misbehave on conventional denaturing gels.

Restriction analysis

Complex constructions such as small DNA polyhedra[5.8,5.12] or Borromean rings[5.45] must be designed carefully from the start, so that proof of construction can be obtained. As noted above, crystallography is not readily available to characterize these species, and molecular weights of 150–800 kD are beyond the range of current NMR capabilities. Unclosed structures, if large enough, have been characterized by AFM[5.46] and cryo-transmission electron microscopy.[5.47,5.48,5.49] For closed molecules, it is possible to characterize the products topologically. The analysis entails the judicious insertion of restriction sites into the molecules. These sites are utilized to break the products down to smaller catenanes that can be synthesized independently, and to which they can be compared. Thus, the cube contains a unique restriction site in each of its edges, as shown in Figure 5-11. The presence of these restriction sites violates, to some extent, the rules discussed in Chapter 2, but there do not seem to have been any negative consequences of this violation. The DNA cube (Chapter 3) was built from a linear triple catenane corresponding to the ultimate left–front–right faces of the target. Figure 5-12 shows that cleavage of the left–front and right–front edges of the cube leads to the top–back–bottom linear triple catenane. Likewise, the cleavage of any of the strands of the Borromean rings leads to two cyclic single strands (see Chapter 4).

5.2.3 Characterization of devices

DNA nanomechanical devices hold great promise in the developing field of nanorobotics. Devices may consist of ligated[5.31] or annealed[5.50,5.51] components. The requirements for characterization of the system are no less stringent for a device than for a static target. In addition, you have to demonstrate that the device responds to the triggers for which it has been designed. Mechanical motion is the hallmark of a device, and this intramolecular motion leads to a change in some distance within the molecule that can be monitored by fluorescence resonance energy transfer (FRET). FRET has been used successfully to

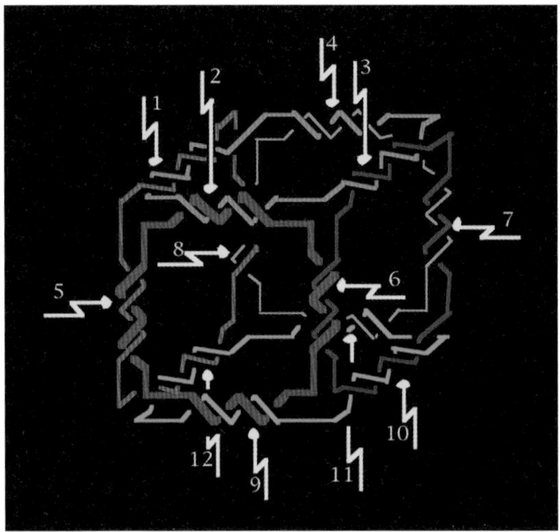

Figure 5-11 *Restriction sites on a DNA molecule with the connectivity of a cube.* The molecule is drawn with all the twisting of each edge shown only in the middle of the edge. Each of the six strands is colored differently. Lightning bolts point to twelve different restriction sites which allow each edge to be cleaved individually. Reprinted by permission from John Wiley & Sons.[5.17]

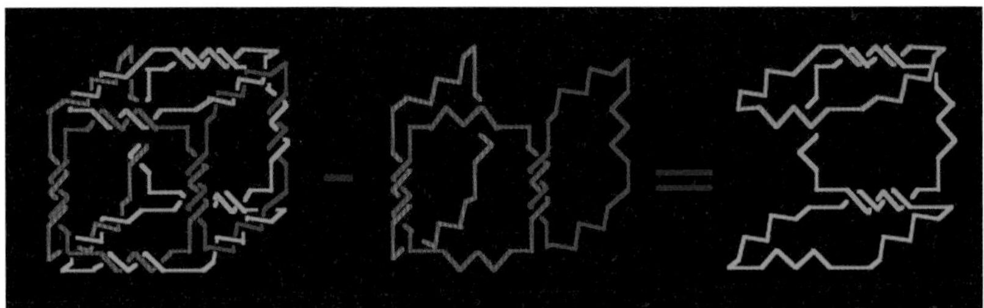

Figure 5-12 *Proof of DNA cube synthesis.* The precursor to the cube is the left–front–right linear triple catenane. Restriction of the front–left and front–right edges of the cube removes it, leaving the top–back–bottom linear triple catenane, demonstrating that all closed strands are present. Reprinted by permission from John Wiley & Sons.[5.43]

demonstrate the action of devices predicated on the B–Z transition[5.31] and on the binding and removal of a specific strand[5.50]. A sequence-specific device based on hybridization topology has also been demonstrated recently by atomic force microscopy. An edge-sharing half-hexagon attached roughly perpendicular to the device provides a 17 nm lever arm whose repositioning is readily visible at the 7–10 nm resolution of the AFM.[5.51] Walkers have been demonstrated by crosslinking,[5.52,5.53] FRET,[5.54] and AFM.[5.55,5.56]

Sometimes a complex device is built which leads to the production of a product; examples include a device that directs the formation of particular polymers[5.57] and a nanomechanical assembly line.[5.55] In this case, assuming that all the intermediate steps have been checked out during the troubleshooting phase of the construction, you can just characterize the product; in the two cases noted here, this was done by sequencing the DNA polymer product and by examining the nanoparticles joined by the assembly line in the TEM. Something that is important to remember is that it is possible to cherry-pick results when analyzing results by AFM or TEM, but you have to be careful to report the percentages of correct products. Along the same lines, and perhaps even more important, is the need to remember that when performing constructions with ensembles of molecules there will be 10^9–10^{15} molecules in your reaction vessel, and you should take into account the presence of all the molecules in the pot when planning your assembly: unless you demonstrate it, you can't guarantee the absence of crosstalk between your reaction components.[5.55]

5.2.4 Characterization of DNA arrays

Periodic arrays

We described above how reporter strands can be used to establish the extent that a system has ligated in one dimension. However, such an approach is of less utility in two dimensions (or three), because it will not reveal the nature or extent of faults very well. Direct observation of the array is necessary. To date, the key technique used is atomic force microscopy (AFM). Periodic arrays of DX molecules,[5.43] TX molecules,[5.22] and conventional[5.4] and bowtie[5.58] DNA parallelograms have all been characterized by this method. As with all microscopy, the investigator can usually find the structure that is sought; consequently, any observations must be challenged by making a molecular-level change that leads to a predictable change in the AFM pattern. For example, DAE molecules can be interspersed with DAE+J molecules (Figure 5-10) whose extra helical domain is oriented out of the plane of the array. For DAE molecules of dimensions ~4 × 16 nm, alternating DAE molecules with DAE+J molecules leads to a striped feature every 32 nm, as seen in Figure 5-13a; if there is only a single DAE+J molecule in every four molecules, then the stripes will be separated by 64 nm, seen in Figure 5-13b.[5.43] To ensure reliable results, changes in molecular design must lead to predictable changes in AFM patterns. Turberfield and his colleagues have used TEM in some cases to characterize 2D arrays.[5.59] X-ray diffraction clearly is required to characterize three-dimensional periodic DNA constructions.[5.60,5.61]

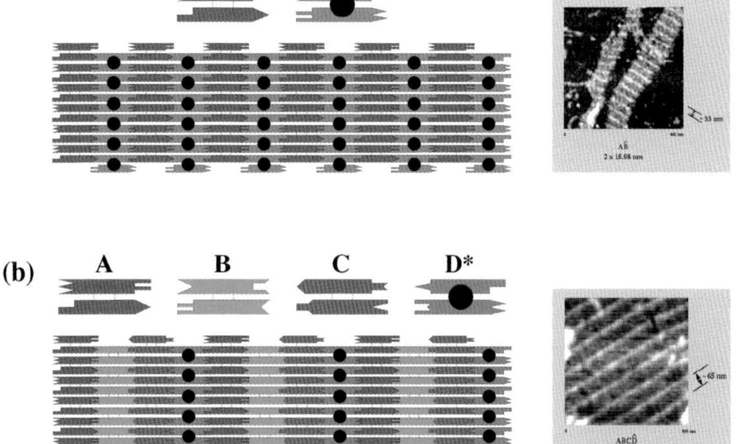

Figure 5-13 *Atomic force microscope (AFM) characterization of a 2D DX lattice.* (a) *A 2-component lattice consisting of a DX molecule and a DX+J molecule.* The DX molecule is labeled A and the DX+J molecle is labeled B*. A filled circle on the DX+J molecule indicates the presence of the extra domain. The sticky ends are indicated geometrically. The horizontal dimensions of these tiles are 16 nm, so there ought to be a stripe-like feature resulting from the extra domain in the lattice, with stripes separated by ~32 nm. The AFM image at right confirms this prediction. (b) *A 4-component lattice consisting of three DX molecules and one DX+J molecule.* The same conventions apply as in (a), so the stripes ought to be separated by ~64 nm, as illustrated in the AFM image at right.

Aperiodic arrays

The founding of experimental DNA-based computation by Adleman[5.62] has led to an interest in other types of DNA self-assembly. Winfree[5.63] pointed out that branched DNA motifs could be used to do logical computation. The idea is that sticky ends can be assigned a logical value, and that their assembly can then be used to perform the computation. We'll talk more about this in Chapter 12. You have to be able to characterize the product, to get the answers. If the answer is a single strand, you can sequence the strand.[5.64,5.65] Algorithmic assembly can also be used to program patterns in two or three dimensions, such as making Sierpinski triangles.[5.66] In 2D, the success can be demonstrated by AFM, but new methods, beyond AFM and crystalline diffraction, will be needed in 3D to confirm the presence of designed aperiodic patterns. It is now time to consider the requirements of 3D systems, and the ways we can approach them.

References

5.1 M.H. Caruthers, Gene Synthesis Machines: DNA Chemistry and Its Uses. *Science* **230**, 281–285 (1985).

5.2 H. Wang, R.J. Di Gate, N.C. Seeman, The Construction of an RNA Knot and Its Role in Demonstrating That E. Coli DNA Topoisomerase III is an RNA Topoisomerase. In *Structure, Motion, Interaction and Expression of Biological Macromolecules*, ed. R.H. Sarma, M.H. Sarma, New York, Adenine Press, pp. 103–116 (1998).

5.3 G. Wu, N. Jonoska, N.C. Seeman, Construction of a DNA Nano-Object Directly Demonstrates Computation. *Biosystems* **98**, 80–84 (2009).

5.4 C. Mao, W. Sun, N.C. Seeman, Designed Two-Dimensional DNA Holliday Junction Arrays Visualized by Atomic Force Microscopy. *J. Am. Chem. Soc.* **121**, 5437–5443 (1999).

5.5 N.C. Seeman, M.F. Maestre, R.-I. Ma, N.R. Kallenbach, Physical Characterization of a Nucleic Acid Junction. In *Progress in Clinical and Biological Research, Vol. 172A: The Molecular Basis of Cancer*, ed. R. Rein, New York, Alan R. Liss, pp. 99–108 (1985).

5.6 A. Podtelezhnikov, C. Mao, N.C. Seeman, A. Vologodskii, Multimerization-Cyclization of DNA Fragments as a Method of Conformational Analysis. *Biophys. J.* **79**, 2692–2704 (2000).

5.7 S.M. Du, H. Wang, Y.-C. Tse-Dinh, N.C. Seeman, Topological Transformations of Synthetic DNA Knots. *Biochem.* **34**, 673–682 (1995).

5.8 R. Williams, *The Geometrical Foundation of Natural Structure*, New York, Dover (1979).

5.9 N.C. Seeman, The Design of Single-Stranded Nucleic Acid Knots. *Mol. Eng.* **2**, 297–307 (1992).

5.10 T-J. Fu, Y.-C. Tse-Dinh, N.C. Seeman, Holliday Junction Crossover Topology. *J. Mol. Biol.* **236**, 91–105 (1994).

5.11 Y. Zhang, N.C. Seeman, A Solid-Support Methodology for the Construction of Geometrical Objects from DNA. *J. Am. Chem. Soc.* **114**, 2656–2663 (1992).

5.12 J. Chen, N.C. Seeman, The Synthesis from DNA of a Molecule with the Connectivity of a Cube. *Nature* **350**, 631–633 (1991).

5.13 N.R. Kallenbach, R.-I. Ma, N.C. Seeman, An Immobile Nucleic Acid Junction Constructed from Oligonucleotides. *Nature* **305**, 829–831 (1983).

5.14 B. Ding, N.C. Seeman, Operation of a DNA Robot Arm Inserted into a 2D DNA Crystalline Substrate. *Science* **314**, 1583–1585 (2006).

5.15 T.-J. Fu, N.C. Seeman, DNA Double Crossover Structures. *Biochem.* **32**, 3211–3220 (1993).

5.16 T.-J. Fu, B. Kemper, N.C. Seeman, Endonuclease VII Cleavage of DNA Double Crossover Molecules. *Biochem.* **33**, 3896–3905 (1994).

5.17 X. Li, X. Yang, J. Qi, N.C. Seeman, Antiparallel DNA Double Crossover Molecules as Components for Nanoconstruction. *J. Am. Chem. Soc.* **118**, 6131–6140 (1996).

5.18 S.M. Du, S. Zhang, N.C. Seeman, DNA Junctions, Antijunctions and Mesojunctions. *Biochem.* **31**, 10955–10963 (1992).

5.19 D. Rodbard, A. Chrambach, Estimation of Molecular Radius, Free Mobility and Valence Using Polyacrylamide Gel Electrophoresis. *Anal. Biochem.* **40**, 95–134 (1971).

5.20 Y. Wang, J.E. Mueller, B. Kemper, N.C. Seeman, The Assembly and Characterization of 5-Arm and 6-Arm DNA Junctions. *Biochem.* **30**, 5667–5674 (1991).

5.21 X. Wang, N.C. Seeman, The Assembly and Characterization of 8-Arm and 12-Arm DNA Branched Junctions. *J. Am. Chem. Soc.* **129**, 8169–8176 (2007).

5.22 T. LaBean, H. Yan, J. Kopatsch, F. Liu, E. Winfree, J.H. Reif, N.C. Seeman, The Construction, Analysis, Ligation and Self-Assembly of DNA Triple Crossover Complexes. *J. Am. Chem. Soc.* **122**, 1848–1860 (2000).

5.23 T. Wang, Ph.D. thesis, New York University (2007).

5.24 J. Chen, N.C. Seeman, The Electrophoretic Properties of a DNA Cube and Its Sub-Structure Catenanes. *Electrophor.* **12**, 607–611 (1991).

5.25 J.P. Cooper, P.J. Hagerman, Gel Electrophoretic Analysis of the Geometry of a 4-Way Junction. *J. Mol. Biol.* **198**, 711–719 (1987).

5.26 D.R. Duckett, A.I.H. Murchie, S. Diekmann, E. von Kitzing, B. Kemper, D.M.J. Lilley, The Structure of the Holliday Junction and its Resolution. *Cell* **55**, 79–89 (1988).

5.27 R. Sha, F. Liu, M.F. Bruist, N.C. Seeman, Parallel Helical Domains in DNA Branched Junctions Containing 5′, 5′ and 3′, 3′ Linkages. *Biochem.* **38**, 2832–2841 (1999).

5.28 J.P. Cooper, P.J. Hagerman, Geometry of a Branched Junction Structure in Solution. *Proc. Nat. Acad. Sci.(USA)* **86**, 7336–7340 (1989).

5.29 A.I.H. Murchie, R.M. Clegg, E. von Kitzing, D.R. Duckett, S. Diekmann, D.M.J. Lilley, Fluorescence-Energy Transfer Shows That the 4-Way DNA Junction is a Right-Handed Cross of Antiparallel Molecules. *Nature* **341**, 763–766 (1989).

5.30 P.S. Eis, D.P. Millar, Conformational Distributions of a 4-Way DNA Junction Revealed by Time-Resolved Fluorescence Resonance Energy-Transfer. *Biochem.* **32**, 13852–13860 (1993).

5.31 C. Mao, W. Sun, Z. Shen, N.C. Seeman, A DNA Nanomechanical Device Based on the B-Z Transition. *Nature* **397**, 144–146 (1999).

5.32 W. Shen, M. Bruist, S. Goodman, N.C. Seeman, A Nanomechanical Device for Measuring the Excess Binding Energy of Proteins that Distort DNA. *Angew. Chem. Int. Ed.* **43**, 4750–4752 (2004).

5.33 H. Gu, W. Yang, N.C. Seeman, A DNA Scissors Device Used to Measure MutS Binding to DNA Mis-Pairs. *J. Am. Chem. Soc.* **132**, 4352–4357 (2010).

5.34 R.-I. Ma, N.R. Kallenbach, R.D. Sheardy, M.L. Petrillo, N.C. Seeman, 3-Arm Nucleic Acid Junctions Are Flexible. *Nucl. Acids Res.* **14**, 9745–9753 (1986).

5.35 S.G. Fischer, L.S. Lerman, Length-Independent Separation of DNA Restriction Fragments in 2-Dimensional Gel Electrophoresis. *Cell* **16**, 191–200 (1979).

5.36 C.H. Spink, L. Ding, Q. Yang, R.D. Sheardy, N.C. Seeman, Thermodynamics of Forming a Parallel Holliday Crossover. *Biophys. J.* **97**, 528–538 (2009).

5.37 L.A. Marky, N.R. Kallenbach, K.A. McDonough, N.C. Seeman, K.J. Breslauer, The Melting Behavior of a Nucleic Acid Junction: A Calorimetric and Spectroscopic Study. *Biopolymers* **26**, 1621–1634 (1987).

5.38 W. Sun, C. Mao, H. Iwasaki, B. Kemper, N.C. Seeman, No Braiding of Holliday Junctions in Positively Supercoiled DNA Molecules. *J. Mol. Biol.* **294**, 683–699 (1999).
5.39 A. Kimball, Q. Guo, M. Lu, N.R. Kallenbach, R.P. Cunningham, N.C. Seeman, T.D. Tullius, Conformational Isomers of Holliday Junctions. *J. Biol. Chem.* **265**, 6544–6547 (1990).
5.40 S.M. Du, S. Zhang, N.C. Seeman, DNA Junctions, Antijunctions and Mesojunctions. *Biochem.* **31**, 10955–10963 (1992).
5.41 B. Balasubramanian, B. Pogozelski, T.D. Tullius, DNA Strand Breaking by the Hydroxyl Radical is Governed by the Accessible Surface Areas of the Hydrogen Atoms of the DNA Backbone. *Proc. Nat. Acad. Sci. (USA)* **95**, 9738–9743 (1998).
5.42 M.L. Petrillo, C.J. Newton, R.P. Cunningham, R.-I. Ma, N.R. Kallenbach, N.C. Seeman. The Ligation and Flexibility of 4-Arm DNA Junctions. *Biopolymers* **27**, 1337–1352 (1988).
5.43 E. Winfree, F. Liu, L. A. Wenzler, N.C. Seeman, Design and Self-Assembly of Two-Dimensional DNA Crystals. *Nature* **394**, 539–544 (1998).
5.44 X. Yang, L.A. Wenzler, J. Qi, X. Li, N.C. Seeman, Ligation of DNA Triangles Containing Double Crossover Molecules. *J. Am. Chem. Soc.* **120**, 9779–9786 (1998).
5.45 C. Mao, W. Sun, N.C. Seeman, Assembly of Borromean Rings from DNA. *Nature* **386**, 137–138 (1997).
5.46 R.P. Goodman, I.A.T. Schaap, C.F. Tardin, C.M. Erben, R.M. Berry, C.F. Schmidt, A.J. Turberfield, Rapid Chiral Assembly of Rigid DNA Building Blocks for Molecular Nanofabrication. *Science* **310**, 1661–1665 (2005).
5.47 Z. Shen, H. Yan, T. Wang, N.C. Seeman, Paranemic Crossover DNA: A Generalized Holliday Structure with Applications in Nanotechnology. *J. Am. Chem. Soc.* **126**, 1666–1674 (2004).
5.48 S.M. Douglas, H. Dietz, T. Liedl, B. Högberg, F. Graf, W.M. Shih, Self-Assembly of DNA into Nanoscale Three-Dimensional Shapes. *Nature* **459**, 414–418 (2009).
5.49 Y. He, T. Ye, M. Su, C. Zhang, A.E. Ribbe, W. Jiang, C.D. Mao, Hierarchical Self-Assembly of DNA into Symmetric Supramolecular Polyhedra. *Nature* **452**, 196–202 (2008).
5.50 B. Yurke, A.J. Turberfield, A.P. Mills, Jr., F.C. Simmel, J.L. Newmann, A DNA-Fuelled Molecular Machine Made of DNA. *Nature* **406**, 605–608 (2000).
5.51 H. Yan, X. Zhang, Z. Shen, N.C. Seeman, A Robust DNA Mechanical Device Controlled by Hybridization Topology. *Nature* **415**, 62–65 (2002).
5.52 W.B. Sherman, N.C. Seeman, A Precisely Controlled DNA Bipedal Walking Device. *Nano Letters* **4**, 1203–1207 (2004).
5.53 T. Omabegho, R. Sha, N.C. Seeman, A Bipedal DNA Brownian Motor with Coordinated Legs. *Science* **324**, 67–71 (2009).
5.54 J.-S. Shin, N.A. Pierce, A Synthetic DNA Walker for Molecular Transport. *J. Am. Chem. Soc.* **126**, 10834–10835 (2004).
5.55 H. Gu, J. Chao, S.J. Xiao, N.C. Seeman, A Proximity-Based Programmable DNA Nanoscale Assembly Line. *Nature* **465**, 202–205 (2010).
5.56 K. Lund, A.J. Manzo, N. Dabby, N. Michelotti, A. Johnson-Buck, J. Nangreave, S. Taylor, R.J. Pei, M.N. Stojanovic, N.G. Walter, E. Winfree, H. Yan, Molecular Robots Guided by Prescriptive Landscapes. *Nature* **465**, 206–210 (2010).

5.57 S. Liao, N.C. Seeman, Translation of DNA Signals into Polymer Assembly Instructions. *Science* **306**, 2072–2074 (2004).

5.58 R. Sha, F. Liu, D.P. Millar, N.C. Seeman, Atomic Force Microscopy of Parallel DNA Branched Junction Arrays. *Chem. Biol.* **7**, 743–751 (2000).

5.59 D.N. Selmi, R.J. Adamson, H. Attrill, A.D. Goddard, R.J.C. Gilbert, A. Watts, A.J. Turberfield, DNA-Templated Protein Arrays for Single-Molecule Imaging. *Nano Letters* **11**, 657–660 (2011).

5.60 J. Zheng, J.J. Birktoft, Y. Chen, T. Wang, R. Sha, P.E. Constantinou, S.L. Ginell, C. Mao, N.C. Seeman, From Molecular to Macroscopic via the Rational Design of a Self-Assembled 3D DNA Crystal. *Nature* **461**, 74–77 (2009).

5.61 T. Wang, R. Sha, J.J. Birktoft, J. Zheng, C. Mao, N.C. Seeman, A DNA Crystal Designed to Contain Two Molecules per Asymmetric Unit. *J. Am. Chem. Soc.* **132**, 15471–15473 (2010).

5.62 L.M. Adleman, Molecular Computation of Solutions to Combinatorial Problems. *Science* **266**, 1021–1024 (1994).

5.63 E. Winfree, Algorithmic Self-Assembly of DNA, Theoretical Motivations and 2D Assembly Experiments. In *Proc. 11th Conversation in Biomolecular Stereodynamics*, ed. R.H. Sarma, M.H. Sarma, New York, Adenine Press, pp. 263–270 (2000).

5.64 C. Mao, T. LaBean, J.H. Reif, N.C. Seeman, Logical Computation Using Algorithmic Self-Assembly of DNA Triple Crossover Molecules. *Nature* **407**, 493–496 (2000); *Erratum: Nature* 408, 750–750 (2000).

5.65 G. Wu, N. Jonoska, N.C. Seeman, Construction of a DNA Nano-Object Directly Demonstrates Computation. *Biosystems* **98**, 80–84 (2009).

5.66 P.W.K. Rothemund, N. Papadakis, E. Winfree, Algorithmic Assembly of DNA Sierpinski Triangles. *PLOS Biol.* **2**, 2041–2054 (2004).

6

A short historical interlude: the search for robust DNA motifs

So far, we have been talking about DNA branched motifs and the various things we could make from them, often by holding them together with sticky ends. Here, we want to talk about motifs and cohesion that work, to a pretty good approximation, in the same way that structures and strong glues work on the macroscopic scale. It may seem that there is almost no need for this chapter, but in fact the lack of robust motifs was the major stumbling block to building periodic arrays and DNA nanomechanical devices in the late 1980s and much of the 1990s. It's worth thinking about. Just to be clear about what we mean, we'll define a rigid component as one that can specify the vectors of DNA double helix axes (and hence the angles between them) within limits of flexibility no greater than those of linear duplex DNA.

Periodic arrays need to be made from components that are fairly rigid. Anyone familiar with crystals knows that the structures derived from them are produced by summing up Fourier series, where the amplitudes are experimentally available and the phases are derived by a variety of methods. Fourier series are periodic sinusoidal functions, sines and cosines. These are functions that correspond to projections of the radius of a circle as it traverses the circular trajectory. It is clear from this relationship that designs that are aimed at making periodic functions must be prevented from cyclizing, primarily by rigidity, so that the cycles do not poison the growth of the lattice. In a similar fashion, robust nanomechanical devices function like their analogs on the macroscopic scale, by changing structural states without undergoing major deformations or multimerization or breakdown as a consequence of thermal noise. Rigidity is a requirement for robustness in nanomechanical devices, although it is possible and sometimes useful to make devices that are not robust.[6.1]

The need to discover robust motifs was apparent fairly early in the history of this field. A 3-arm junction with sticky ends was designed and purchased. The

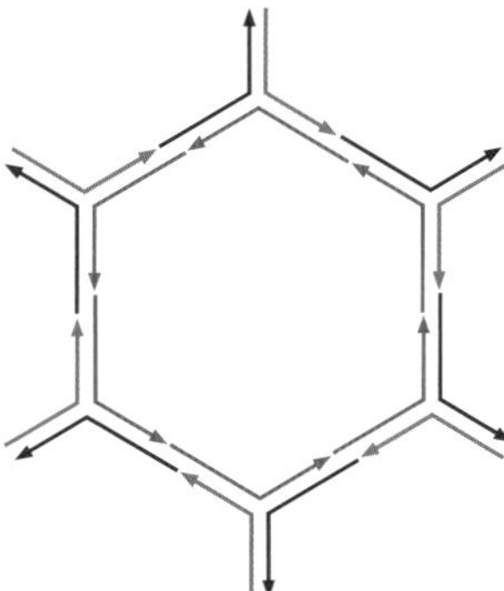

Figure 6-1 *An unrealized ideal for the 3-arm junction.* A 3-arm junction is shown with red and blue strands containing complementary sticky ends. Oligomerizing a 3-arm junction with a single sticky-end pair certainly does not result in the hexagonal arrangement drawn.

naïve notion behind this junction was that six 3-arm junctions would self-assemble to look like a hexagon, as shown in Figure 6-1. There were three different strands, shown as red, blue, and green. The blue–green arm was blunt and unphosphorylated, but the blue–red and green–red arms contained complementary sticky ends. Both the red and the green strands were phosphorylated, so that ligation could occur.[6.2] There are two turns between the vertices of the hexagon, so the red strand behaves as a reporter strand. When ligated together, it was believed that a hexamer of the red strand would be the dominant product. Figure 6-2 shows the details of a ligation-closure experiment that was performed using reporter strands, and demonstrates that this expectation was not met at all. The 3-arm junction appears to be sufficiently flexible that any cycle from trimer to hexamer can be obtained. The hexamer should not be taken as an upper bound; ligation of branched molecules is so inefficient that it is hard to get more molecules to be ligated into a cyclic arrangement, although longer molecules have been reported.[6.3,6.4]

Bulged 3-arm junctions[6.5] are somewhat less flexible than simple 3-arm junctions. An extra two thymidines on one of the strands allows the two arms 3' to that strand to stack on one another. This may be a sequence-specific effect, with different arms doing the stacking.[6.6] The impact of a TT bulge is shown in Figure 6-3. The percentage of cyclization through each of the strands is shown as two numbers, with and without a bulge.[6.4] The results are not overwhelming, but the 78% linearity of strand 1 was strong enough to suggest another ligation-closure experiment be tried to see if triangles made from bulged 3-arm junctions were rigid. The design of the bulged triangle ligation-closure experiment is shown in Figure 6-4.

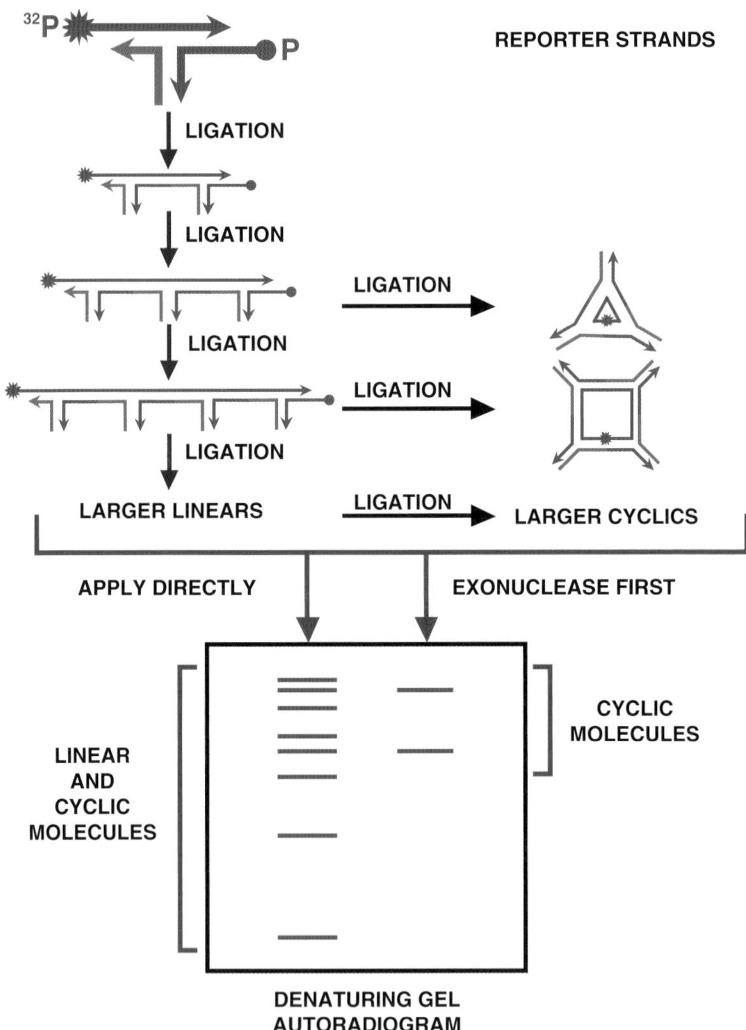

Figure 6-2 *The results of oligomerizing 3-arm junctions.* The blue strand contains a phosphate group, and the red strand contains a phosphate group containing radioactive ^{32}P. When oligomerized, the red strand acts as a reporter strand, indicating the structure from which it is derived, linear or cyclic, as indicated. When applied to a denaturing gel, only the radioactive strand is visible. If applied directly (left side of the schematic gel), all bands are visible, both linear and cyclic. However, if first treated with exonuclease, only the cyclic strands survive, so the different components can be distinguished.

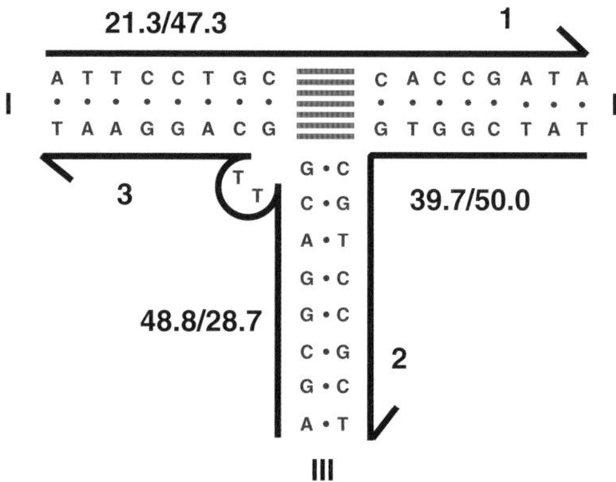

Figure 6-3 *Cyclization percentages in bulged 3-arm branched junctions.* Strand 1 cyclizes only 21.3% of the time when the bulge is in strand 3, vs. 47.3% when the bulge is lacking. This suggests that the addition of 12 backbone bonds to strand 3 allows domain I to stack on domain II.[6.4]

Figure 6-4a shows that two different kinds of bulged DNA triangles were used in the experiment. The I-type triangles (blue and left) have the bulges on their inner strands, and the O-type triangles (red and right) have the bulges on their outer strands. Figure 6-4b shows a strand representation of these two triangles. The need for a reporter strand to assay the products means that it is necessary to alternate these triangles. This reporter strand is shown in red in Figure 6-4b, and also in Figure 6-4c, which indicates the target of assembling these triangles, a hexagonal arrangement of triangles. Thus, if these triangles were to form the type of robust structure that was sought, a circular molecule with the length corresponding to the trajectory shown in red (a cyclic hexamer) would be the result of the experiment. Unfortunately, once the catenated reporter strand was isolated, its length was found to correspond to a tetramer. Thus, the bulged-junction triangle system did not satisfy the requirements of a rigid component.[6.7]

This result sounds somewhat discouraging, but there was a backup plan available were the bulged triangles to fail, as they did. The 80% stacking of two arms of the bulged junction was clearly not a strong basis on which to predicate robust motifs. However, the notion presented itself that perhaps this system could be backed up by reinforcing the stacking arms with another helix adjacent to it. In other words, make a hybrid of a bulged 3-arm junction and a double-crossover (DX) molecule (Chapter 3). Thus, the DX+J (shown in Figure 5-10 as a DAE+J motif) might provide the necessary stiffness needed to make robust branched DNA motifs. Clearly, it was of interest to examine the

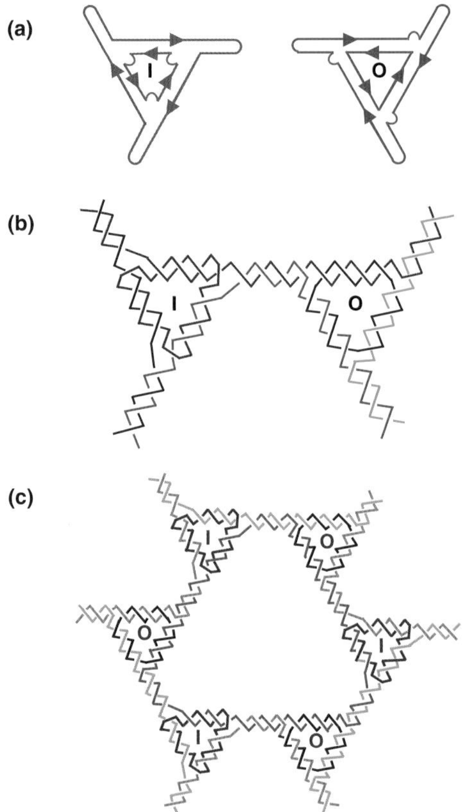

Figure 6-4 *Bulged-junction triangles built to yield hexagons.* (a) *Simple ladder-like representations of the triangles.* Note that in the triangle labeled I, the bulges are on the inner strand; they are on the outside strand in the triangle labeled O. (b) *Helical representation of triangle I ligated to triangle O.* Note that the red strand is a reporter strand. (c) *The unrealized goal of bulged-junction triangles ligated to form a hexagon.* The alternation of the I and O triangles ought to produce a hexagon, if the bulged junctions are sufficiently rigid. The data do not support this notion. Reprinted by permission from the American Chemical Society.[6.7]

stiffness properties of the DX motif by itself. This too proved to be rigid, providing the basis for both periodic matter designed from DNA and robust DNA-based nanomechanical devices.[6.3]

The experiments that demonstrated the stiffness of both the DX and the DX+J motifs are shown in Figures 6-5 and 6-6, respectively. These gels are autoradiograms showing ligation-closure experiments, where the reporter strand of a DAE molecule has been labeled. There are several components to each of these gels, with the most important lanes in their center sections. Let's

Lanes	1	2	3	4	5	6	7	8
	JY 21		DX				DX with nicks	
Ligase	+	+	+	+	+	+	+	+
Hha I	-	-	-	-	+	+	-	-
Exo I + III	-	+	-	+	-	+	-	+

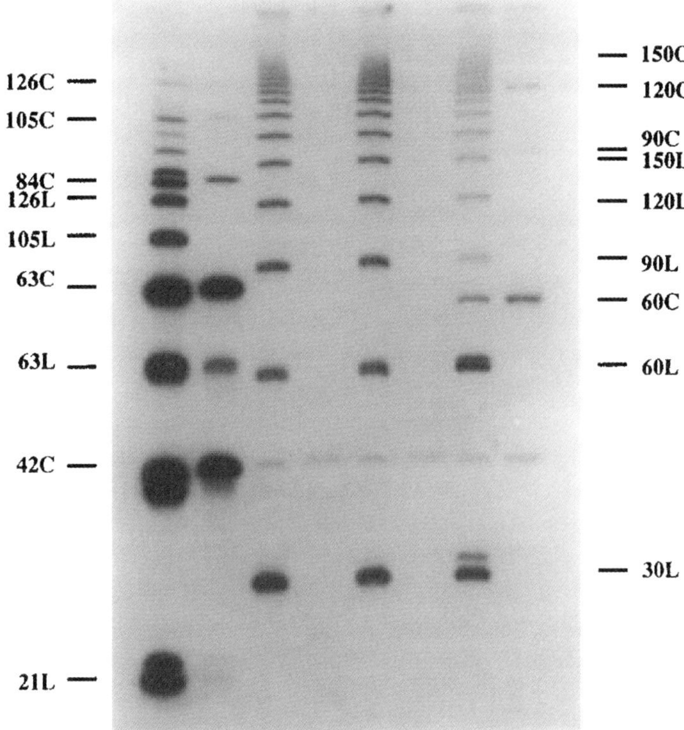

Figure 6-5 *Denaturing gel of the ligation of DX molecules.* Bands labeled L are linear and bands labeled C are cyclic. The lanes containing JY21 act as markers, demonstrating how this conventional 3-arm branch junction cyclizes readily. As noted in Figure 6-2, treatment with exonucleases reveals which bands are cyclic. There are no cyclic molecules in the ligated DX lanes. Hha I digestion is needed to simplify the complex catenanes that would result from cyclization, were any present. Note the cyclic fiducial marker of length 42 that has been added to each lane to make sure that no error has been committed in the exonuclease digestion of the DX molecules. At right, unligated DX molecules cyclize somewhat, as expected. Reprinted by permission from the American Chemical Society.[5.17]

Lanes	1	2	3	4	5	6	7	8
	JY 21		DX + Junction				DX with nicks	
Ligase	+	+	+	+	+	+	+	+
Hha I	-	-	-	-	+	+	-	-
Exo I + III	-	+	-	+	-	+	-	+

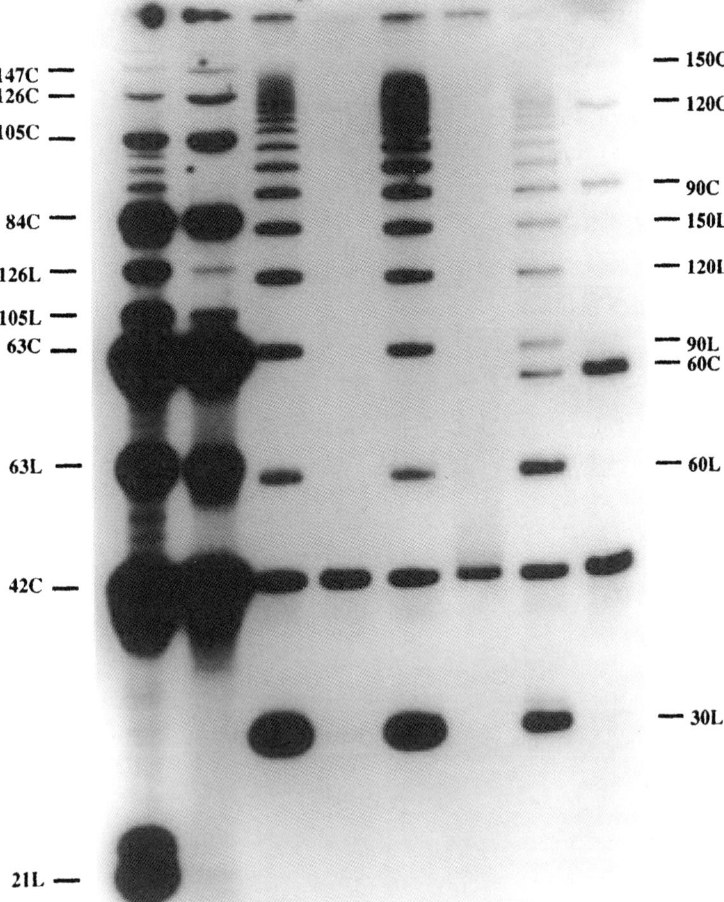

Figure 6-6 *Denaturing gel of the ligation of DX+J molecules.* The conventions of Figure 6-5 pertain to this gel as well. The results are similar, although the intensity of the fiducial band of size 42C is much stronger. Reprinted by permission from the American Chemical Society.[5.17]

look at Figure 6-5 first. On the far left is a junction like the one shown in Figure 6-1, a 3-arm branched junction. The first lane of this section contains a complete ligation series, including cyclic and linear molecules. The second lane contains the same materials, but they have been digested by exonucleases, so the only bands left derive from circular molecules. You can see that quite a bit of the material has cyclized. The central section of the gel contains the DX molecule, with a label in the reporter strand. Lanes 3 and 5 show the ligation reactions. True cyclic molecules in this system are complex catenanes, so they require release by restriction enzyme Hha I (lane 5). Lanes 4 and 6 show the results of treating these materials with exonucleases, to release cyclic materials. It is clear that virtually no cyclic materials are present, attesting to the rigidity of the DX molecules. Lanes 7 and 8 show the same DAE molecule, but it has been nicked twice in the domain that does not contain the reporter strand; thus it is now just a linear DNA molecule containing two successive junctions. A series of cyclic bands is present. A key component of this gel is the faint band of a 42-mer cyclic molecule. Equal amounts of this molecule have been added to each lane so that we can know that equal amounts of material have added to the lanes containing cyclic material. This shows that there is not just a loading error when no other bands are seen in these lanes.

Figure 6-6 is the analogous gel for a DAE+J molecule. The left and right portions of the gel are the same as in Figure 6-5. The central part of the gel now contains ligation and/or digestion of the DAE+J molecule, rather than treatment of a DAE molecule. Again, it is seen that virtually no cyclic molecules are formed by the ligation reaction. The 42C marker band is much more prominent here than in Figure 6-5, just because the sample was more heavily labeled. From the combination of Figures 6-5 and 6-6, we know that DX molecules are robust molecules, molecules that can be treated as fairly rigid motifs. Later analysis of the DAE molecule has shown that its persistence length, a measure of its stiffness, is roughly twice that of linear duplex DNA.[6.8]

The discovery that the DX molecule was a rigid motif was a key breakpoint in the history of structural DNA nanotechnology. To this writer's knowledge, it was the last time that rigid motifs were sought by gels reporting ligation-closure experiments. Once the DX and DX+J motifs were found to be robust, serious attempts were made to make 2D arrays. The analysis of 2D arrays has been performed almost exclusively by means of atomic force microscopy (AFM). From then on, it was easier to try to make a 1D or 2D array from a new motif and to examine it by AFM to see whether it behaved in a robust fashion. Through such experiments, many new motifs have been discovered. Many are related to the DX motif, such as the TX motif (three parallel helix axes),[6.9] the PATX motif,[6.10] the 6-helix bundle (6HB),[6.11] the paranemic crossover motif (PX),[6.12] and various other motifs that are characterized by the presence of helix axes that fused into a parallel array. We shall talk about these and others in Chapter 7. We are now ready to tackle the self-assembly of periodic material to an acceptably high resolution.

References

6.1 B. Yurke, A.J. Turberfield, A.P. Mills, Jr., F.C. Simmel, J.L. Newmann, A DNA-Fuelled Molecular Machine Made of DNA. *Nature* **406**, 605–608 (2000).

6.2 R.-I. Ma, N.R. Kallenbach, R.D. Sheardy, M.L. Petrillo, N.C. Seeman, 3-Arm Nucleic Acid Junctions Are Flexible. *Nucl. Acids Res.* **14**, 9745–9753 (1986).

6.3 X. Li, X. Yang, J. Qi, N.C. Seeman, Antiparallel DNA Double Crossover Molecules as Components for Nanoconstruction. *J. Am. Chem. Soc.* **118**, 6131–6140 (1996).

6.4 B. Liu, N.B. Leontis, N.C. Seeman, Bulged 3-Arm DNA Branched Junctions as Components for Nanoconstruction. *Nanobiology* **3**, 177–188 (1994).

6.5 N.B. Leontis, W. Kwok, J.S. Newman, Stability and Structure of 3-Way DNA Junctions Containing Unpaired Nucleotides. *Nucl. Acids Res.* **19**, 759–766 (1991).

6.6 M.S. Yang, D.P. Millar, Conformational Flexibility of Three-Way DNA Junctions Containing Unpaired Nucleotides. *Biochem.* **35**, 7959–7967 (1996).

6.7 J. Qi, X. Li, X. Yang, N.C. Seeman, The Ligation of Triangles Built from Bulged Three-Arm DNA Branched Junctions. *J. Am. Chem. Soc.* **118**, 6121–6130 (1996).

6.8 P. Sa-Ardyen, A.V. Vologodskii, N.C. Seeman, The Flexibility of DNA Double Crossover Molecules. *Biophys. J.* **84**, 3829–3837 (2003).

6.9 T. LaBean, H. Yan, J. Kopatsch, F. Liu, E. Winfree, J.H. Reif, N.C. Seeman, The Construction, Analysis, Ligation and Self-Assembly of DNA Triple Crossover Complexes. *J. Am. Chem. Soc.* **122**, 1848–1860 (2000).

6.10 W. Liu, X. Wang, T. Wang, R. Sha, N.C. Seeman, A PX DNA Triangle Oligomerized Using a Novel Three-Domain Motif. *Nano Letters* **8**, 317–322 (2008).

6.11 F. Mathieu, S. Liao, C. Mao, J. Kopatsch, T. Wang, N.C. Seeman, Six-Helix Bundles Designed from DNA. *Nano Letters* **5**, 661–665 (2005).

6.12 T. Wang, Ph.D. thesis, New York University (2007).

7
Combining DNA motifs into larger multi-component constructs

With the preamble of the previous chapters, we are now ready to talk about making things from DNA that are larger than individual units like the knots or catenanes or polyhedra discussed in Chapters 3 and 4. The initial goal of structural DNA nanotechnology was to self-assemble crystals, usually thought of as repeating units in three dimensions. However, the notion of periodicity is not limited to 3D; it can also apply to 2D and 1D. In fact one of the most influential ideas of the early twentieth century was Schrödinger's discussion of Delbrück's suggestion that genetic material, thought correctly at the time to be one-dimensional, was an aperiodic crystal.[7.1] As we all know, that's a pretty good description of the naturally occurring DNA double helix, remarkably good for the early 1940s. In this chapter, we'll talk about one-dimensional, two-dimensional, and three-dimensional arrangements of DNA motifs. It's a little hard to separate the different types of crystals, so we will discuss them when particular points relevant to those dimensions arise, rather than separating topics by dimensionality. For example, you might think of tubular structures as one-dimensional or two-dimensional, depending on whether you are thinking about their tubular axes or about the surface arrangement of their components.

As you might imagine from all the discussion in Chapter 4, topology comes to play a major role when we concatenate DNA motifs. The basic quantum of DNA topology is the half-turn of double helical DNA; we emphasized this point when we discussed converting knot structures to single-stranded DNA constructs. Sometimes we call the half-turn a "unit tangle." Figure 7-1 is a slightly larger version of Figure 1-8, which showed four 4-arm junctions being assembled into a quadrilateral, with lots of sticky ends left over to make something bigger. Besides the number of junctions, the difference in Figure 7-1 is that the twisting of the DNA duplex is included in the picture. Figure 7-1a shows what the structure looks like when there is an even number

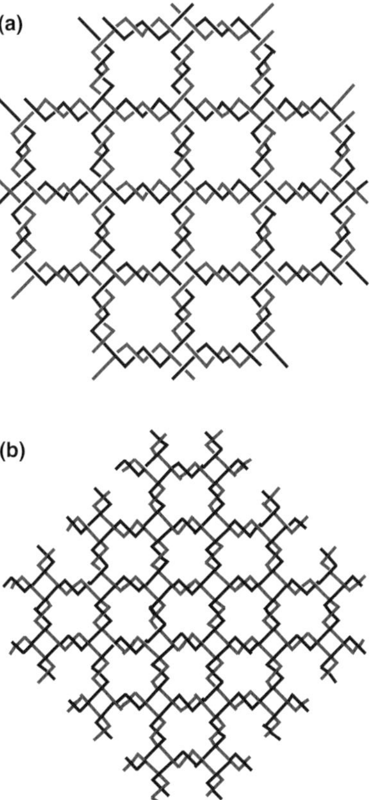

Figure 7-1 *Topological consequences of DNA constructs.* (a) *A square-like array where the vertices are separated by an even number of double helical half-turns.* The resulting 2D lattice is a chain-mail-like arrangement where the red strands are linked to the blue strands. The separation of vertices is two turns, so each edge represents two links. (b) *The same array where the vertices are separated by an odd number of double helical half-turns.* In this case, the red and blue strands form an interwoven pattern, with the blue strands extending from lower left to upper right, while the red strands extend from lower right to upper left.

of double helical half-turns (in this case four) between vertices. The resulting structure is akin to chain mail, with a red circle linked twice (because there are two turns between vertices) to each of four blue strands that flank it; likewise, the blue circles are linked twice to each of the four red strands that flank them. Figure 7-1b shows what the structure looks like when there is an odd number of half-turns (three in this case) between vertices. What we see is a woven pattern, where the blue strands all go from lower left to upper right, while the red strands all go from lower right to upper left. This is an unusual type of weave, because all the nodes have the same sign (negative).

The first reported construction containing multiple components was a 2D array made from DX tiles.[7.2] We discussed this structure briefly in Chapter 5, when we described the use of AFM in structural DNA nanotechnology. Let's look a little more closely at these assemblies, shown in Figure 5-13. The schematic drawing there does not really describe the 2D DX arrangement in sufficient detail. There are five possible DX arrangements, but here we are only talking about DX molecules where the crossovers take place between strands of opposite polarity, called DAE and DAO molecules. As shown in Figure 3-8, there are three kinds of DX molecules where the crossovers take place between strands of the same polarity, but they are not usually used for nanoconstruction, because of the multimerization shown in the gel in Figure 5-6.

As seen in Figures 3-8 and 5-10, DAE molecules and DAO molecules differ from each other by the number of half-turns between crossovers. Those with an even number of half-turns are termed DAE molecules and those with an odd number of half-turns are termed DAO molecules. However, when making a 2D arrangement, we also have to think about whether there are an even or odd number of half-turns between crossovers in adjacent units. Thus, there are four possible topologies for the arrangement shown schematically in Figure 5-13. These different topologies are shown in Figure 7-2; thus, there is an arrangement that comes from separating DAE molecules by an even number of half-turns (DAE-E) and one that comes from separating DAE molecules by an odd number of half-turns (DAE-O). Likewise, there is an arrangement in which DAO molecules are separated by an even number of half-turns (DAO-E) and one where DAO molecules are separated by an odd number of half-turns (DAO-O). Don't forget that we are really talking about separating the crossovers in adjacent molecules by this number of half-turns, not the molecules themselves.

The DAE-E arrangement is characterized, in principle, by a series of cyclic molecules, such as the magenta molecules at the centers of the DAE molecules, and the red, green, light-blue, and dark-blue cycles (only the red and dark-blue cycles are shown complete in Figure 7-2). The DAE-O system is quite different. In addition to the small magenta and purple cycles, there are infinite zigzag light-blue and green strands oriented vertically in the drawing, and infinite red and dark-blue strands oriented horizontally in the drawing. The red and dark-blue strands are analogous to the reporter strands discussed in Chapters 5 and 6, although the light-blue and green strands could, in principle, serve the same role. The interlacings of the strands in the DAO-E and DAO-O systems are

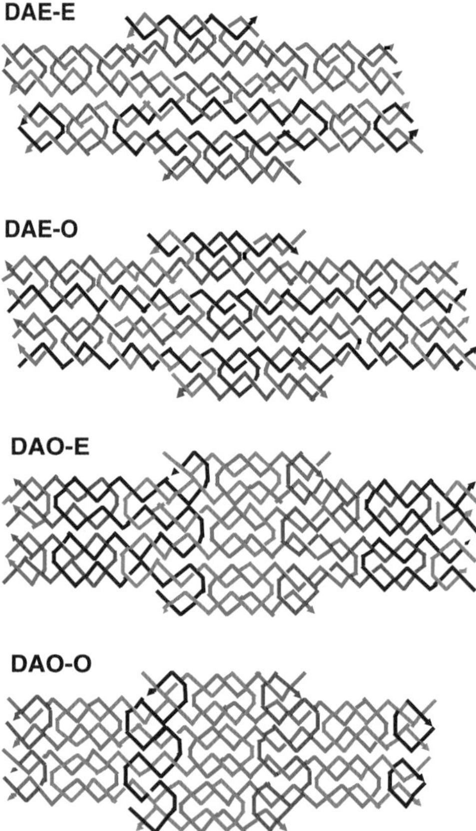

Figure 7-2 *Two-dimensional DX lattices.* These drawings represent DX molecules joined into a 2D lattice with separations of even or odd numbers of half-turns. From top to bottom are shown DAE molecules separated by even numbers of half-turns (DAE-E) and then odd numbers of half-turns (DAE-O), DAO molecules separated by even numbers of half-turns (DAO-E) and then odd numbers of half-turns (DAO-O). DAE-E contains only cyclic strands, but DAE-O contains both horizontal and vertical reporter strands, although the vertical reporters are zigzag strands. Both DAO lattices contain zigzag vertical reporter strands, but no horizontal reporter strands.[5.43]

somewhat different, but their structures are similar: infinite strands zigzagging vertically in each arrangement.

At the time that the 2D construction of DX molecules was first contemplated, it was not clear how the assembly would be demonstrated. The two groups working together on this project chose different lattices, but it was clear that one of them should be the DAE-O lattice, because, in principle, it provided reporter strands in both directions. As a practical matter, it is not possible to ligate every nick in a 2D lattice, so the reporter strands were regarded as potential supplementary support. The key data had to be provided by direct visualization, which turned out to be from atomic force microscopy.[7.3] It is useful to have a marker to know that such an assembly is happening as designed. As discussed in Chapter 5 (see Figure 5-13), it is possible to design the assemblies with features that are spaced differently, so as to demonstrate that the array is forming according to the design.

The microscopy caveat. The key rule of all microscopy (AFM, TEM, dissecting scopes, etc.) seems to be that if you seek a particular image, you will find it. This means that all imaging must be buttressed by controls, statistics, and challenges. An example of challenges to a 1D construction is shown in Figure 7-3.[7.3] Three experiments are shown, from top to bottom. Figure 7-3a shows a 3-turn DNA triangle, one of whose edges is a DX molecule. The domain including red strands contains 4.5 turns of DNA from sticky end to sticky end. The long strand seen in the 1030 nm image is very unlikely with the 4-nucleotide sticky end used in this construction in the absence of ligation. The ligation was aided a lot because the sticky end was generated by restriction, leading to complete phosphorylation of the 5′ termini. When these molecules are ligated together, one expects a repeat of 9 turns (~32 nm) of DNA from triangle point to triangle point on the same side of the continuous duplex. The AFM image on the right with the 188 nm zoom shows about six points in each direction, about what is expected. That's good news, but it just can't be believed without further evidence. Figure 7-3b shows a second experiment, where a DAE-type DX molecule is inserted between the triangles. This molecule contains 4 turns of DNA between its sticky ends, so the unit of the DX molecule and the triangle consists of 8.5 turns of DNA, and the repeat on one side of the ligated image should be around 17 turns (60.5 nm), nearly double the repeat in the first experiment. The 188 nm zoom here shows a repeat of roughly that length. As a further challenge to the first result, Figure 7-3c shows a variation on the experiment in Figure 7-3b. The separation of the sticky ends in the DX molecule is now 4.5 turns, leading to all the triangles being limited to one side of the 1D array, with a separation similar to the first experiment, about 32 nm. This result is seen clearly in the 188 nm zoom at the right of the panel.

Modifications of 2D self-assembled arrays. Patterns derived from arrays that contain the DX+J motif can be modified in various ways.[7.4] Figure 7-4 shows an array similar to that in Figure 5-13. However, in this case there are two DX+J motifs, B′ and D*. The filled circles representing the hairpins of the DX+J components have been colored differently, to indicate differences in their

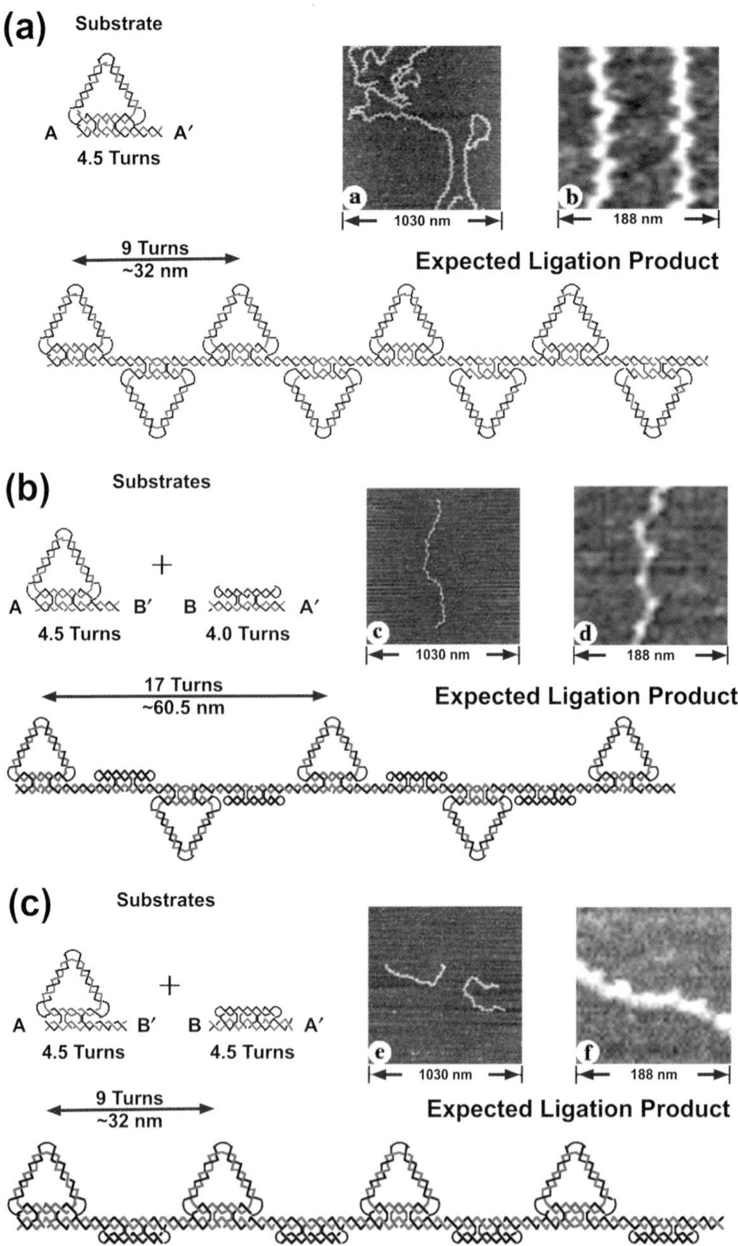

Figure 7-3 *Ligation of DNA triangles containing a DX domain.* (a) *Only the triangle is ligated.* Sticky ends A and its complement A′ are separated by 4.5 turns, leading to a zigzag pattern. At right, a zigzag molecule is visible and the zoom reveals that the points of the triangles are separated by the expected 9-turn separation of ~32 nm. (b) *A triangle and a DX molecule are ligated in an alternating pattern.* The DX adds 4 turns to the separation, maintaining the zigzag pattern but almost doubling the separation. At right, the zoom reveals that the separation of points is nearly double. (c) *The triangle and a 4.5-turn DX are ligated in an alternating pattern.* The extra half-turn in the DX now eliminates the zigzag pattern, and puts all the triangles on the same side of the continuous shaft. This is visible at right, where the original separation is seen, but one side is smooth. Reprinted by permission from the American Chemical Society.[5.44]

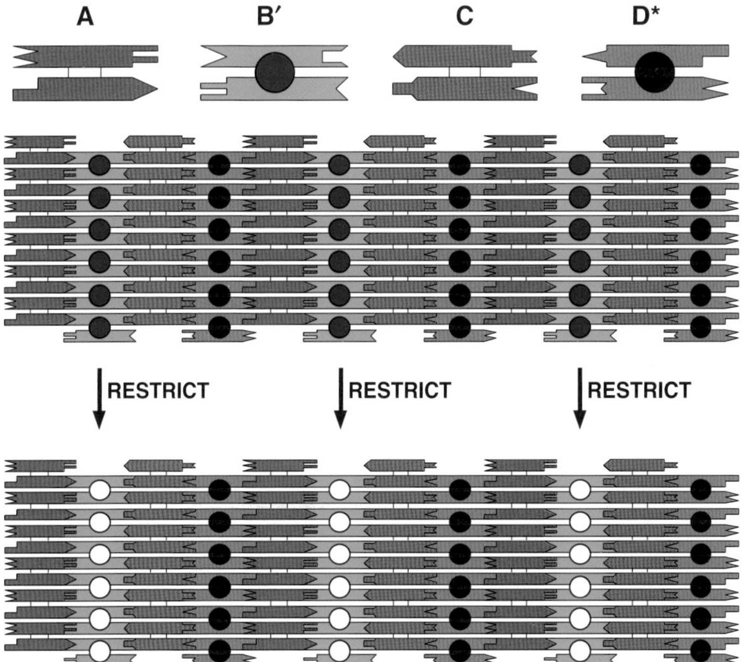

Figure 7-4 *Modification of the surface features of a 2D DNA array by restriction.* The DX and DX+J tiles in this array are 16 nm wide, leading to a stripe every 32 nm. The blue feature in the B′ DX+J tile contains a restriction site that is not present in the D* tile. When treated with the restriction enzyme, the blue feature is removed, leading to an array with a 64 nm stripe. Reprinted by permission from the American Chemical Society.[7.4]

sequences. The blue hairpin of B′ contains a restriction site, but the black hairpin of D* lacks this restriction site. The separation of the stripes generated by the blue and black hairpins is 32 nm. However, if we treat the array with the restriction enzyme that cleaves the blue hairpin, the resulting pattern has separations of 64 nm, because the blue hairpins have been removed.

Figure 7-5 shows that we can perform the inverse operation. In this diagram, the A, B, and D* tiles are the same as the ones in Figure 7-4. However, the B° tile lacks enough of a hairpin to produce a pattern; it is only a short duplex tailed by a sticky end, denoted by a white circle. The pattern is thus a set of stripes separated by 64 nanometers. It is possible to add a longer hairpin fragment and convert this pattern to a 32-nanometer stripe. This can be done if the sticky end contains about six nucleotides. However, if we want only to anneal the additional hairpin, without ligation, the sticky end should be about 12 nucleotides long.

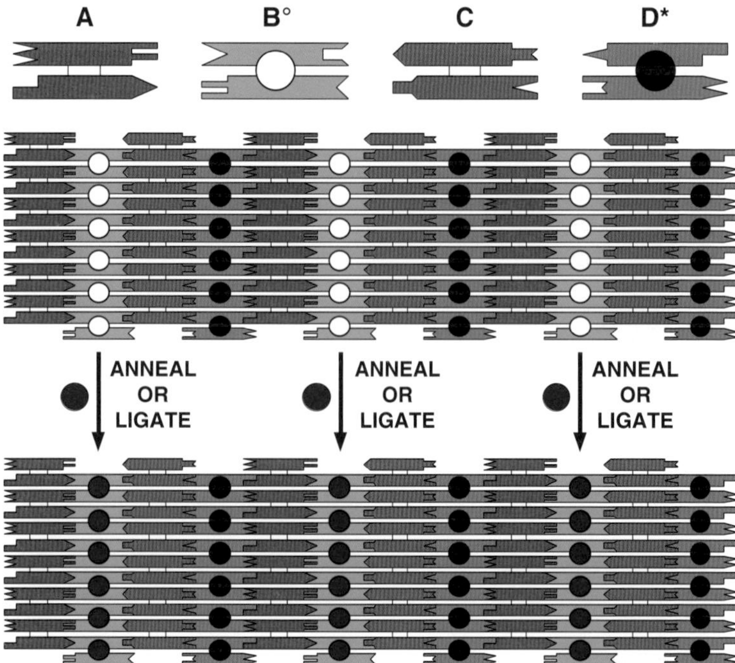

Figure 7-5 *Modification of the surface features of a 2D DNA array by ligation or annealing.* This image shows the opposite modification to that in Figure 7-4. Starting with a 64 nm stripe, the array is converted to one with a 32 nm stripe. The white circles represent a sticky end to which a hairpin (represented by a blue circle) is ligated, or just annealed non-covalently. Reprinted by permission from the American Chemical Society.[7.4]

The pattern-generating feature does not need to be something as passive as a simple hairpin. It can also be something more sophisticated, like a deoxyribozyme (DNAzyme).[7.5] Figure 7-6 shows a diagram similar to Figures 7-4 and 7-5. However, the $\hat{B}$ tile now contains a DNAzyme that is represented by the green filled circle. This DNAzyme autocatalyzes its own cleavage in the presence of Cu^{2+}. In the absence of Cu^{2+} cations, the pattern is the 32 nm stripe we have discussed before. When Cu^{2+} is added to the solution, the pattern changes to the 64 nm stripe. This conversion is shown in AFM images in Figure 7-7, where the white scale bars represent 100 nm.

Adding TX molecules to the system. The construction of 2D lattices from DX components presaged the construction of other lattices made from similar motifs. For example, the TX molecule (Figure 3-7) was also used as the basis for a 2D lattice.[7.6] However, the TX (and wider molecules) need not be connected to fill all the space. Figure 7-8 shows an example of this, where

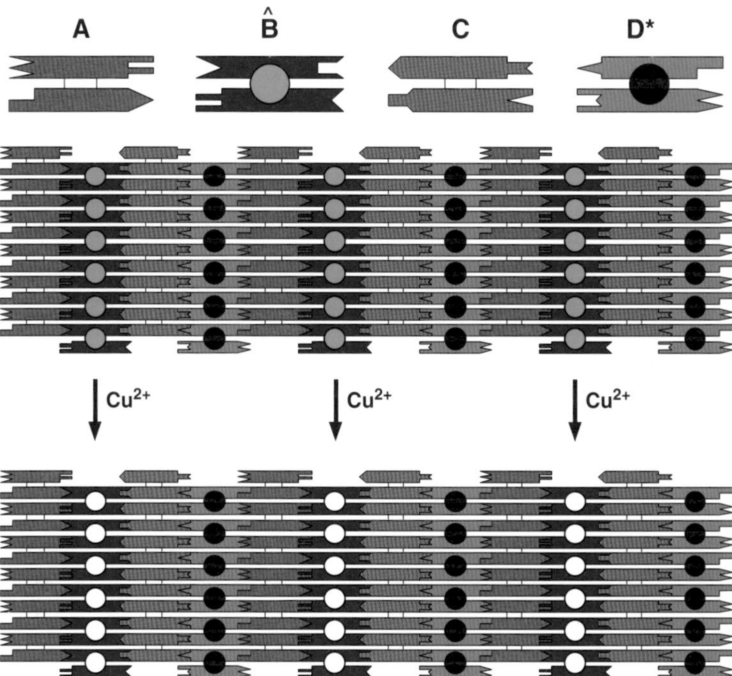

Figure 7-6 *Modification of a DNAzyme-containing array by auto-digestion.* In this case, the green circle represents an auto-digesting DNAzyme whose presence leads to a stripe. In the presence of Cu^{2+} cations, the feature auto-digests, leaving only the hairpin on the D* tile, and the stripe spacing decreases from 32 to 64 nm, as in Figure 7-4. Reprinted with permission from the American Chemical Society.[7.5]

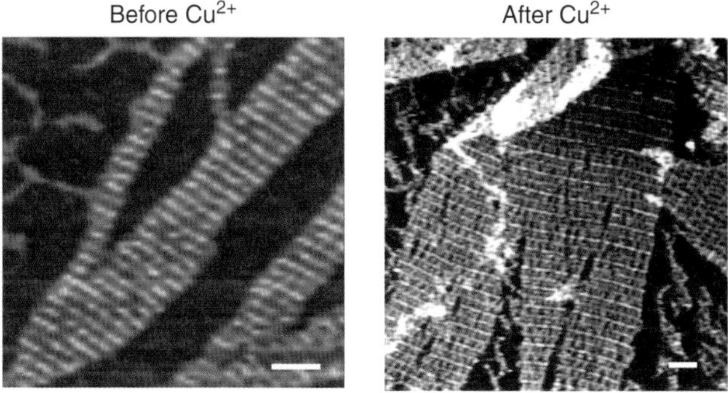

Figure 7-7 *AFM images of the modification shown in Figure 7-6.* The left panel shows the 32-nm-spacing stripe present before addition of Cu^{2+}, and the right panel shows the 64-nm-spacing stripe present afterwards. Reprinted with permission from the American Chemical Society.[7.5]

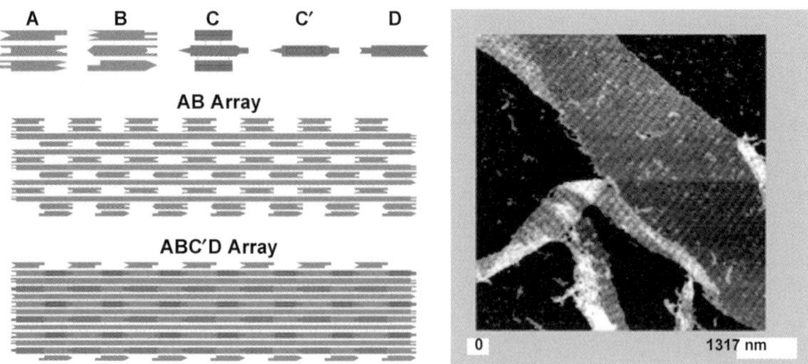

Figure 7-8 *A 2D TX array with four components.* The four components are shown at the top: the gold and blue TX molecules, A and B respectively, that form a 2D array AB with gaps as shown in the middle; the blue-and-red C tile that is re-phased to be nearly perpendicular to the plane formed by A and B and binds to the middle domain of the gold A tiles; and a pink duplex, D, that binds to the middle domain of the blue B tiles, filling the gap there. The C tile is called C′ when it is rotated. The ABC′D array is shown at the bottom. At right is an AFM image of the ABC′D array, showing the striped features caused by the C′ tiles. Reprinted by permission from the American Chemical Society.[5.22]

the AB array is connected in a 1–3 fashion. This leaves open spaces, so that it is possible to add more components to the arrangement. Two different components have been added to the ABC′D array. D is a simple duplex, which is inserted between the middle domains of the B tiles. The C tile is another TX tile, so it can't be simply inserted between the A tiles in the same way. However, if it is re-phased by three nucleotides, about 102°, it can be fit into that slot. This rotated tile is what is meant by the C′ designation. The AFM pattern that results from this 4-tile array is shown on the right of the figure.

We didn't really discuss it earlier, but Figure 1-8 is a simplification and an idealization. In fact, the 4-arm junction does not really look like the simple crossroads we drew there, except in the absence of Mg^{2+} cations.[7.7] Figure 7-9 tells the real story of the 4-arm junction in the presence of Mg^{2+}, and it reminds us of an important lesson. The figure at the upper left shows a 4-arm junction as we have been idealizing it. The strands are indicated by Arabic numerals, the arms are labeled by Roman numerals, and the arrowheads represent the 3′ ends of the strands. A nominal four-fold symmetry symbol is at the center of the junction. The top middle panel closes the ends of the arms and selects colors that will represent the ultimate stacking pattern, red and blue. The top right panel rotates the image, and shows the way that the structure is going to close, thereby destroying the four-fold symmetry. The lower panels show the actual structure, and the remaining symmetry of the system. They also show us two

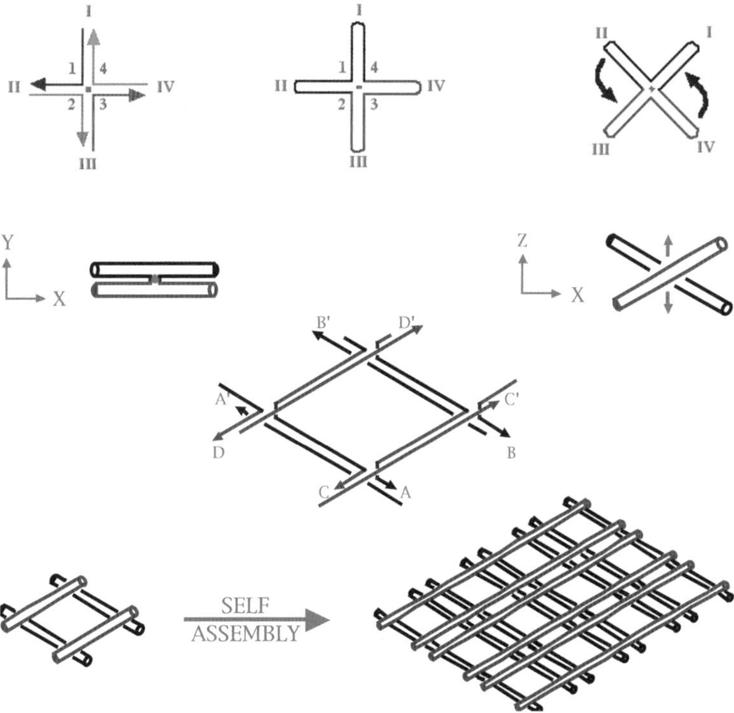

Figure 7-9 *An array formed from 4-arm junction tiles.* The top row shows, from left to right, a 4-arm junction, its domain structure (two blue domains and two red domains), and the way it stacks, as indicated by arrows, to form a single blue domain and a single red domain. This 2-domain structure is shown in two views in the central row, with rotation about the horizontal direction indicated. Note the 60° rotation between domains, visible in the view at right. The strand structure of four of these domains is shown, along with sticky ends, in the drawing below this row. At bottom is a simple domain picture of this same image, and an indication of how it forms a 2D lattice. Reprinted by permission from the American Chemical Society.[5.4]

axes each of the coordinate system. On the left of the second row, we can see that the blue strand arms have stacked on each other, as have the red arms. On the right of the second row, the structure has been rotated 90° about the X axis, showing that the stacking domains are twisted by 60°.[7.8] The lesson is evident here: if a nucleic acid structure can increase its stacking by undergoing a structural transition, it will do so. It took the author many years to realize the importance of this rule.

The strand structure of an arrangement of four of these molecules is shown in the central portion of Figure 7-9. The parallelogram-like structure of this group

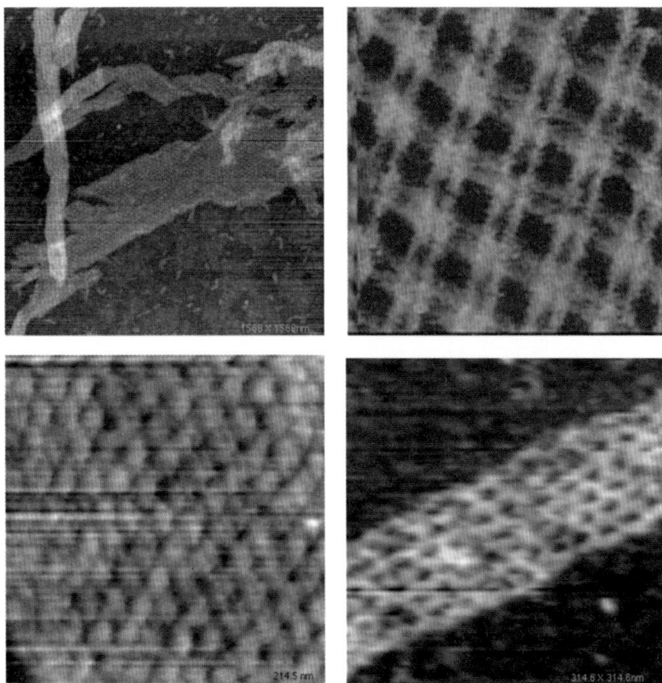

Figure 7-10 *AFM images of the 2D array shown in Figure 7-9.* At upper left is a long view of such an array, and a zoom is shown at lower left. The array contains 6-turn edges, so the large separations shown are four turns (~14 nm) and the short ones are two turns (~7 nm). The same array is shown at high resolution at the upper right. A larger array (8 turns, 6-turn squares, and 2-turn inter-square spacings) is shown at lower right. The short spacings are not visible. Reprinted by permission from the American Chemical Society.[5.4]

of strands is evident.[7.9] The bottom row of Figure 7-9 suggests that it is possible to take these parallelograms and organize them to produce a parallelogram lattice of DNA. Figure 7-10 shows AFM images of such a parallelogram lattice. The dimension of each of the parallelogram repeats is six turns of DNA. The overhangs are about one turn, so that the large cavities are 4 × 4 turns (~13 × 13 nm), and the small cavity is about 2 × 2 turns (~7 × 7 nm). In the pictures at the upper left and lower right of Figure 7-10, the 2-turn separations are not resolved. The 2-turn separation is well resolved in the image in the lower left. In an exceptionally well-resolved image, the 2 × 2-turn separation is evident in the image on the upper right. The angle between helices is seen to be ~63° in the image on the lower left, but it is closer to 85° in the image on the upper right (made under different conditions).

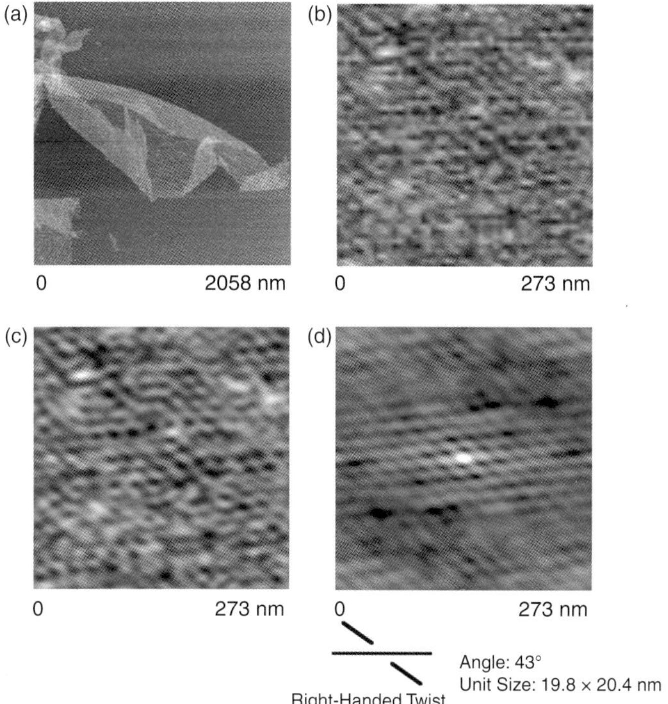

Figure 7-11 *Analysis of a 2D array of junctions distorted by a hydrogen bond.* (a) *A long-view AFM image of the array;* (b) *a zoom;* (c) *a Fourier-smoothed image;* (d) *the autocorrelation function of (c).* As schematized below the images, the angle is indeed 43°, as suggested by the crystallographic analysis. Reprinted by permission from the American Chemical Society.[7.12]

2D arrays as an analytical tool. In addition to being a route to 2D arrays, this type of parallelogram system can be used as an analytical tool to examine features of Holliday junctions. For example, the first crystal structure of a Holliday junction[7.10] found an inter-helical angle of about 40° between the helix axes, differing markedly from the ~60° angles found both in solution by gel mobility and fluorescence experiments[7.8,7.11] and in earlier 2D arrays.[7.9] Two explanations were immediately evident: (1) this was an artifact induced by crystal packing; or (2) a close contact seen in the crystal structure could be interpreted as a sequence-specific hydrogen bond causing the change in inter-helical angle. To distinguish the two possibilities, a parallelogram array was constructed using the junction-flanking sequence used in making the crystal. Figure 7-11 shows the results of this analysis. Panel a is an unzoomed image, and panel b is a zoom; panel c is a Fourier-smoothed image, and panel d is the autocorrelation function of panel c. From this image, it is clear that the angle is around 43°, as summarized by the schematic at the bottom of the drawing.[7.12] Thus, the sequence-specific hydrogen bonding found in the crystal structure appears to persist in solution, distorting the junction structure from its conventional value.

One-dimensional arrays of parallelogram structures have also been used analytically. They have confirmed gel electrophoretic data suggesting that the bowtie junction is more like a parallel structure than an antiparallel structure.[7.13] The bowtie structure is shown in Figure 7-12, where it is compared with the conventional junction structure. The key difference is that the junction portion of the structure contains a 5′, 5′ linkage on one strand and a 3′, 3′ linkage on the strand opposite it; the other two strands are normal. Figure 7-12a shows the angle conventions, and Figure 7-12b shows what the structures of the normal and bowtie junctions look like. Figure 7-12c shows the structures of the juxtapositions in both parallel and antiparallel versions of the structures. Note that the terms "parallel" and "antiparallel" are just broad indicators of the range of the inter-helical structure, with the antiparallel structure for the normal junction being 60° away from the ideal. It's hard to get too much further away, but the bowtie junction does it by being 70° away from the parallel ideal, as we'll see. The same sort of 2D lattice that showed that the normal junction has an angle of about 60° away from parallel was constructed for the bowtie junction, and it is visible in Figure 7-13; an angle of about –70° is seen. However, this doesn't answer the question shown at the top of Figure 7-14: which angle are we seeing? The answer is found by using the system shown at the bottom of Figure 7-14. Four parallelograms are constructed: I, II, III, and IV. I and II are designed to be the fulcrums of V-shaped arrays, and III and IV are designed to be extenders of those parallelograms. If the parallelograms look like the gray parallelograms, Figure 7-14 shows they will be shaped one way, and if they look like the red parallelograms, they will be shaped differently. The top left of Figure 7-15 shows the consequences for a V-shaped array if the fulcrum parallelogram I is shaped like the gray or red parallelograms. Similarly, the right top of 7-15 shows the consequences for a V-shaped array with parallelogram II as the fulcrum. Below each of these schematics are AFM images of the actual V-shaped arrays. It is clear that the bowtie junction is shaped like the gray parallelograms.

To return to two-dimensional arrays for a moment, we might ask what other systems can be used to build them. We have already looked at DX and TX tiles, and we have just looked at junction parallelograms. Many other systems failed. For example, we considered bulged junctions in Chapter 6. However, what happens if bulged junctions are combined with DX tiles, to make the structure

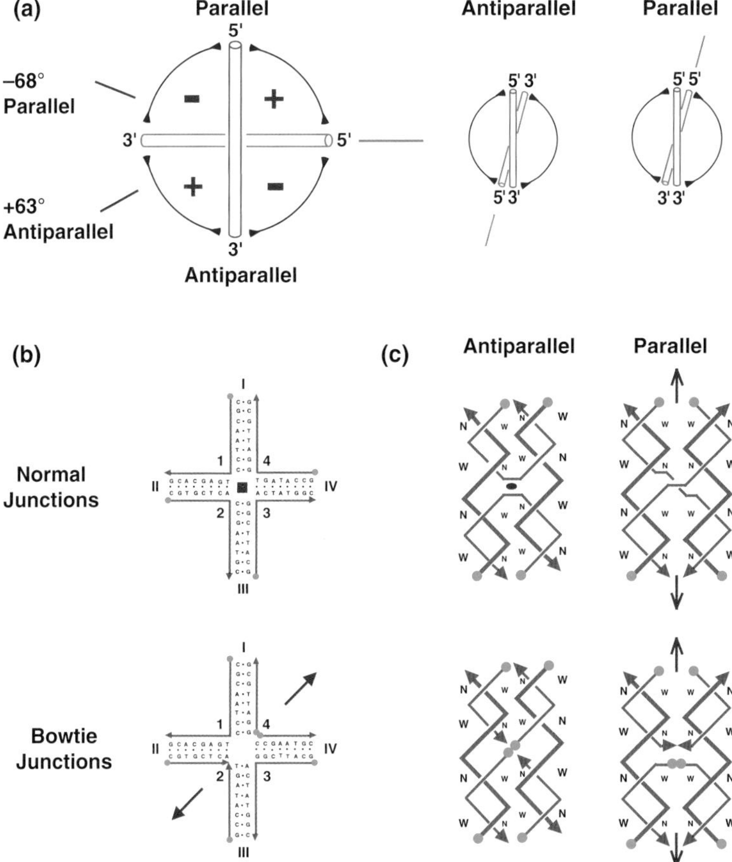

Figure 7-12 *Structural features of bowtie junctions.* (a) *Ranges of the domains.* The circle represents the positions available for the orientations of the helical strands in each domain. The actual positions are indicated at the far left. Idealized parallel and antiparallel positions are shown at the right. (b) *Images of normal and bowtie junctions.* Strand numbers are shown in Arabic numerals and arm numbers in Roman numerals. 5′ ends are shown as filled green circles, and 3′ ends are shown as red arrowheads. Apparent symmetry elements are shown: a four-fold axis for the normal junction and a two-fold axis for the bowtie junction. (c) *Strand structures of stacked junctions.* Symmetry elements in the antiparallel and parallel conformations are shown. Major grooves are denoted by W (wide) and minor grooves by N (narrow). Two-fold symmetry elements are indicated. Reprinted by permission from the American Chemical Society.[5.27]

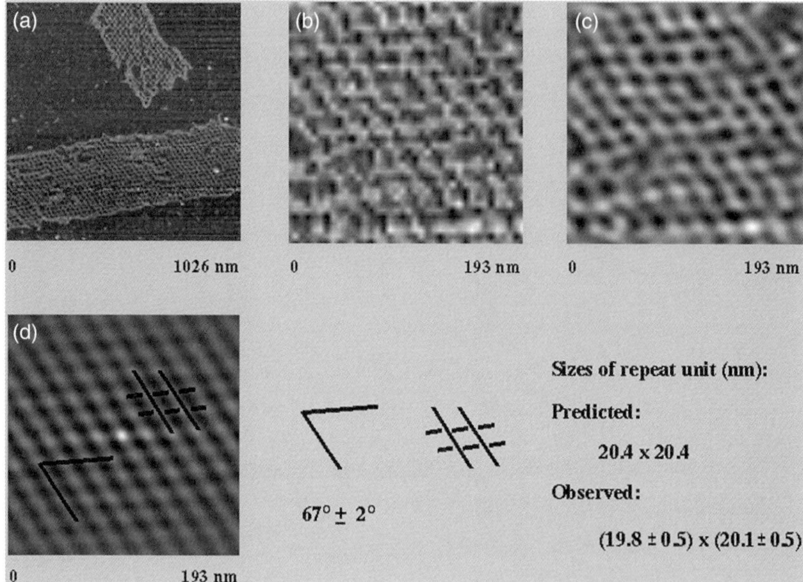

Figure 7-13 *AFM analysis of bowtie junction arrays.* Panels a, b, c, and d represent the same treatments of the data as shown for conventional junctions in Figure 7-11. The 67° angle is noted, but the parallel or antiparallel nature of the junctions is not clear from the imaging. Reprinted by permission from the American Chemical Society.[5.27]

shown in Figure 7-16a? That structure proved successful in making a 2D lattice.[7.14] There were two different things that were changed between the bulged-junction triangle of Figure 6-4 and the DX triangle structure of Figure 7-16a: the structure was twice as thick, and there were two sticky ends at the extremities of each triangle. It was easy to test whether two sticky ends was the key element, because removing one sticky end from each triangle is simple to do. When the assembly was performed with just single cohesion, rather than double cohesion, no array formed. Thus, the key to the formation of 2D arrays from triangles seems to be double cohesion. Are there other instances where double cohesion might help array formation?

Examples of the strength of double cohesion. You know when anybody asks a rhetorical question like that, the answer is always "yes." A convenient and increasingly popular motif in structural DNA nanotechnology is the 6-helix bundle (6HB).[7.15] If we consider the DNA double helix to be 10.5-fold, then seven nucleotides corresponds to 2/3 of a turn, and 14 nucleotides corresponds to 4/3 of a turn. Thus, placing DX molecules around a cyclic path with connections every 7 or 14 nucleotides leads to a 6HB motif, shown in

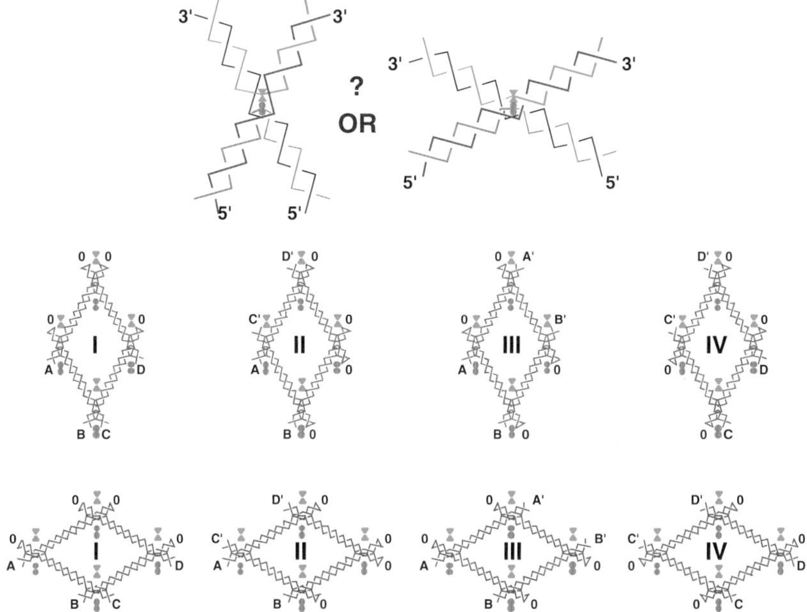

Figure 7-14 *Schematics of a system to analyze the parallel or antiparallel nature of bowtie junctions by AFM.* The top panel demonstrates the question being asked of the junctions: which conformation is it, parallel (left) or antiparallel (right)? The second row shows the conformation of a series of gray rhombuses if "parallel" is the correct answer. 0 represents closed hairpins and A, B, C, and D represent sticky ends, with A′, B′, C′, and D′ representing their complements. Rhombuses I and II represent apices of V-shaped arrays, while III and IV represent rhombuses to extend the sizes of the V so it is visible in the AFM. The bottom row shows the same rhombuses in red if "antiparallel" is the right answer. Reprinted by permission from the American Chemical Society.[5.27]

Figure 7-17a. The red hexagons in the picture are considered to be in front of the blue hexagons. This leads to double cohesion of a series of 6-helix bundles, producing a 2D array. AFM images of 6HB arrays are shown in Figures 7-17b, 7-17c, and 7-17d. Figure 7-17e shows the autocorrelation pattern of a 2D 6HB array.[7.8] Although it is easy to imagine 3D arrays made from 6HB arrays, the crystals grown from the motif have not diffracted. It should come as no surprise that 1D arrays are readily produced from 6HB motifs.

Another motif that has proved valuable in 2D, but has not produced 3D crystals, is the 3D-DX triangle motif.[7.16] This is based on the Mao group's tensegrity triangle (see below), except that it contains DX motifs rather than individual double helices. A selection of 2D arrays is shown in Figure 7-18. Figure 7-18a illustrates a schematic of the motif itself. The over-and-under

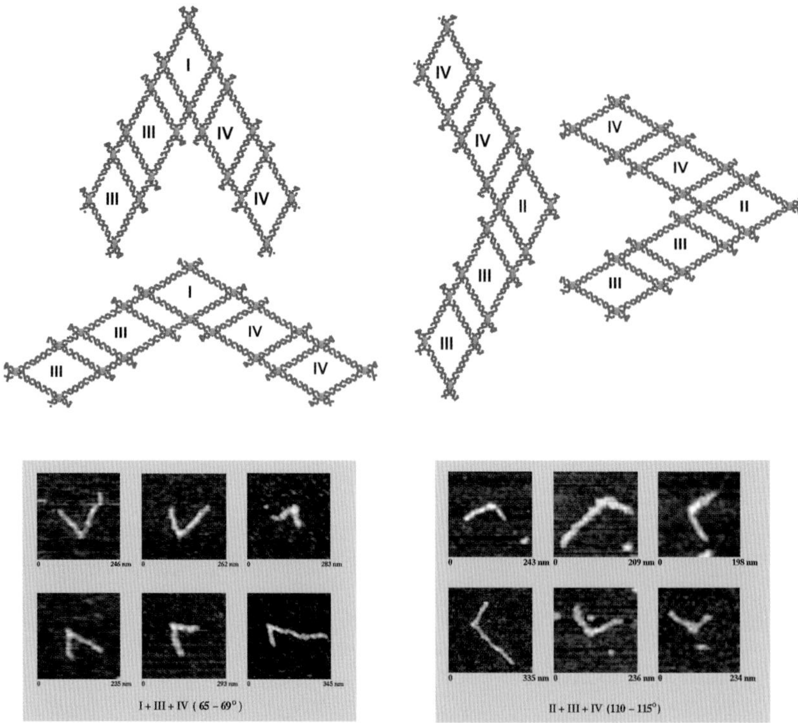

Figure 7-15 *AFM images of the experiment schematized in Figure 7-14.* The left side shows the schematic V-structures in gray and red that would result from using rhombus I as the apex of the V. It is clear that the parallel (gray) V is the correct answer, because all the structures display an acute angle. The right side shows the same experiment, but now using rhombus II as the apex of the V. Again the gray answer is closest to the experimental data, because the structures all display an obtuse angle. Reprinted by permission from the American Chemical Society.[5.27]

pattern of each edge is seen in each of the three DX domains of the molecule. The nearly horizontal DX at the top is above the adjacent DX on the left and below the one on the right. Figures 7-18b, 7-18c, and 7-18d illustrate three independent periodic sections of the 3D-DX motif with even numbers of half-turns between all crossovers. Note the extensive curling in all three panels, although there is a nice flat image as well, in Figure 7-18c. By contrast, Figures 7-18e, 7-18f, and 7-18g illustrate three independent periodic sections of the 3D-DX motif with alternating even and odd numbers of half-turns between crossovers. These images are in the same order as in panels b–d. Note the lack of curling relative to panels b–d. Figures 7-18h through 7-18l show 2-tile independent sections and autocorrelation functions. Panels h, j, and

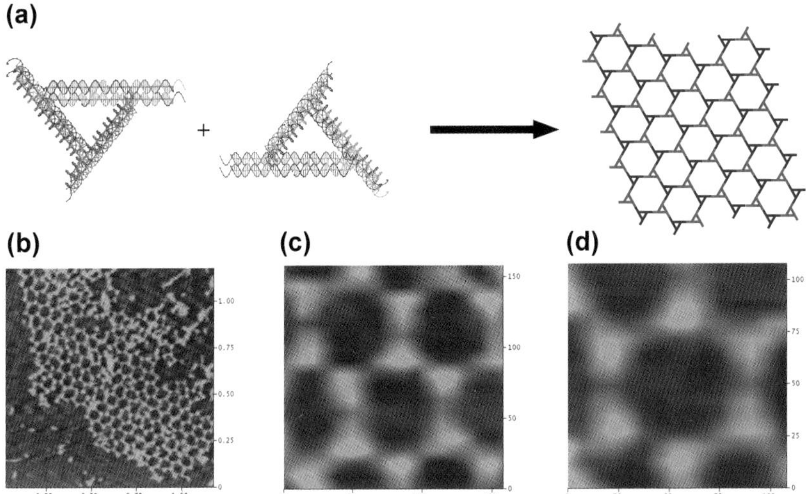

Figure 7-16 *DX bulged-junction triangles.* (a) *A schematic showing how two DX bulged-junction triangles can form a pseudo-hexagonal lattice.* Two molecular representations of the triangles are shown, alternating as red and blue triangles in the lattice structure at right. (b) *A 2D lattice of DX bulged-junction triangles.* The formation of the lattice is evident. (c) and (d) *Successive zooms of the image in (b).* The dimensions of the triangles are evident from the images.[7.14]

l are 2-tile images in the same order as in the previous panels. Panels i and k are the autocorrelation functions of the images in panels h and j, respectively. As we will see later, this powerful motif has been used to organize metallic nanoparticles.

Other motifs have been connected by double cohesion. The parallelogram arrays discussed above are not very robust when the edges are made long. The images shown in Figures 7-9 and 7-10 correspond to what is known as a $(4 + 2) \times (4 + 2)$ structure, where there are four double helical turns between vertices within the parallelogram and two between vertices from parallelogram to parallelogram. It is hard to go much further than this in dimension, although small $(6 + 2) \times (6 + 2)$ and $(4 + 4) \times (4 + 4)$ arrays can be seen in Figures 7-19a and 7-19b, respectively. However, using double cohesion and DX motifs (schematic in Figure 7-19c), it is possible to make an $(8 + 4) \times (8 + 4)$ array, as shown in panels d–g of Figure 7-19. Figure 7-19h shows a failure to produce this array using single cohesion and double helices.[7.16]

An unusual motif related to the tensegrity triangle and the 3D-DX motif is the skewed TX motif. Schematics of it and its components are shown in different views in Figure 7-20a. Its 2D array is held together by double

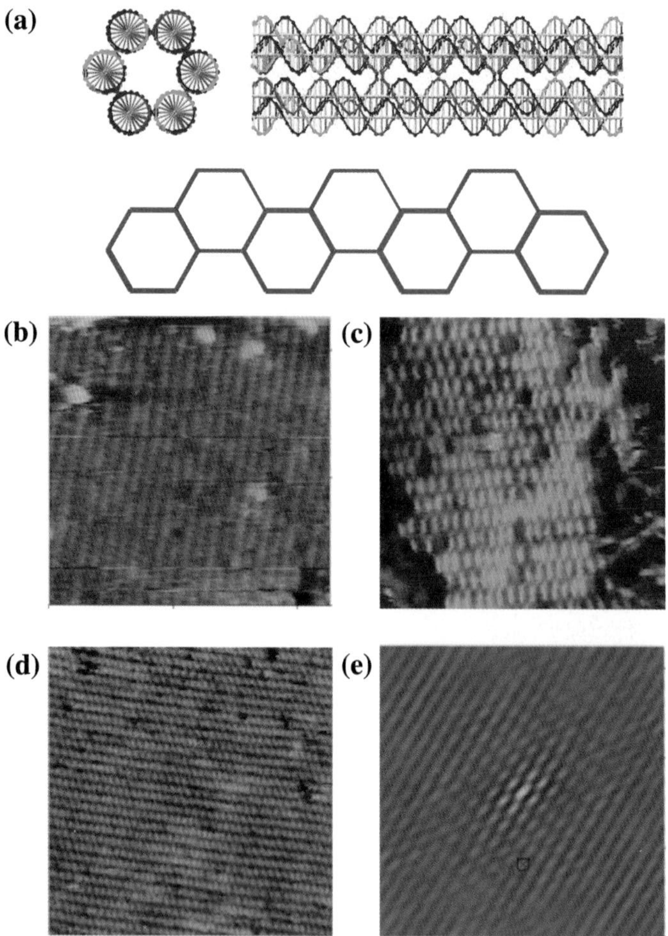

Figure 7-17 *2D arrays formed by 6-helix bundles.* (a) *Schematics.* Top and side views of the 6-helix bundle are shown. Below that, the motif is schematized as a hexagon. The red hexagons are nearer the reader than the blue ones. Only the helices that connect them in this fashion contain sticky ends. (b), (c), (d) *AFM images of the arrays formed.* (e) *The autocorrelation function of the array.*[7.16]

cohesion, and forms quite nicely. Figures 7-20b through 7-20d show 2D arrays formed by using one pair of blunt ends and two pairs of sticky ends to form the arrays. This motif also fails to produce diffracting 3D crystals. Nevertheless, it should be clear from the foregoing that double cohesion is a very potent method for producing cohesion in structural DNA nanotechnology.[7.16] We will encounter it again, and we should regard it as a standard component of the toolbox that increases the strength of DNA intermolecular interactions.

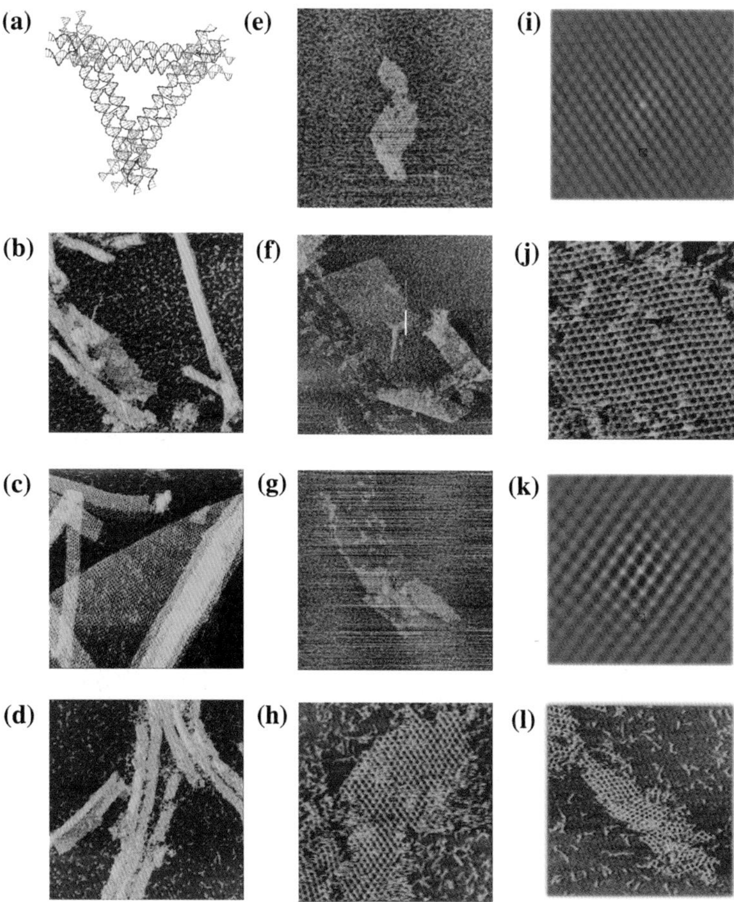

Figure 7-18 *2D arrays of 3D-DX triangles.* These are tensegrity triangles whose elements are DX molecules rather than single helices. The triangles span 3-space, but they only have sticky ends on two of their DX domains. (a) *A molecular diagram.* (b), (c), (d) *Arrays with even numbers of half-turns between crossovers.* (e), (f), (g) *Arrays with alternating even and odd numbers of half-turns between crossovers.* (h), (j), (l) *Arrays formed from two different triangles.* (i) *Autocorrelation function of (h).* (k) *Autocorrelation function of (j).*[7.16]

So what about 3D? In Chapter 1 we talked about it, but we haven't discussed it yet. The first thing to realize is that the bar is a little higher when we talk about 3D. In the ordinary investigator's hands, AFM on a good day (a good tip and good luck) has a resolution of about 7–10 nm. Looking back at Figure 7-10, the picture on the lower left shows two double helices separated by 7 nm. The remarkably high-quality picture on the upper right shows some features that

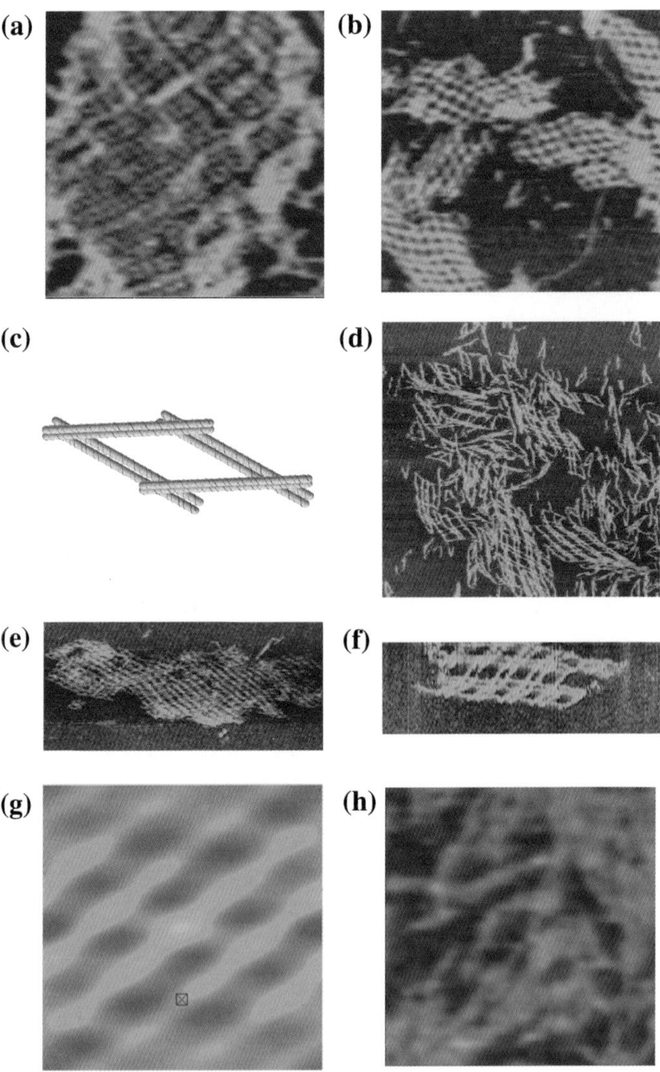

Figure 7-19 *Parallelogram arrays with double cohesion.* (a), (b) *Poorly formed (6+2) × (6+2) and (4+4) × (4+4) arrays.* (c) *Schematic of (8+4) × (8+4) DX molecule.* (d), (e), (f), (g) *(8+4) × (8+4) DX arrays.* (h) *Failure to produce (8+4) × (8+4) array with single helices.*[7.16]

may be of even higher resolution. However, the other images in Figure 7-10 are clearly of lower resolution, not resolving the two nearest double helices at all. The technique for looking at 3D is X-ray crystallography, not AFM. Although there are many technical issues involved with this method, it is certainly

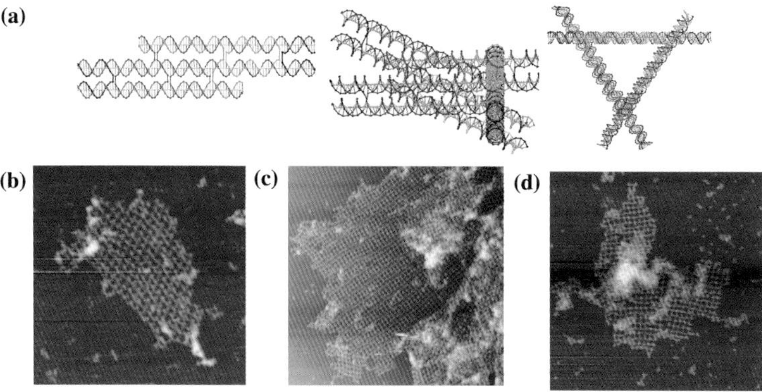

Figure 7-20 *Skewed TX motif arrays.* (a) *Schematic images of the motif.* (b), (c), (d) *2D arrays formed from the motif.* This motif uses double cohesion to form 2D arrays.[7.16]

possible to resolve atoms by it. Resolution of 1 Å or better is not uncommon for crystallographic determinations with small molecules. The limits on the resolution of a crystallographic experiment are a consequence of crystal quality, not the limitations of the technique. The downside of this high-resolution capability is that one is looking at finer details, and resolution that would be remarkable for AFM, say 10 Å (1 nm), is unacceptable or marginally acceptable for X-ray crystallography. In a typical crystallographic determination involving an unknown structure, the molecular boundaries are only marginally distsinct,[7.17] and at least 4 Å resolution is needed to trace the backbone of a nucleic acid structure.[7.18] Fortunately, in the case of a designed structure, there are no issues of molecular boundaries, so one can get away with a resolution of 10 Å if the structure is known to have formed according to plan, and the phasing is good.

3D structures. For the first designed 3D DNA structure, it was important to make sure that the structure was derived in an unbiased fashion. This is done by using a model-independent method known as isomorphous replacement, rather than by using a method known as molecular replacement, which requires a model. The crystals were based on a tensegrity triangle,[7.19] diffracted to 4 Å resolution initially,[7.20] and now diffracted to a little better than 3 Å. The helix axes of the tensegrity triangle span 3-space, so the motif is a good one for building a 3D crystal. Before getting to the crystal I'll talk about below, there were a lot of failures. The two characteristics of the failure crystals were long sticky ends and a fundamentally 2D motif that was tricked into making 3D arrays. For example, a combination of two TX molecules, joined by a short length of duplex that had a non-half-integral twist between junctions, was a dismal failure.

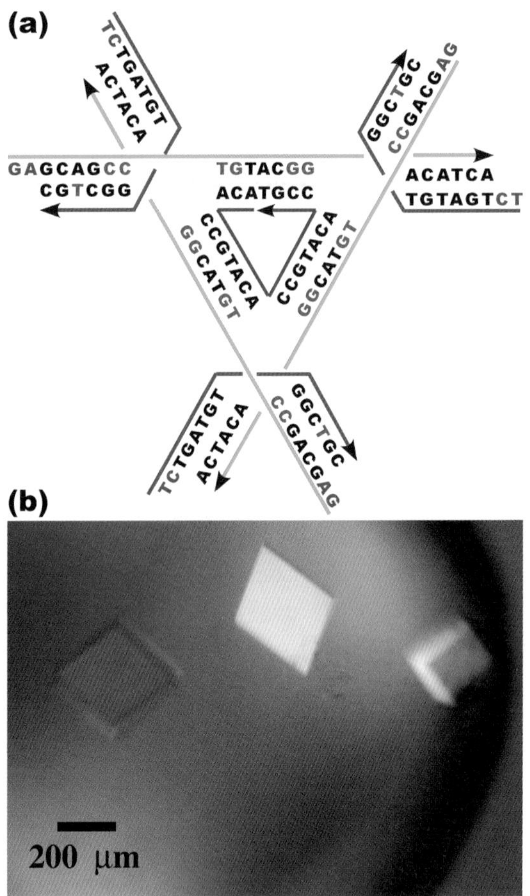

Figure 7-21 *The 2-turn tensegrity triangle.* (a) *Schematic view of the triangle and its sequence.* Note that the sequence has been symmetrized by three-fold rotation. (b) *Crystals produced by self-assembly of the motif.* These crystals are macroscopic and are visible to the naked eye.[5,60]

The sticky ends that worked to provide the best resolution were only two nucleotides long. There are many arguments about the importance of reversibility (the opportunity for the system not to get locked into an unwanted structure) in crystal formation, but it is unclear to this author how that bears on short sticky ends, considering that the system was three-fold symmetrized. The sequence and pictures of the tensegrity triangle crystals are shown in Figure 7-21. Note the three-fold symmetry of the sequence, and the rhombohedral nature of the crystals. These crystals are macroscopic, and can be seen with the naked eye. The molecular structure of the tensegrity triangle is shown in

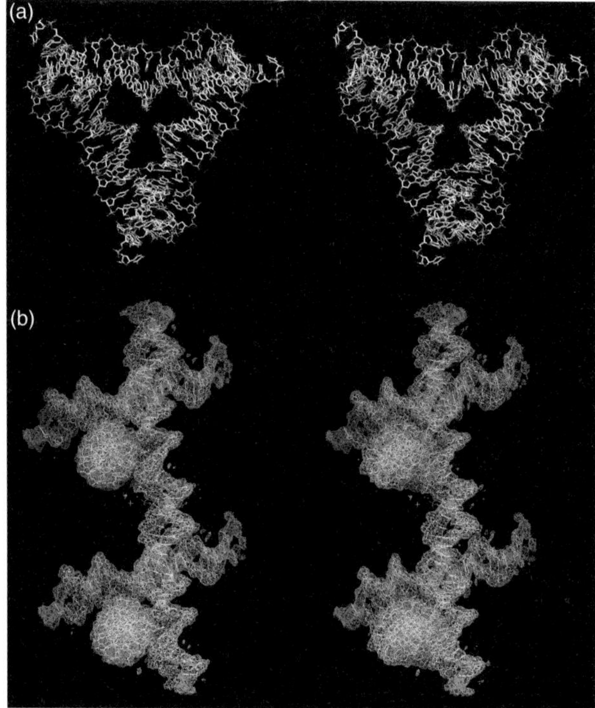

Figure 7-22 *Stereographic images of the crystal structure of the triangle to 4 Å resolution.* (a) The upper image is the individual triangle. (b) The lower image shows the connection between two triangles (with magenta highlighting of the region). The electron density of the molecule has been superimposed on the molecular structure in blue.[5,60]

stereoscopic projection in Figure 7-22a. The view is down the three-fold axis of the crystal. The over-and-under nature of the structure can be seen readily in this view. Figure 7-22b is another stereoscopic projection that shows that the intermolecular interactions of the triangles occur through the sticky ends.

The nature of the designed crystal structure can be appreciated through the stereoscopic projections shown in Figure 7-23. Figure 7-23a illustrates a triangle at its center, and its six nearest-neighbor triangles. The base pairs in the directions of each of the helix axes have been colored differently. Thus, there is the red direction, the green direction, and the yellow direction; the directions of the three helix axes are linearly independent, thereby spanning 3-space. The covalent crossovers between the three helices are not readily apparent from this view, but they are present nevertheless: these are triangles, molecules containing 63 nucleotide pairs, and they migrate accordingly on a

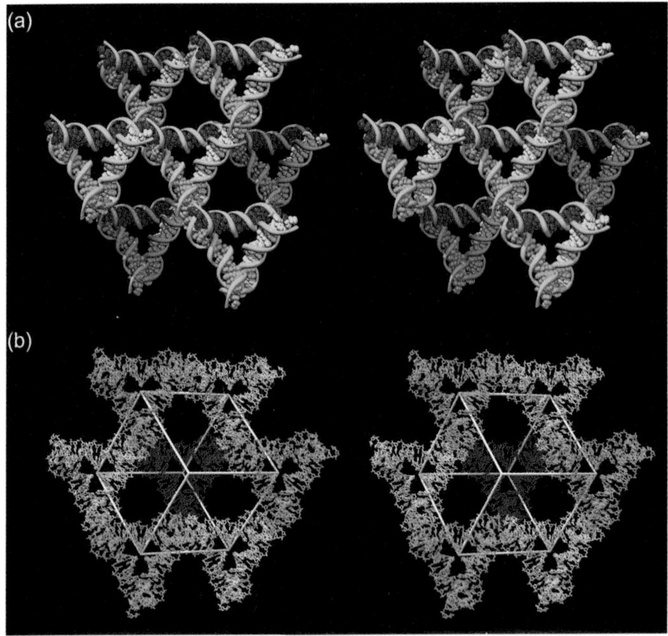

Figure 7-23 *Stereographic images of the crystal structure.* (a) *The surroundings of a triangle.* Each triangle is joined to six other triangles by sticky-ended cohesion. The schematic shows that the three directions defined by the helix axes span 3-space. The red direction goes from rear to front, as does the green direction and the yellow direction. (b) *Eight triangles surround the vertices of a rhombohedron.* The red triangle sits on the rear vertex of a rhombohedron. It is connected to the three yellow triangles that flank vertices nearer to the viewer. The yellow triangles are connected to the green triangles that flank vertices nearer yet to the viewer. The front vertex has been left vacant for clarity, but would contain another red triangle.[5.60]

non-denaturing polyacrylamide gel. Although most macromolecular crystal structures represent the packing of globular shapes, this is more of a rod-like structure.

Another way to think about this structure is that it forms a rhombohedron; this a cube-like structure in which one of the body diagonals has been stretched or shrunk, leaving the polyhedron with only a single three-fold axis, rather than four of them. This aspect of the crystal structure is seen in the stereoscopic projection of Figure 7-23b, which is a view down the three-fold axis. The vertices of the rhombohedron are shown, and the center of a triangle drawn in red flanking the rear vertex. The red triangle is connected by sticky ends to each of the yellow triangles that flank it and whose centers lie in a place closer to

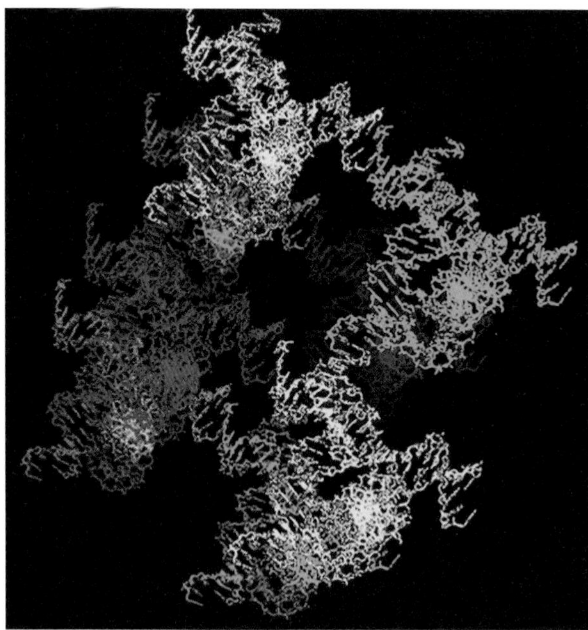

Figure 7-24 *Eight triangles that form the rhombohedron.* The 2-nucleotide sticky-ended connections between the triangles in the face closest to the viewer are particularly clear.

you. The yellow triangles are in turn connected to three green triangles, even closer to you. A second red triangle lying above the first one and closer yet to you has been omitted for clarity. Thus, the eight triangles flank a rhombohedral cavity, which is seen in Figure 7-24. Figure 7-25 is an artistic rendering of the same view as seen in Figure 7-23b, but which shows the view in the context of the whole crystal structure. The same three layers are seen (dim, bright, brightest), but now you can see how the whole crystal is made up of these units. The volume of this cavity is about 100 nm^3.

Other crystals have been designed using the same strategy.[7.20] For example, Figure 7-26 shows the crystal structures of 3-turn and 4-turn tensegrity triangle lattices. These are not significantly different from the lattices designed from 2-turn tensegrity triangles. The volumes of their cavities are about 370 nm^3 and 1000 nm^3, respectively. Their resolutions are about 6.5 Å and 10 Å, respectively, and the structures have been determined by molecular replacement techniques.

Multiple molecules per asymmetric unit. The ability to control the unit cell contents by means of sticky ends has also enabled the construction of crystals with more than a single crystallographic repeat. Figure 7-27 shows two slightly

Figure 7-25 *A view of the crystal structure showing how the rhombohedra fill space.* The faintest triangles are at the rear, then there is a plane of triangles nearer the viewer, and then another plane closer yet. This is analogous to Figure 7-23b, but now it is evident how the crystal structure is formed in the vertical and horizontal directions by many such interactions. Reprinted with permission from David Goodsell.

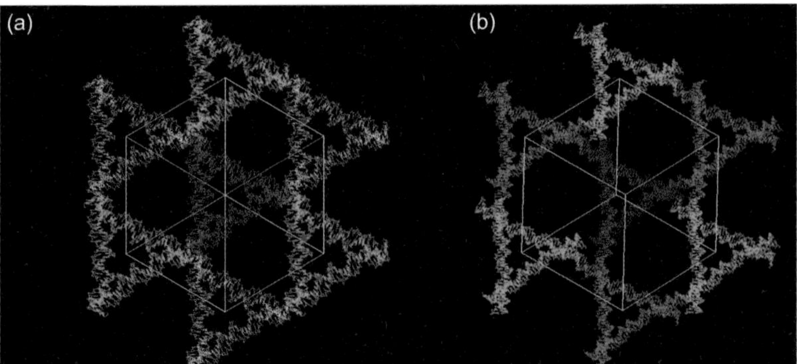

Figure 7-26 *Larger crystals.* (a) *A crystal structure constructed from triangles containing three turns per edge.* (b) *A crystal structure constructed from triangles containing four turns per edge.* These crystals are not significantly different from the two-turns-per-edge crystal.

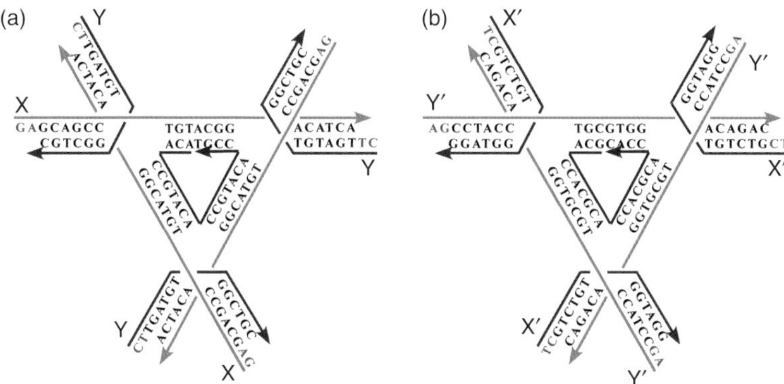

Figure 7-27 *Triangles that form a crystal with two independent molecules per asymmetric unit of the unit cell.* The A triangle and the B triangle are independent molecules, but inspection of the sticky ends shows that they have been programmed to alternate within the crystal. Reprinted by permission from the American Chemical Society.[5.61]

different 2-turn tensegrity triangles. They do not associate with themselves but each species only associates with the other species, as expected from the sticky end design.[7.21] The two molecules are designated A and B. Figure 7-28a is another stereoscopic picture, and it is analogous to Figure 7-23a; it shows the organization of the two molecules within the crystal. The green and red molecules represent the two different species of the tensegrity triangles. You can see readily from the alternation of red and green molecules that each species of triangle is only bound to the other species. In a similar fashion, Figure 7-28b is analogous to Figure 7-23b, in that it emphasizes the cavities within the crystals. You can see that in this drawing four corners of the rhombohedron are flanked by red triangles, and the other corners are flanked by green triangles; each body diagonal and each edge has a red triangle on one end and a green triangle on the other end. It is clear that a red tetrahedron and a green tetrahedron intersect to form the rhombohedron. The tetrahedra are not Platonic tetrahedra, because their intersection forms a rhombohedron rather than a cube. The cavity at the center of the rhombohedron is visible, and it is the same size as the cavity shown in Figure 7-23b.

The ability to control the contents of the unit cell gives us the ability to control some of the properties of the crystals. An example shown in Figure 7-29 illustrates nine different species of crystals. The central picture shows a control crystal, but all the others have dye molecules attached to one or both of the triangle molecules. The top row shows crystals where Cy3, which is a pink dye, has been attached covalently to the A molecule (left), to the B molecule (right),

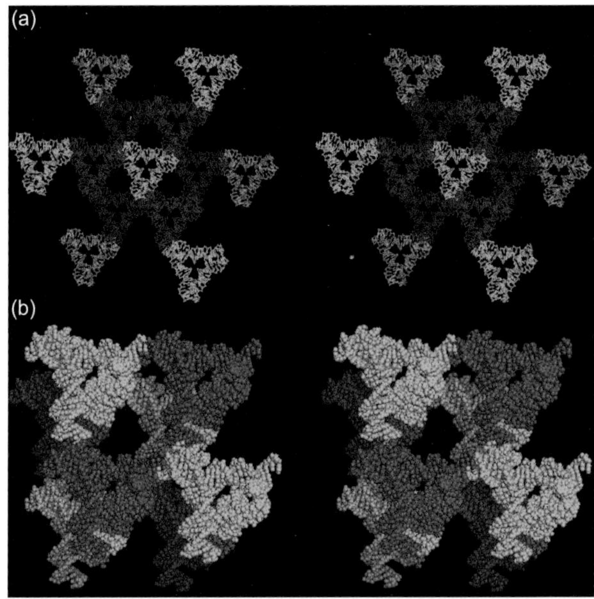

Figure 7-28 *A crystal programmed to contain two molecules per asymmetric unit.* Both panels are stereographic images of the crystal structure. One molecule is drawn in red and the other in green. (a) *The surroundings of an individual triangle.* The green triangle in the center is connected to six red triangles, just as in Figure 7-23a. The next-nearest neighbors in the same direction (only) are shown in green. (b) *The rhombohedron formed by four molecules of each kind.* Eight molecules are shown. The opposite vertices along each edge contain different species triangles. Reprinted by permission from the American Chemical Society.[5.61]

or to both (center). Similarly, the bottom row shows crystals where Cy5, a blue dye, has been attached to the A molecule (left), to the B molecule (right), or to both (center). The middle row shows crystals where Cy3 has been attached to the A molecule and Cy5 to the B molecule (left) or they have been attached in the opposite order (right). It is clear that the presence of the dyes controls the colors of the crystals, either when they are attached as one color (top and bottom rows) or when they are mixed (middle row).

The ultimate goal of 3D self-assembly of robust DNA motifs is to organize other species, ranging from nanoelectronics to potential drug targets. Although nanoelectronics probably does not require ultra-high-resolution crystals (< 1 nm), potential drug targets must be very well organized (~3 Å or better) for this form of crystallization to be of use. At this time, we do not know why the best self-assembled crystals only diffract to 3 Å resolution. Potential candidates for the culprits in this effort include the nature of the freezing

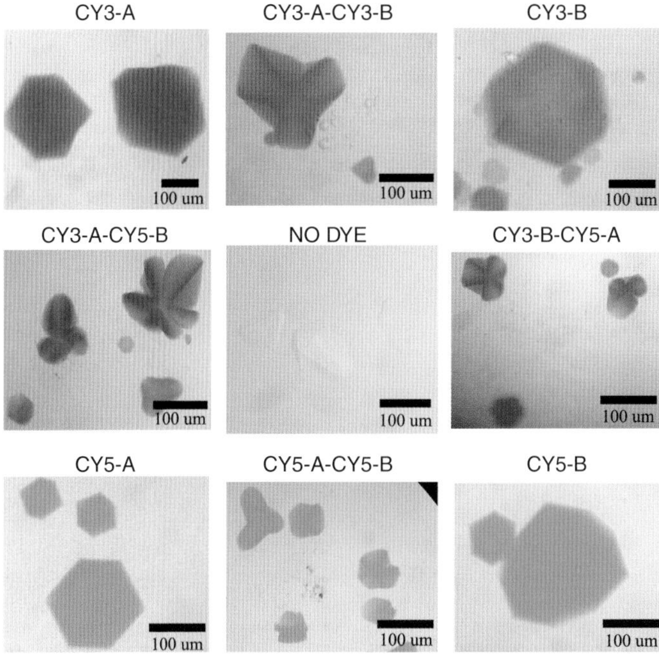

Figure 7-29 *Controlling the color: a macroscopic property of a crystal containing two independent molecules.* Two dyes have been employed: Cy3, a pink dye, and Cy5, a blue dye. In the top row, the A molecule (left), the B molecule (right), or both (center) have had Cy3 attached to it. The crystals are consequently colored pink. In the bottom row, the A molecule (left), the B molecule (right), or both (center) have had Cy5 attached to it. The crystals are now colored blue. The left image in the middle row has Cy3 attached to the A molecule and Cy5 attached to the B molecule, whereas the order has been reversed on the right. The crystals are colored purple, as expected. The central image contains control crystals to which no dye has been attached. Thus, a macroscopic property has been programmed using a microscopic chemical attachment. Reprinted by permission from the American Chemical Society.[5.61]

process for stick-like motifs, the inherent floppiness of DNA, and the need of the components of a lattice to find their own free-energy minimum. It is impossible to exclude mundane factors such as the details of the crystallization medium. The inherent impurity of synthetic DNA has been excluded, but the other factors are still undergoing active investigation.

We have achieved our major goal in self-assembly: the arrangement of DNA motifs in periodic arrays that diffract X-rays to adequate resolution. Now it is time to think about using these constructs made from robust components in systems that are not frozen forever.

References

7.1 E. Schrödinger, *What is Life?*, Cambridge, Cambridge University Press (1944).

7.2 E. Winfree, F. Liu, L. A. Wenzler, N.C. Seeman, Design and Self-Assembly of Two-Dimensional DNA Crystals. *Nature* **394**, 539–544 (1998).

7.3 X. Yang, L.A. Wenzler, J. Qi, X. Li, N.C. Seeman, Ligation of DNA Triangles Containing Double Crossover Molecules. *J. Am. Chem. Soc.* **120**, 9779–9786 (1998).

7.4 F. Liu, R. Sha, N.C. Seeman, Modifying the Surface Features of Two-Dimensional DNA Crystals. *J. Am. Chem. Soc.* **121**, 917–922 (1999).

7.5 A.V. Garibotti, S.M. Knudsen, A.D. Ellington, N.C. Seeman, Functional DNAzymes Organized into 2D Arrays. *Nano Letters* **6**, 1505–1507 (2006).

7.6 T. LaBean, H. Yan, J. Kopatsch, F. Liu, E. Winfree, J.H. Reif, N.C. Seeman, The Construction, Analysis, Ligation and Self-Assembly of DNA Triple Crossover Complexes. *J. Am. Chem. Soc.* **122**, 1848–1860 (2000).

7.7 R.M. Clegg, A.I.H. Murchie, D.M.J. Lilley, The Solution Structure of the 4-Way DNA Junction at Low-Salt Conditions: A Fluorescence Resonance Energy Transfer Analysis. *Biophys. J.* **66**, 99–109 (1994).

7.8 A.I.H. Murchie, R.M. Clegg, E. von Kitzing, D.R. Duckett, S. Diekmann, D.M.J. Lilley, Fluorescence-Energy Transfer Shows That the 4-Way DNA Junction is a Right-Handed Cross of Antiparallel Molecules. *Nature* **341**, 763–766 (1989).

7.9 C. Mao, W. Sun, N.C. Seeman, Designed Two-Dimensional DNA Holliday Junction Arrays Visualized by Atomic Force Microscopy. *J. Am. Chem. Soc.* **121**, 5437–5443 (1999).

7.10 B.F. Eichman, J.M. Vargason, B.H.M. Mooers, P.S. Ho, The Holliday Junction in an Inverted Repeat DNA Sequence: Sequence Effects on the Structure of Four-Way Junctions. *Proc. Nat. Acad. Sci. (USA)* **97**, 3971–3976 (2000).

7.11 A. Edmondson, *A Fuller Explanation*, Boston, Birkhauser (1987).

7.12 R. Sha, F. Liu, N.C. Seeman, Atomic Force Measurement of the Interdomain Angle in Symmetric Holliday Junctions. *Biochem.* **41**, 5950–5955 (2002).

7.13 R. Sha, F. Liu, D.P. Millar, N.C. Seeman, Atomic Force Microscopy of Parallel DNA Branched Junction Arrays. *Chem. Biol.* **7**, 743–751 (2000).

7.14 B. Ding, R. Sha, N.C. Seeman, Pseudohexagonal 2D DNA Crystals from Double Crossover Cohesion. *J. Am. Chem. Soc.* **126**, 10230–10231 (2004).

7.15 F. Mathieu, S. Liao, C. Mao, J. Kopatsch, T. Wang, N.C. Seeman, Six-Helix Bundles Designed from DNA. *Nano Letters* **5**, 661–665 (2005).

7.16 P.E. Constantinou, T. Wang, J. Kopatsch, L.B. Israel, X. Zhang, B. Ding, W.B. Sherman, X. Wang, J. Zheng, R. Sha, N. C. Seeman, Double Cohesion in Structural DNA Nanotechnology. *Org. Biomol. Chem.* **4**, 3414–3419 (2006).

7.17 S.H. Kim, G.J. Quigley, F.L. Suddath, A. McPherson, D. Sneden, J.J. Kim, J. Weinzierl, P. Blattman, A. Rich, 3 Dimensional Structure of Yeast Phenylalanine Transfer RNA: The Shape of the Molecule at 5.5 Å Resolution. *Proc. Nat. Acad. Sci. (USA)* **69**, 3746–3750 (1972).

7.18 S.H. Kim, G.J. Quigley, F.L. Suddath, A. McPherson, D. Sneden, J.J. Kim, J. Weinzierl, A. Rich, 3 Dimensional Structure of Yeast Phenylalanine Transfer RNA: Folding of the Polynucleotide Chain. *Science* **179**, 285–288 (1973).

7.19 D. Liu, M.S. Wang, Z.X. Deng, R. Walulu, C. Mao, Tensegrity: Construction of Rigid DNA Triangles from Flexible Four-Arm DNA Junctions. *J. Am. Chem. Soc.* **126**, 2324–2325 (2004).

7.20 J. Zheng, J.J. Birktoft, Y. Chen, T. Wang, R. Sha, P.E. Constantinou, S.L. Ginell, C. Mao, N.C. Seeman, From Molecular to Macroscopic via the Rational Design of a Self-Assembled 3D DNA Crystal. *Nature* **461**, 74–77 (2009).

7.21 T. Wang, R. Sha, J.J. Birktoft, J. Zheng, C. Mao, N.C. Seeman, A DNA Crystal Designed to Contain Two Molecules per Asymmetric Unit. *J. Am. Chem. Soc.* **132**, 15471–15473 (2010).

8
DNA nanomechanical devices

We have been talking so far about static DNA structures: we make them with some expectation of their 3D geometry, or perhaps of their topology, and then they sit there, perhaps changing shape as a result of thermal fluctuations. However, it is certainly possible to do more with DNA than just make static species. It is also possible to make molecules that can change their shapes, producing multi-state devices that can produce mechanical action on the nanoscale. The key DNA devices operate on two different principles: either DNA structural transitions, or programmed sequence-dependent devices whose states can be addressed or created individually. DNA structural transitions can be based on the contents of the solution, but they are only individually addressable by nuanced chemistry, such as the formation of the various knots shown in Figure 4-4, or by the action of protein molecules.

Robust devices. As all of you reading this book are aware, things can happen in chemical systems that do not happen on the macroscopic scale. If we are making a DNA-based nanomechanical device, we term it a "robust" device if all of the molecules undergo the same transition, looking the same after the transition as they did before. If the strands of its framework can recombine and perhaps form another species (a dimer or a breakdown product), then we would say that the device is not robust. Exchange of strands while in an intermediate state is another example of non-robust behavior. Sometimes the conditions will define the robustness of the device. If the device is based on a DNA structural transition, a partial transition of the ensemble of molecules will be the result if the transition trigger is not present in an overwhelming quantity; such conditions could be corrected, if necessary. The basic notion of a robust device is that it behaves like a macroscopic device made of simple machines: a lever is either up or down, a torsional element is either rotated or not, a movement occurs or it doesn't, throughout the ensemble of molecules in the system.

Shape-shifters

A key category of devices can be thought of as *shape-shifters*, molecules that have different structures under different conditions. Clearly a molecule at equilibrium will have a fixed shape or ensemble of shapes, and they won't change. Something must upset this equilibrium if the molecule is to change shape. Fortunately, there are many ways to change the shape of DNA. Historically, the first DNA molecules to be made to change shape underwent structural transitions. There are a number of different structural transitions that have been used to change the shape of DNA.

Ironically, the first structural transition used was an unwieldy and somewhat obscure transition, branch migration (branch migration is shown in Figure 2-9).[8.1] Most of the DNA in cells is negatively supercoiled, meaning that it is under-linked for the number of nucleotide pairs that it contains. For example, if a cyclic DNA molecule contains 10 000 nucleotide pairs, and we say for simplicity that there are 10 nucleotide pairs per Watson–Crick turn, we would expect 1000 Watson–Crick turns of the double helix, resulting in one strand of the DNA to be linked to the other strand 1000 times. If by the expenditure of energy we were able to achieve a state where the DNA was linked only 950 times, then we would say that the DNA is 5% negatively supercoiled. The DNA is stressed in this high-energy state, and it may undergo a variety of sequence-dependent transitions to lower its energy. As explained by many authors (e.g., see reference 8.2), removing nucleotide pairs from the central double helix by extruding a cruciform is one way to do this. When an intercalating dye is added to DNA, its character changes in those sites where the intercalator slips in between nucleotide pairs: the twist decreases, thereby decreasing the stress on negatively supercoiled DNA (think about it – it sounds a little counterintuitive the first time).

Figure 8-1 illustrates a device based on this notion.[8.1] The construct shown is a small circle of DNA that contains a cruciform attached at its top. Only the four red nucleotide pairs contain the sequence symmetry that permits branch migration. All of the rest are base-paired, but in such a way that they won't change partners. If an intercalating dye is added, the twist in the main circle will decrease. Branch migration enables the four nucleotide pairs in the side-arms of the cruciform to enter the central circle and increase the twist, thereby relaxing the system. Removal of the dye has the opposite impact on the system. This is not a very useful or programmable device, but it was the first deliberate nanomechanical construct built from DNA.

A robust shape-shifting device. Figure 8-2 shows a much more convenient and controllable shape-shifting device.[8.3] The device consists of two DX molecules (shown with red and blue strands) that flank a stretch of

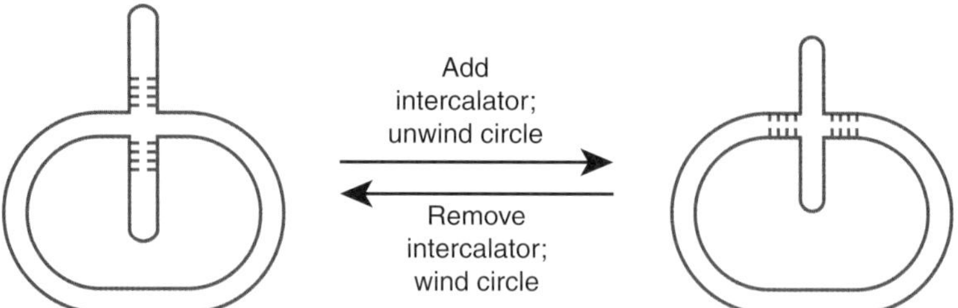

Figure 8-1 *A primitive DNA nanomechanical device based on branch migration.* A tetra-mobile junction is placed in a small DNA circle. In the presence of an intercalator, the circle is unwound, pulling the mobile bases into the circle. When the intercalator is removed, the mobile bases are extruded from the circle. Reprinted by permission from John Wiley & Sons.[8.1]

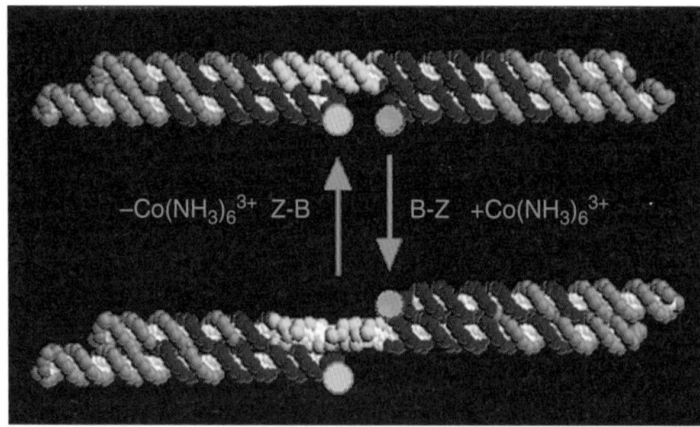

Figure 8-2 *A device based on the B–Z transition.* Two DX molecules, consisting of red and blue strands, flank a stretch of proto-Z-DNA, drawn in yellow. At top, all parts of the molecule are in the B-form. Upon the addition of $Co(NH_3)_6^{3+}$, the proto-Z-DNA undergoes a transition from B-DNA to Z-DNA. With 20 nucleotide pairs in the proto-Z-DNA region, the domain on the lower right is rotated 3.5 turns to the upper right. The pair of FRET dyes shown as pink and green circles report the net half-turn motion of that domain.[5.31]

proto-Z-DNA (shown with yellow strands). This is a sequence that can undergo the transition from right-handed B-DNA to left-handed Z-DNA.[8.4] The device consists of one long shaft and two domains that lie below it in the top panel of Figure 8-2. The B–Z transition has two components: a sequence capable of undergoing it (most typically, but not exclusively,[8.5] a sequence of alternating CG) and an agent to promote it. In the absence of the promoting agent, all the DNA will be in the B-form, as illustrated in Figure 8-2. Here, the promoting agent is $Co(NH_3)_6^{3+}$. When it is added to the solution, the proto-Z-DNA undergoes the B–Z transition over a range of about 3.5 turns,

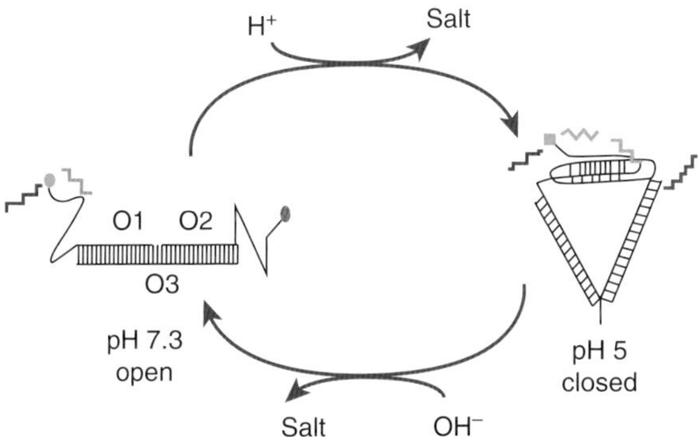

Figure 8-3 *An I-motif device for reporting the pH of its locale.* At high pH (left) the device is open, but at acid pH the device is closed, forming the I-motif and leading to a FRET signal. Reprinted by permission from John Wiley & Sons.[8.6]

corresponding to the yellow nucleotide pairs. The result of this change is shown in the lower panel of the figure. Note that the structure in the yellow portion of the diagram is quite different from the B-DNA structure in the rest of the molecule. When this occurs, the domain on the bottom right is rotated up to the other side of the shaft, as shown in the lower panel. How do we know this? Both the upper and lower panels of the diagram contain a green circle and a pink circle. These are respectively the donor and acceptor dyes of a FRET pair. Their distance increases after the transition, and this results in a decreased fluorescence transfer from the donor dye to the acceptor dye. This device is a robust device, and it is reversible, although not conveniently so. Nevertheless, this was the first robust device. Note that, in the absence of the stiff domains provided by the DX motifs, it would not be possible to use FRET reliably to establish that the device had changed shape.

Other devices have been constructed that are based on DNA structural transitions. For example, the I-motif of DNA, involving stretches of cytosine (Figure 2-7), is promoted by lowering the pH. Figure 8-3 shows a nanodevice developed by Yamuna Krishnan[8.6] and her colleagues for monitoring the pH in various parts of the cell. The system contains three strands, two of which contain oligo-dC-containing stretches. The strand labeled O1 in the diagram contains a short single-stranded region to promote flexibility. When the pH is above 7, the two ends are far apart. However, when the pH is lowered, the two parts come together to form the I-motif. The two ends of these strands contain dyes, which do not interact at higher pH but display a significant FRET signal when the device is switched to acidic conditions.

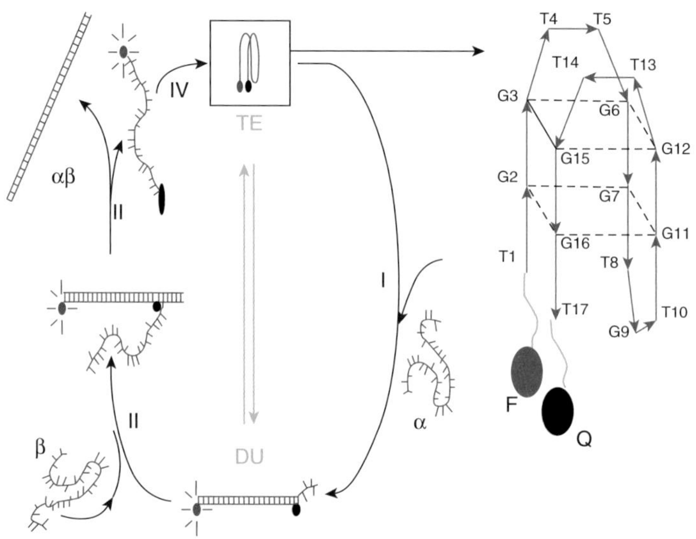

Figure 8-4 *A device based on the G4 structure.* A single-stranded system contains the G4 structure, in which a dye is brought into proximity with a quencher. When the complement to the strand is added, the G4 structure is disrupted and the quencher is separated from the dye, producing a fluorescent signal. Reprinted by permission from the American Chemical Society.[8.7]

If a device can be made from the I-motif, you might think that a device could also be made from the tetra-G motif. Of course, that's right. An example of a system developed by Weihong Tan's laboratory[8.7] is shown in Figure 8-4. A single-stranded G4 structure is shown at the right of the picture. The two ellipses attached to it are a dye and a quencher, so the system will not fluoresce in this state. When the complement to the strand (labeled α) is added, the DNA prefers to be in the Watson–Crick base-paired state, so the G4 pairing disappears. When this happens, the dye is separated from its quencher, leading to fluorescence. Removal of strand α by strand β completes the machine cycle, and leads to the G4 structure reappearing. How does β remove α from the G-rich strand? That's a subject for a new section.

We have seen a variety of DNA structural transitions used to make DNA nanodevices. Given that there are other structural transitions of DNA, other devices of this sort may be possible. None of them has really used the key property of DNA, which is the programmability of structure or behavior as a consequence of sequence. Surely it must be possible to utilize the sequence-specific interactions to control a device. The importance of being able to do this is clear from the device based on the B–Z transition. There are only so many

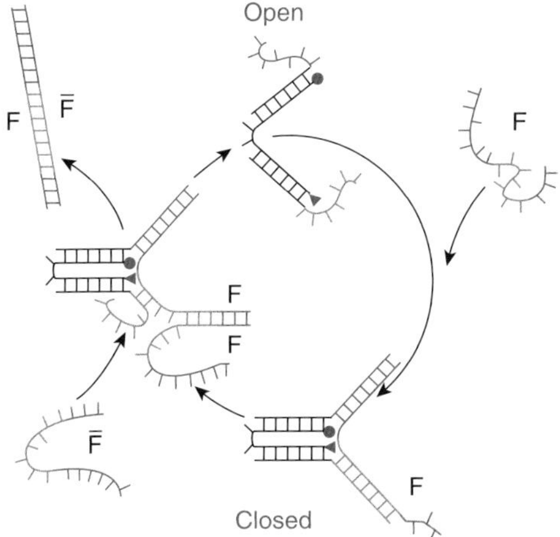

Figure 8-5 *A sequence-driven DNA tweezers.* Starting from the top, the "open" state of the tweezers, strand F brings the two arms of the tweezers together in the "closed" state. Addition of the F̄ fuel strand removes F from the tweezers and opens the tweezers again. The cycle is possible only because of the "toeholds" on the F strand, so that there are more base pairs at each step.[5.50]

different Z-forming sequences that can be addressed independently, without overlap in the conditions. Whether the number is three, as suggested or by Figure 4-4, or larger, it is certainly limited. It is important to be able to address as many devices as possible, so as to create a large multitude of states.

Sequence-specific devices. The first device that was sequence-specific was a tweezers-like machine developed by Bernie Yurke and his colleagues.[8.8] The device and its machine cycle are shown in Figure 8-5. Let's start at the top with the state labeled "Open." Three strands are seen, and the triangle and the circle are the usual dye components, where the circle represents a fluorescein derivative and the triangle is a rhodamine derivative. There is a long strand and two short strands, as we noted above with the I-motif device. However, the short strands have extensions on them, one shown as blue and one as green. In the open state, the dyes are too far apart to interact. They are brought into proximity by the strand at the upper right, labeled F. Part of F is complementary to the blue unpaired section and part of F is complementary to the green unpaired section. When F binds to the rest of the device, it brings the two double helical domains together to form a structure that is like the inside of a DAO molecule (see Figure 3-8). This is the state labeled "Closed" in Figure 8-5; this tweezers-like action is detected by the FRET between the two dyes.

The really important part of this work is the next step. F consists of more than just the blue and green sections. It also has a red section, which is called a "toehold." The whole system works because of strand $\bar{F}$, whose sequence consists of complements to the red portion, to the green portion, and to the blue portion of strand F. When strand $\bar{F}$ is added to the solution, its toehold binds to the toehold. Then, through single-stranded branch migration, it works its way through the green and blue sections of F, displacing it from the rest of the tweezers. A duplex of F and $\bar{F}$ is formed, and this is waste for the system. The open state is restored. It is key to realize that the toehold-mediated strand removal is isothermal, requiring nothing more than the addition of the $\bar{F}$ strand (often called the "fuel" strand) to reset the system. Virtually all enzyme-free nanomechanical device systems are based on toehold-mediated strand removal. Although we described it earlier, it should be clear that the G4 device of Figure 8-4 was based on this principle: that's how strand β removed strand α.

The PX-JX_2 device. One problem with the tweezers device as reported was that it was not robust. A certain amount of dimerization occurred during the transitions. A robust rotational device based on isothermal toehold-based strand removal was developed shortly thereafter.[8,9] This device was based on the PX state of DNA and one of its topoisomers, the JX_2 state. These two structures are shown in Figure 8-6. The PX state, shown on the left, looks like a turn and a half of a blue duplex wound around a turn and a half of a red duplex, resulting in two double helical domains, one on the left and one on the right. In fact, the structure is a little more subtle than that. If you look closely at the domain on the left, you see, going from top to bottom, a red half-turn of DNA, then a red–blue half-turn, then a blue half-turn, and then a blue–red half-turn, followed again by a red half-turn. The other domain has just the opposite coloring. The top has been labeled with a red A and a blue B, in line with the colors of the strands. The bottom has been labeled in the same way with a blue C on the left and a red D on the right. The other motif, JX_2, is a topoisomer of PX. It lacks a pair of crossovers in the middle of the structure. This difference results in the wrapping consisting of a single turn, rather than a turn and a half. The red D is now on the left, and the blue C is on the right. Thus, there is a half-turn difference between the two structures.

Although Figure 8-6 shows the structural basis of the two states of the PX-JX_2 device, it doesn't show how it works. This is shown in Figure 8-7. The PX structure is shown on the left, but there have been some changes from Figure 8-6. Although it is not important, the red strands have been joined by hairpin loops, and so have the blue strands, so there is just one red strand and just one blue strand. The key difference is that the red strand and the blue strand have each been interrupted by a green strand. The green strands are called "set" strands, because they set the state of the device. The green strand interrupting the red strand has a horizontal segment on its 5′ end, and the green strand

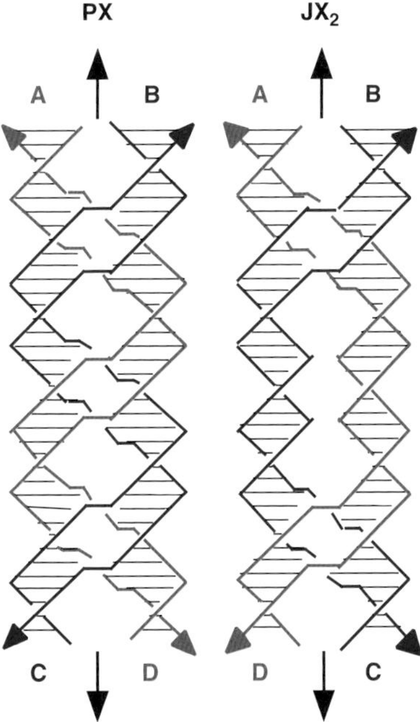

Figure 8-6 *The PX and JX_2 structures.* The PX molecule looks like a blue double helix and a red double helix wrapped around each other, but closer inspection reveals that there are half-turn regions where the red and blue strands actually pair. The JX_2 molecule lacks two crossover sites present in the PX molecule. Consequently, although the tops of the two molecules are the same, their bottoms are rotated by a half-turn about the central two-fold axis, as shown by the positions of the C and D labels in the two molecules.[5.51]

interrupting the blue strand has a horizontal segment on its 3′ end. These horizontal segments are toeholds. Other than the toeholds, the green strands are just like the red and blue strand segments that they have replaced, performing the crossovers that generate the PX state. Just as in the tweezers device, the green strands can be removed by adding their complete complements, including the toeholds. This is shown in process I, which shows the green strands being removed by other green strands ("unset" or "fuel" strands) with black dots on their 5′ ends. These dots represent biotin groups, and the addition of magnetic streptavidin beads enables removal of the duplex from solution. When the green strands have been removed, the result is a naked frame, which is shown at the top of the figure. The large single-stranded regions where the green strands have been removed result in the frame being a poorly defined intermediate in the 3D structural sense. Process II shows the addition of

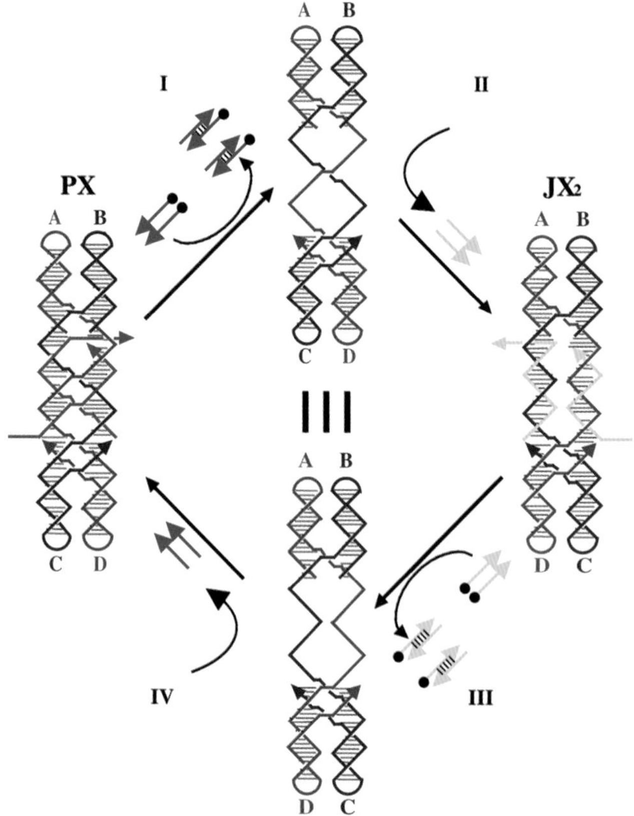

Figure 8-7 *The machine cycle of the PX-JX_2 device.* Starting from the left, the PX state is maintained by two green "set" strands containing horizontal toeholds. Addition of the complete complements of the green strands ("unset" or "fuel" strands) pulls them out, leaving a naked frame at the top. The green duplex contains biotin groups (black dots) that can be removed from solution by magnetic streptavidin beads. Addition of the yellow set strands puts the device in the JX_2 state. Addition of the unset strands to this state again leaves a naked frame (at the bottom), so it is ready to return to the PX state with the addition of more green set strands.[5.51]

yellow set strands that convert the poorly structured frame to a well-defined JX_2 state. The yellow strands form simple double helices to produce the JX_2 state, rather than the crossovers that are formed by the green strands to generate the PX state. Note that, just as in Figure 8-6, the tops of both well-structured states of the device are identical, but the bottoms have been rotated by a half-turn. The machine cycle is completed on the bottom half of Figure 8-7. In process III, a pair of yellow unset strands removes the yellow set strands to reveal the naked

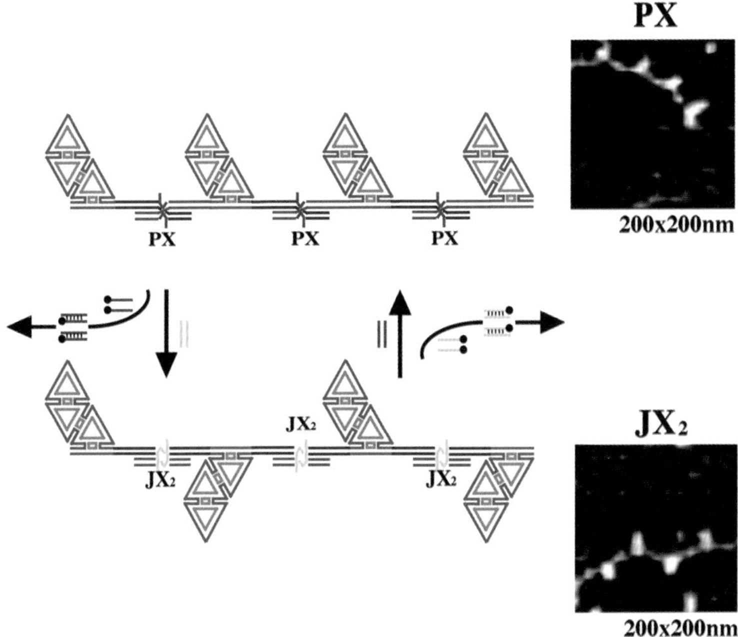

Figure 8-8 *A system for using the AFM to demonstrate the operation of the PX-JX$_2$ device.* A series of trapezoidal half-hexagons are connected by the device. In the green-strand PX state, the trapezoids are parallel, but they form a zigzag in the yellow-strand JX$_2$ state. AFM images at the right demonstrate both states.[5.51]

frame again. Addition of the green set strands in process IV completes the cycle to return the system to the PX state.

Figure 8-8 illustrates a system whereby the action of the PX-JX$_2$ device can be visualized directly in the AFM. DNA half-hexagon trapezoids are connected by the device. In the PX state, all the trapezoids are parallel to each other, but in the JX$_2$ state they form a zigzag pattern. Thus, conversion from one state to another should produce two different patterns. The trapezoids are large enough to be seen in the AFM. The structures corresponding to the two different states are shown in Figure 8-8. The differences between the parallel PX arrangement and the zigzag JX$_2$ arrangement are evident in the AFM images that are shown.

Multiple states. This same basic system can be used to produce a nanomechanical device with three robust states, not just two.[8.10] This device is shown in Figure 8-9. Figure 8-9a shows the three distinct states that can be formed by the system. We are already familiar with the PX and JX$_2$ states, shown with green and yellow set strands, just as they are drawn in Figures 8-7 and 8-8, but a

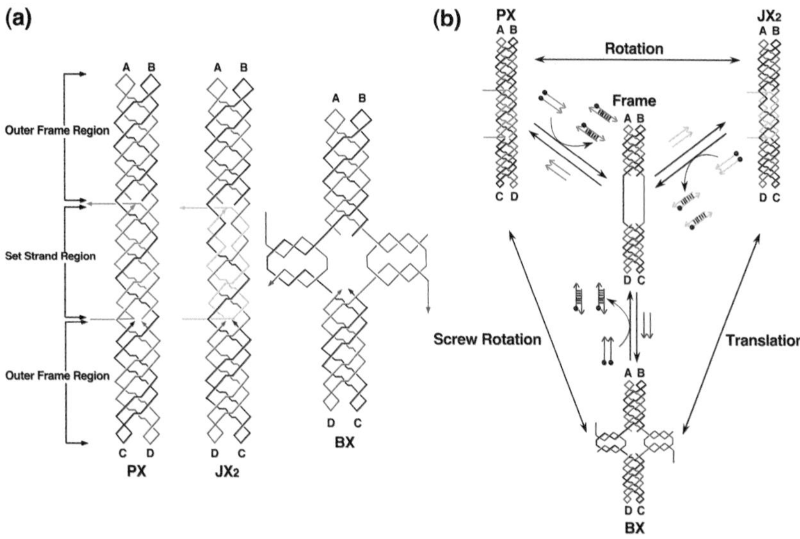

Figure 8-9 *A three-state device.* (a) *The three states: PX, JX_2, and BX.* This is similar to the states shown in Figure 8-6, except that the cruciform BX state is new. (b) *Forming the three different states.* The unstructured naked-frame intermediate of the PX-JX_2 device is the intermediate here as well. However, a new set of set strands, drawn in purple, can contract the intermediate to form the BX state. Thus, the transition from the PX to the JX_2 state is still rotation, but the transition from the JX_2 state to the BX state is translation, and the transition of the BX state to the PX state is a screw rotation.[8.10]

new state has been added, which we have termed the BX state. It is drawn with purple set strands that cause it to form an X-like structure. Closer inspection of the three molecules in Figure 8-9a shows that the C and D hairpins in the BX molecule are just like those in the JX_2 molecule, so the BX state corresponds to a contraction of the JX_2 state. Figure 8-9b shows the transitions that can be undergone by this device. At the center of the diagram is the poorly structured frame, which can act as a branching point for the state of the system; all final states are the result of adding set strands to the frame. The top of the diagram shows the same transitions discussed earlier, the rotational transition between the PX state and the JX_2 state. The bottom-right of the diagram shows the contraction between the JX_2 state and the BX state. This contraction is a simple translational motion. The bottom left part of the diagram shows the motion that occurs between the BX and PX states. This is a two-fold screw rotation, consisting of a rotation and a translation. This is the same kind of motion that one undergoes when moving along the stairs of a helical (usually called spiral) staircase, or along the nucleotides in the DNA double helical backbone. Of

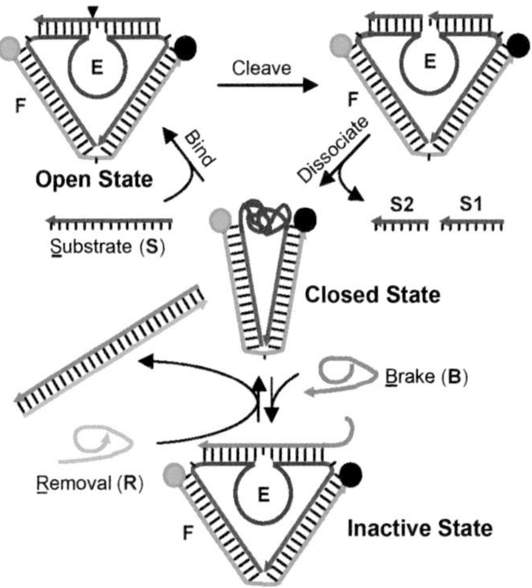

Figure 8-10 *A DNAzyme-driven machine.* The DNAzyme is labeled E. It cleaves an RNA strand, which is the substrate (S). The machine opens when the substrate binds. The two products do not pair strongly enough to bind after cleavage, leading to the single-stranded DNAzyme collapsing, and allowing the machine to close. The lower part of the diagram shows an uncleavable "brake," which binds instead of the substrate. However, the device can be readied for action if the brake is removed via the removal strand (R) binding to it through its toehold. Reprinted by permission from the American Chemical Society.[8.11]

course, those motions usually have a very small rotational component, and here it is a half-turn, but the principle is the same: rotation combined with translation.

Autonomous devices. All of the devices described above are clocked devices: the experimenter changes something about the environment of the device and some structural feature of the device itself is altered. It is evident that it would be desirable to create machines that run without experimenter intervention. This has now been done in a few instances. In one case, Mao and colleagues have built a device based on a DNAzyme that cleaves RNA.[8.11] The advanced version of this device is shown in Figure 8-10. In the open state (upper left), the device binds a strand of RNA and it opens. When the RNA is cleaved (upper right), the two cleavage products are sufficiently short that they dissociate from the device; when the products dissociate, the device closes. If there is another molecule of the substrate in the solution, it too can be bound, and the system can go through another round. The system shown in Figure 8-10 has another level of sophistication. A DNA brake that cannot be cleaved by the DNAzyme can be added to the solution, as shown at the bottom of the drawing.

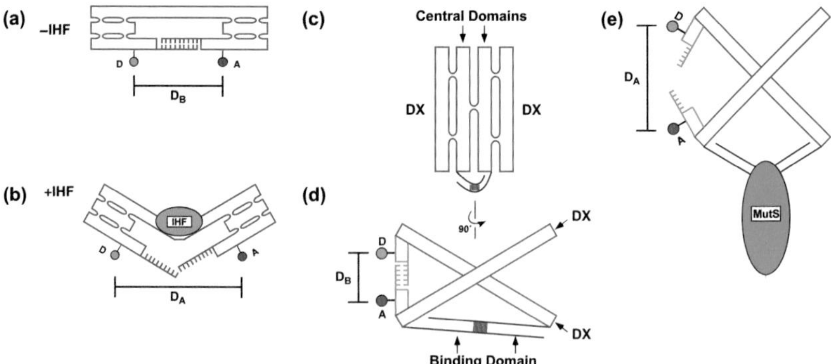

Figure 8-11 *DNA devices that measure the binding of DNA-distorting proteins to DNA.* FRET pairs to monitor the binding are shown as D and A; distances before and after binding are indicated as D_B and D_A, respectively. (a) and (b) *The binding of integration host factor (IHF) to a DNA device.* When IHF binds, it breaks a sticky end. The strength of the sticky end can be titrated to measure the binding constant of IHF to its sequence. (c), (d), and (e) *The binding of MutS to a scissors-like device.* (c) and (d) are two orthogonal views of the scissors. (e) shows the distortion of the device and the separation of the sticky ends by the binding of MutS. Reprinted by permission from the American Chemical Society.[5.33]

The machine can be activated by removing the brake, using the standard toehold approach.

Another category of device is one that can be driven simply by using a specific sequence or sequence feature of DNA. Such a device requires a recognition element, such as a protein. There may be many such devices, such as those based on restriction enzymes[8.12] or nucleic acid-based catalysts, such as ribozymes or DNAzymes. Such devices are a step closer to biology, in that they utilize molecular species whose design is currently at least challenging for molecular designers, and is usually not possible at all except by homology, i.e., by changing a small part of an existing protein molecule or by mixing and matching known binding and catalytic segments of nucleic acid-based catalysts.

DNA–protein devices. However, there is a class of protein-recognition devices that can be used to *explain* biological systems, and I'd like to focus on them for a bit. The first such device was developed to estimate the thermodynamics of IHF binding.[8.13] This device is illustrated in panels a and b of Figure 8-11. Figure 8-11a shows the design of the device. It consists of two TX components, one on the left and one on the right. They are connected by a shaft connecting the top domains and containing the binding site for the integration host factor (IHF) protein. This protein is a component of the recombination machinery of the bacterial cell, and one of its key features is that it bends the DNA to which it binds. When IHF binds, it bends the DNA by a large amount – much more than illustrated in the cartoon shown in Figure 8-11b. The actual binding event can be monitored by the pair of FRET dyes shown as a green circle (labeled D for donor) and a red circle (shown as A for acceptor). The bottom two domains are connected by a sticky end of potentially variable strength. Figure 8-11b shows that the sticky-ended interaction is broken when IHF binds to its recognition site in the top domain. It is possible to measure the strength with which IHF binds to its recognition sequence by titrating the strength of the sticky end, which is readily estimated.[8.14] Thus, a DNA-based device that is activated by a protein can be used for analysis of the binding interaction, so long as there is a physical signal (change in separation of the FRET dyes, for example) to report the binding event.

It turns out that the IHF-driven device is not entirely general. It depends on the position on the DNA to which the protein binds. A more general device based on a scissors-like motion has been developed to analyze the binding of MutS protein. MutS is a DNA-repair protein that recognizes mis-paired or unpaired sites in DNA. When it binds to such a site, it bends the DNA. MutS is a large protein that would not bind to the first device without distorting it; hence, the scissors was developed.[8.15] Figure 8-11c shows that this device consists of a pair of DX domains connected by a single crossover site in the middle. This single crossover site acts as the fulcrum of the scissors. A blue section, representing the DNA duplex to which MutS will bind, is shown at the bottom of this panel; the magenta portion of the duplex contains the lesion that MutS will recognize. The transition from Figure 8-11c to Figure 8-11d consists of a 90° rotation about the vertical. Figure 8-11d shows the device looking more like one would expect a scissors to look. The variable-length sticky end, flanked by a pair of FRET dyes, is visible on the left. This part of the device is very similar to the IHF-binding device. When MutS binds to its binding domain, it closes the blades of the scissors, as illustrated in Figure 8-11e. In the same fashion as the IHF device, the ability of MutS to close the blades and alter the FRET signal is a function of how strongly it binds to the binding site. The different possible lesions that are recognized by MutS are characterized by different binding energies, and these are in turn reflected in the strength of the sticky end that MutS is able to separate.[8.15]

Walkers

We decided to call the devices discussed above shape-shifters: they change their robust structures in response to an external stimulus, which may be a change in conditions or the presence of a new activating molecule in solution. There is a second category of DNA-based nanomechanical devices that we can call *walkers*. These differ from shape-shifters in that they change their position relative to an external sidewalk as a result of a structural alteration. Thus, they can move from one place to another. The sidewalk is a key element of the walker: it would just flail around if two of its legs were not able to attach to a sidewalk and then to change their positions of attachment. It is key that the walker never ceases to be attached to the sidewalk, at least in a single position. The enzyme-requiring device of reference 8.12 is indeed such a walker.

The very first protein-independent walker was extremely simple, moving just like an inchworm. Its sidewalk was also simple, the ends of a TX molecule.[8.16] This system is shown in Figure 8-12. The TX molecule is shown as three doubly connected rectangles at the bottom of each of the six panels. The essence of this walker is that it consists of two double helical segments of DNA, drawn largely in ochre, that are connected by a flexible linker. Its feet are labeled “Foot 1” and “Foot 2”; the feet can rest on positions on the sidewalk that are labeled “Foothold A,” “Foothold B,” and “Foothold C,” although C is unoccupied in the first panel. Figure 8-12a shows the walker attached at its first position. The attachment of Foot 1 to Foothold A is achieved by a set strand labeled “Set 1A” and the attachment of Foot 2 to Foothold B is achieved by a set strand labeled “Set 2B.” Strand Set 1A holds Foot 1 to Foothold A, and strand Set 2B holds Foot 2 to Foothold B. Both of these strands contain toeholds (drawn as horizontal extensions). Figure 8-12b shows the addition of an unset (or fuel) strand to the system, and Figure 8-12c shows that strand Set 2B has been removed from the system. The small blue circle represents a biotin group that can be used to remove the duplex of strands Set 2B and Unset 2B from the system. At this point, Foot 2 has been released from the sidewalk, although Foot 1 remains attached to the sidewalk. Figure 8-12d shows that a new strand, Set 2C, has been added to the system, binding Foot 2 to Foothold C. Thus, the walker has extended itself, much as an inchworm might. Figure 8-12e shows that Set 1A has been removed by its unset strand, analogous to the state in Figure 8-12c. However, in this state Foot 2 is securing the walker to the sidewalk, while Foot 1 is dangling loose. The final panel shows that a new set strand, Set 1B, has secured Foot 1 to Foothold B, completing the walk. This walker was shown to be able to walk in both directions. The little red bottoms of the feet represent psoralen groups that have been used to crosslink the walker to its sidewalk in various positions, thereby demonstrating that the walk has taken place. Shortly after this walker was published, a second walker was published that used a leg-over-leg motion, and was demonstrated by FRET data rather than by crosslinking.[8.17]

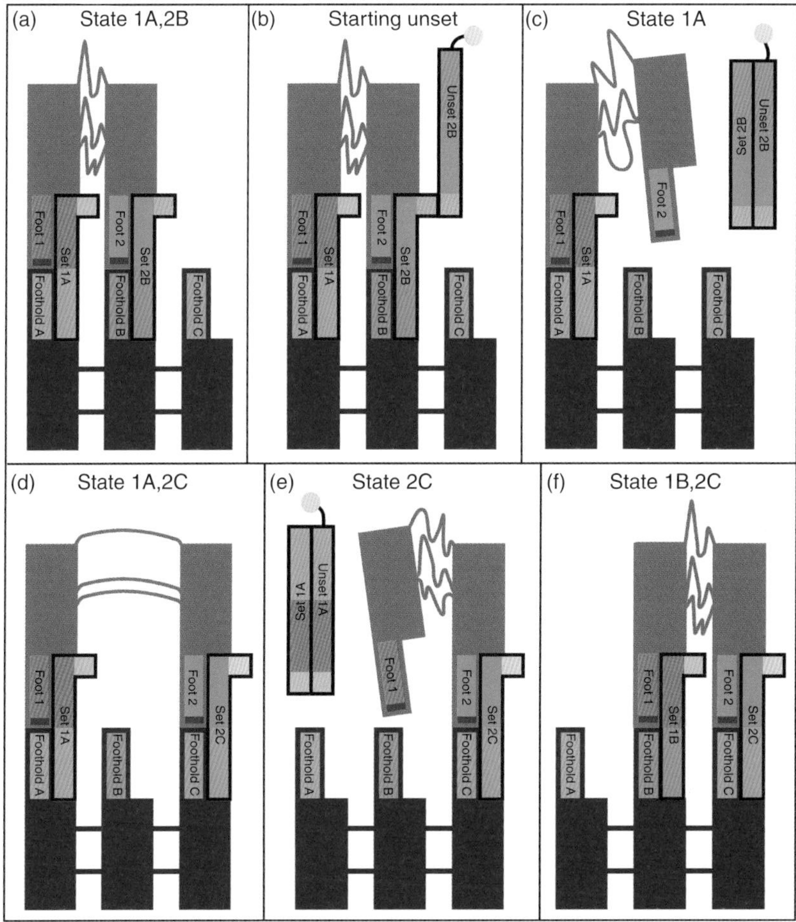

Figure 8-12 *A bipedal walker that walks on a TX sidewalk.* The walker is shown attached by two legs to the left domains of the TX molecule. The next two panels show the release of the right leg using toehold methods. The fourth panel shows the flexible linker extended as a new strand has tied the right leg to a new position on the far right. The fifth panel shows the rear leg released, and the last panel shows it tied down to the middle domain of the TX molecule. Reprinted by permission from the American Chemical Society.[5.52]

The walkers described above are clocked devices. They do not function without the intervention of some sort of operator. It is possible to devise walkers that will function as cascade devices, so long as they do not run out of either fuel or track. The basis for such a device is shown in Figure 8-13a.[8.18] The walker shown there is a single strand of DNA with two solid circles in its

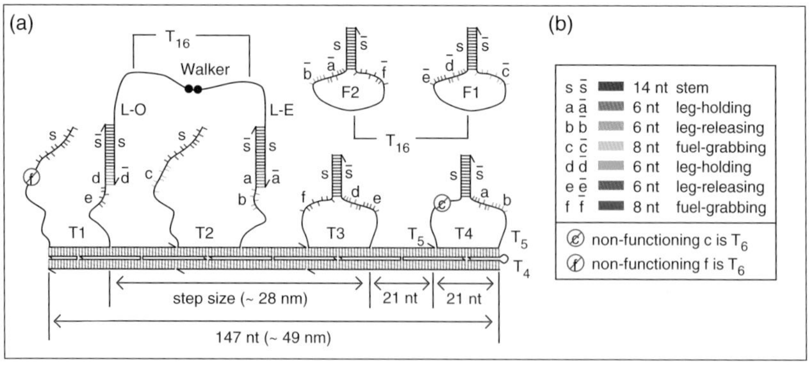

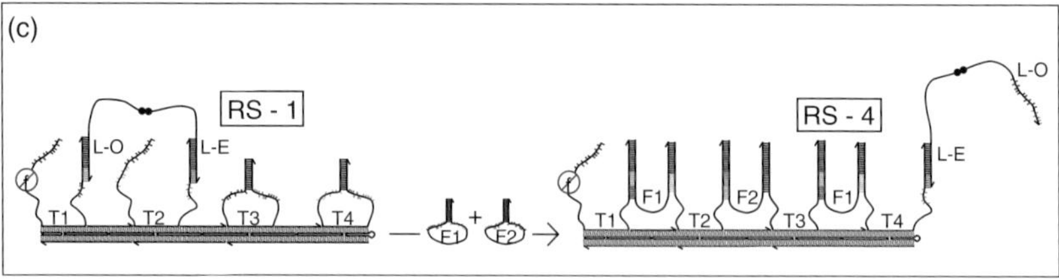

Figure 8-13 *A cascade walker.* (a) *The initial position of the walker.* The strand is linked 5′, 5′ in the middle, and its 3′ ends are paired with two right strands of the T1 (L-O) and T2 (L-E) stem-loops, which are opened. The stem-loops T3 and T4 are intact, and the solution contains fuels F1 and F2. The sidewalk is a long DX molecule. (b) *The special codes for the colored regions.* (c) *The burnt-bridges reaction whereby the two fuel molecules bind successively to the stem-loops, powering the walker to the right.* Reprinted by permission from the AAAS.[5.53]

center. Those circles represent a 5′, 5′ linkage, so that it is symmetric, and has two 3′ ends. The two legs are called L-O (leg-odd) and L-E (leg-even). This walker walks on a track that is a DX molecule containing a series of four (in this case) stem-loop structures, named T1, T2, T3, and T4. T1 and T2 are already open, but T3 and T4 are closed. The step size here is much longer than in the inchworm walker. A variety of functions are needed in each of the stem-loops, which are color-coded in Figure 8-13b. These include leg-holding, leg-releasing, and fuel-grabbing. The system is powered by two different fuel molecules that alternate in supplying the increased amount of base-pairing that powers the device at every step. Figure 8-13c shows that this is a "burnt-bridges" system, whereby the motion of the walker uses up the track as the system proceeds from the first resting state (RS-1) to the final resting state (RS-4).

The way that this walker actually does its walking is shown in Figure 8-14. The system is initiated at step 1 by the fuel strand F1 being attacked by the left

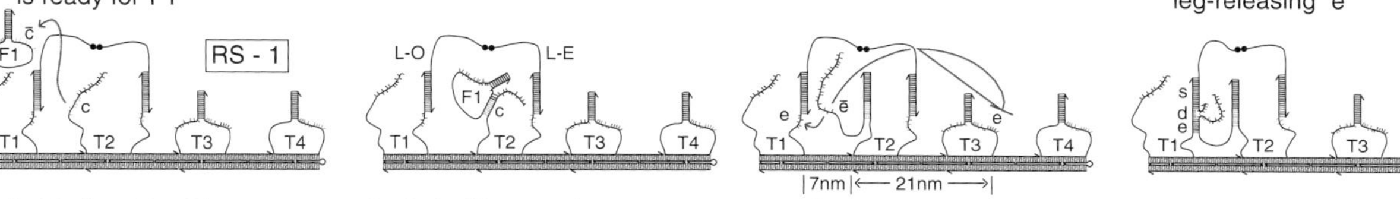
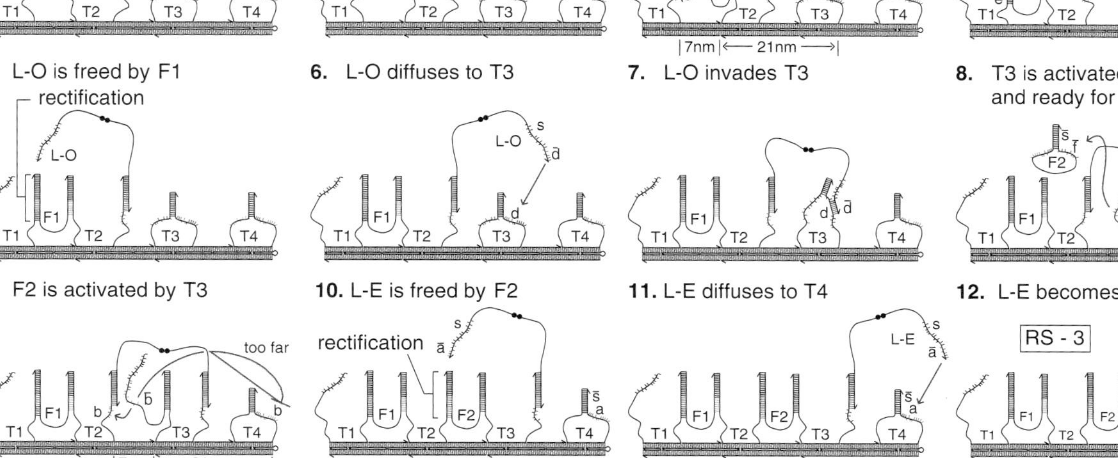

Figure 8-14 *All the steps of the cascade walker in Figure 8-13.* It has gone through more than a cycle, and could, in principle, go further. Reprinted by permission from the AAAS.[5.53]

strand of T2, which it invades in step 2. The invasion separates its two strands (step 3), and it invades and begins to release the walker in step 4. This step is completed in step 5, with L-O being released. In step 6, L-O diffuses over to T3, which it invades in step 7. In step 8, the T3 stem has been separated into two separate strands and can now be attacked by fuel strand F2. This is called Resting State 2 (RS-2). Steps 9–12 show the same events with F2 and L-E that F1 and L-O underwent in the previous steps, leading to Resting State 3 (RS-3). The walk is not complete, though, because now another molecule of F1 can invade T4, leading to the release of L-O, as shown in RS-4 in Figure 8-13. The step-yield has been estimated by crosslinking analysis (analysis of the whole ensemble of molecules in the vessel) to be about 74%. This level of efficiency is not overwhelming, but the principle has moved workers to advance in further directions.

References

8.1 X. Yang, A. Vologodskii, B. Liu, B. Kemper, N.C. Seeman, Torsional Control of Double Stranded DNA Branch Migration. *Biopolymers* **45**, 69–83 (1998).

8.2 A.D. Bates, A. Maxwell, *DNA Topology*, Oxford, IRL Press (1993).

8.3 C. Mao, W. Sun, Z. Shen, N.C. Seeman, A DNA Nanomechanical Device Based on the B-Z Transition. *Nature* **397**, 144–146 (1999).

8.4 A. Rich, A. Nordheim, A.H.-J. Wang, The Chemistry and Biology of Left-Handed Z-DNA. *Ann. Rev. Biochem.* **53**, 791–846 (1984).

8.5 P.S. Ho, M.R. Ellison, G.J. Quigley, A. Rich, A Computer-Aided Thermodynamic Approach for Predicting the Formation of Z-DNA in Naturally Occurring Sequences. *EMBO J.* **5**, 2737–2744 (1986).

8.6 S. Modi, C. Nizak, S. Surana, S. Halder, Y. Krishnan, Two DNA Machines Map pH Changes Along Intersecting Endocytic Pathways in the Same Cell. *Nature Nanotech.* **6**, 459–467 (2013).

8.7 J.J. Li, W. Tan, A Single DNA Molecule Nanomotor. *Nano Letters* **2**, 315–318 (2002).

8.8 B. Yurke, A.J. Turberfield, A.P. Mills, Jr., F.C. Simmel, J.L. Newmann, A DNA-Fuelled Molecular Machine Made of DNA. *Nature* **406**, 605–608 (2000).

8.9 H. Yan, X. Zhang, Z. Shen, N.C. Seeman, A Robust DNA Mechanical Device Controlled by Hybridization Topology. *Nature* **415**, 62–65 (2002).

8.10 B. Chakraborty, R. Sha, N.C. Seeman, A DNA-Based Nanomechanical Device with Three Robust States. *Proc. Nat. Acad. Sci. (USA)* **105**, 17245–17249 (2008).

8.11 Y. Chen, C. Mao, Putting a Brake on an Autonomous DNA Nanomotor. *J. Am. Chem. Soc.* **126**, 8626–8627 (2004).

8.12 P. Yin, H. Yan, X.G. Daniell, A.J. Turberfield, J.H. Reif, A Unidirectional DNA Walker that Moves Autonomously Along a Track. *Angew. Chemie Int. Ed.* **43**, 4901–4911 (2004).

8.13 W. Shen, M. Bruist, S. Goodman, N.C. Seeman, A Nanomechanical Device for Measuring the Excess Binding Energy of Proteins that Distort DNA. *Angew. Chem. Int. Ed.* **43**, 4750–4752 (2004).

8.14 J. SantaLucia, Jr., A Unified View of Polymer, Dumbbell and Oligonucleotide DNA Nearest-Neighbor Thermodynamics. *Proc. Nat. Acad. Sci. (USA)* **95**, 1460–1465 (1998).

8.15 H. Gu, W. Yang, N.C. Seeman, A DNA Scissors Device Used to Measure MutS Binding to DNA Mis-Pairs. *J. Am. Chem. Soc.* **132**, 4352–4357 (2010).

8.16 W.B. Sherman, N.C. Seeman, A Precisely Controlled DNA Bipedal Walking Device. *Nano Letters* **4**, 1203–1207 (2004).

8.17 J.-S. Shin, N.A. Pierce, A Synthetic DNA Walker for Molecular Transport. *J. Am. Chem. Soc.* **126**, 10834–10835 (2004).

8.18 T. Omabegho, R. Sha, N.C. Seeman, A Bipedal DNA Brownian Motor with Coordinated Legs. *Science* **324**, 67–71 (2009).

9

DNA origami and DNA bricks

One of the central concepts in structural science is the notion of *resolution*. What we're talking about here is the extent of detail or precision with which we wish to work. Resolution is usually thought of as the detail level of analysis, particularly optical analysis, but it can also refer to the detail level of construction. For example, when I was a "small molecule" crystallographer, it was useful to have a mental picture of my structures where the rulings were approximately every 10 picometers. My structures themselves were built up from Fourier components whose crests were separated by about 80 picometers (their nominal resolution) or more, but I was able to find meaningful differences between things such as bond lengths that were about 10 picometers different from each other. It is usually more useful to think of macromolecular crystals (resolution typically 200–400 picometers) more crudely, say in a world where the rulings are around 50 picometers apart. Resolution should fit the subject appropriately: if we were building a house with 5 × 10 × 15 cm bricks, this type of thinking would be far too detailed; we would be wasting out time thinking in a world with rulings closer than about a centimeter. Biology is replete with phenomena that take place on multiple distance scales, everywhere from the sub-nanometer scale (e.g., simple enzymatic reactions) to the nanometer scale (e.g., transcription, translation, recombination, and other aspects of nucleic acid metabolism) to the micron scale (cellular phenomena) to the macroscopic scale (organs like muscles and nerves).

DNA origami. In the aspects of structural DNA nanotechnology we have discussed so far, our basic unit has been the DNA double helix, with a diameter of 2 nm. 2D DX arrays should in principle have a repeat perpendicular to the helix axis of about 4–5 nm, but we usually find that the actual repeat is about 6 nm. Thus, these structures are usually designable to within ±1 nm. If we loosen up our design criteria a bit, say to 6 nm, it is possible to design larger structures readily. This is the technique known as DNA origami, first published

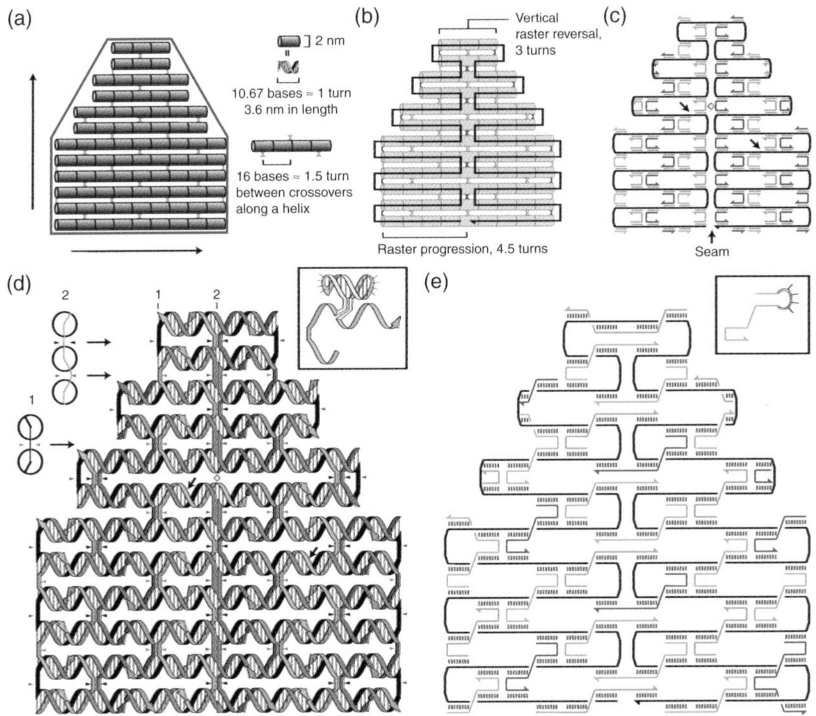

Figure 9-1 *Designing with DNA origami.* (a), (b), and (c) show the principles of designing a hexagon with two right angles from DNA origami. The design is shown in red in (a) and the trajectory of the scaffold strand is shown as a zigzag line in (b). (c) illustrates the notion of holding the scaffold strands together with staple strands. The helical details of the structure are shown in (d). For more details, see reference 9.1. Reprinted by permission from Macmillan Publishers Ltd.[2.29]

by Paul Rothemund in 2006.[9.1] He took the single-stranded form of the M13 virus, ~7500 nt, which is commercially available, and combined it with around 250 "staple" strands to get it to fold into a series of shapes containing parallel helix axes. These shapes are all based on extended versions of the DAO motif. The basic notion is illustrated in Figure 9-1, where a hexagon with two right angles is shown. The basic idea is that the short strands can fold the long strand into the target shape. There are a number of aspects of the physical structure of DNA to which the designer must adhere, of course, such as the helicity and the positions of the major and minor grooves. Several software packages are available to design DNA origami (e.g., see reference 9.2). One of the major

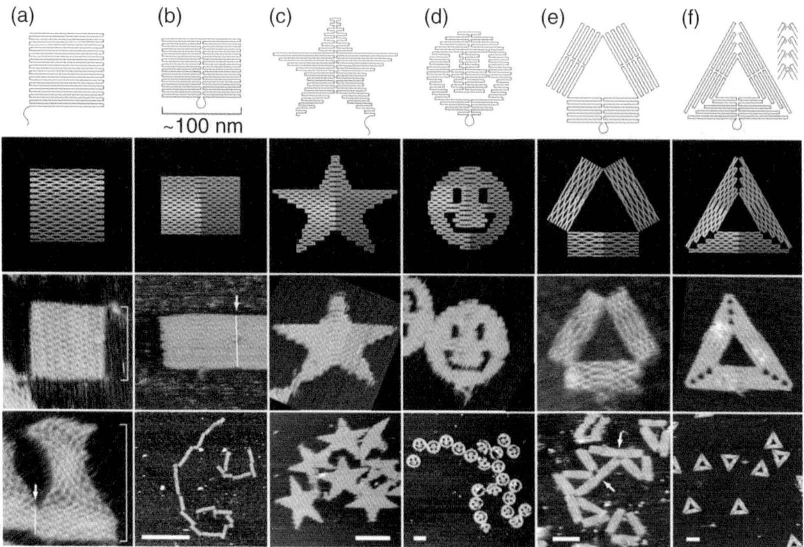

Figure 9-2 *Structures built from DNA origami.* Ideal targets are shown in the top two rows. AFM zooms are shown in the next row, and far views are seen in the bottom row. The versatility of the technique is visible in this collection of images. Reprinted by permission from Macmillan Publishers Ltd.[2.29]

advantages of using origami at a pixilation of 6 nm or so is that the large amount of stoichiometry testing and strand purification needed to assemble the same area can largely be eliminated. It remains necessary to purify those strands that will interact with external DNA species (see Chapter 10), but otherwise, to build the bulk of the area, purification seems to be eliminable, and stoichiometry determination is usually neglected in favor of using an excess of staple strands.

Figure 9-2 shows a variety of shapes that were initially designed and self-assembled using the DNA origami method. The square, rectangle, star, smiley face, and trigonal arrangements have all been made readily, with high yield. It should be pointed out that there are no guarantees that every staple strand is present in every origami molecule. The method is loose enough that we can't afford to care, and nobody has reported a labeled lane gel like the one in Figure 5-5 to establish whether one or more of the strands is systematically missing from any of the numerous origami constructions that have been published. The characterization of origami constructions is usually limited to "cherry-picking" in AFM or TEM experiments. The ensembles shown in Figure 9-2 are among the larger samplings that have been reported in the

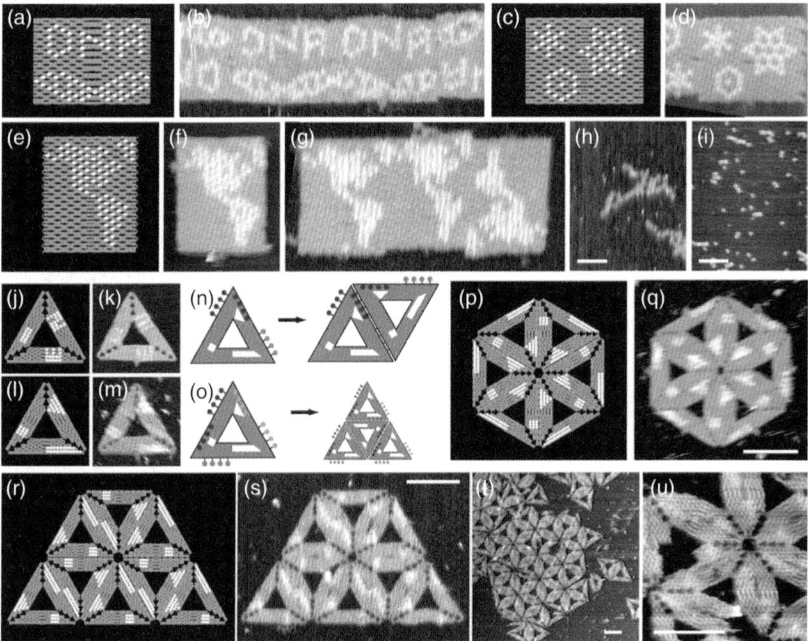

Figure 9-3 *Decoration of an origami surface to produce patterns.* By adding transverse hairpins (see panel d in Figure 9-1), images are created. This is arguably the greatest importance of DNA origami, the addressability of ~8000 nm^2 in about 100–200 positions. Reprinted by permission from Macmillan Publishers Ltd.[2.29]

literature, although cryo-EM examinations require many identical species (e.g., see reference 9.3).

The most important thing about DNA origami is not that DNA strands can be folded into smiley faces or the other arrangements shown in Figure 9-2. Rather, origami gives a way to address something approaching 10 000 nm^2 at about 200 or more loci. In Chapter 10 we will discuss pre-origami work to build a 2D lattice with eight different TX molecules. Getting all of the stoichiometries right and getting all the strands purified in that study was extremely difficult, and in the end the standard M13 DNA origami gives about three times the area of addressable space. Figure 9-3 shows how origami can be used to produce patterns of remarkable complexity. The spelling of "DNA" and the triple hexagonal pattern would both be extremely difficult to do with groups of individual tiles. The intricate map of the western hemisphere shows the extent to which one can go in a simple pattern.

The notion of DNA origami was something that evolved. The first system to combine a long strand with helper strands was a bar-code system shown in Figure 9-4. The bar code has a strand that spans five tiles, and the tiles are selected by the sequence of the bar code.[9.4] Different long strands can select different bar codes, depending upon whether the tile contains a pattern marker or not. An immediate predecessor to DNA origami[9.5] is shown in Figure 9-5. This is an octahedron that largely folds on itself, using the concept of PX-cohesion (see Figure 3-12).[9.6] However, only seven of the 12 edges consist of PX DNA. The other five edges are DX (DAE) motifs, and they are formed by the five cyclic blue strands visible in the image.

The strength of DNA origami is just beginning to be realized. Only a few laboratories have the resources to do the necessary experiments to optimize sequences for particular target structures. We'll review a few of the dramatic experiments that have been done with this interesting technique below, but this field is moving very quickly. Once you realize that origami is such a great way to increase the addressable area (or volume) of a DNA construct, you are suddenly frustrated by the fact that it is so small. We would all like to address square millimeters or centimeters with the precision of origami, but the typical M13-based origami tiles are only ~100 × 100 nm in dimension. Attempts to increase the area available have centered on trying to make lattices of origami. Small hexagonal arrangements are seen in Figure 9-3. The most successful approach to building 2D origami lattices has been to orient the helices in the directions of the unit cell.[9.7] Figure 9-6a shows this strategy, with helices pointed perpendicular to each other. Figure 9-6b shows a zoom of a portion of a 3 micron × 5 micron origami lattice. It should be noted that this was also the successful strategy employed in Chapter 7 to make 3D lattices: propagation of the lattice in linearly independent directions that are parallel to helix axes.

The 6-helix bundle (6HB) motif is shown in Figure 7-17a, and the rest of Figure 7-17 shows analyses of 2D arrays formed from 6HB motifs. Of course, 6HB motifs can also be stacked end-to-end to produce 1D arrays.[9.8] The Shih laboratory has made a 6HB about 1 micron long by combining two 6HB motifs designed from DNA origami. The purpose of these 6HB arrays is to produce long rods that can share a solution with membrane proteins without sticking to them. When the proteins collide with the long 6HB origami units, their rotational trajectories are slightly perturbed, leading to their partial alignment. This partial alignment can be used in NMR experiments to establish the 3D structures of the proteins.[9.9]

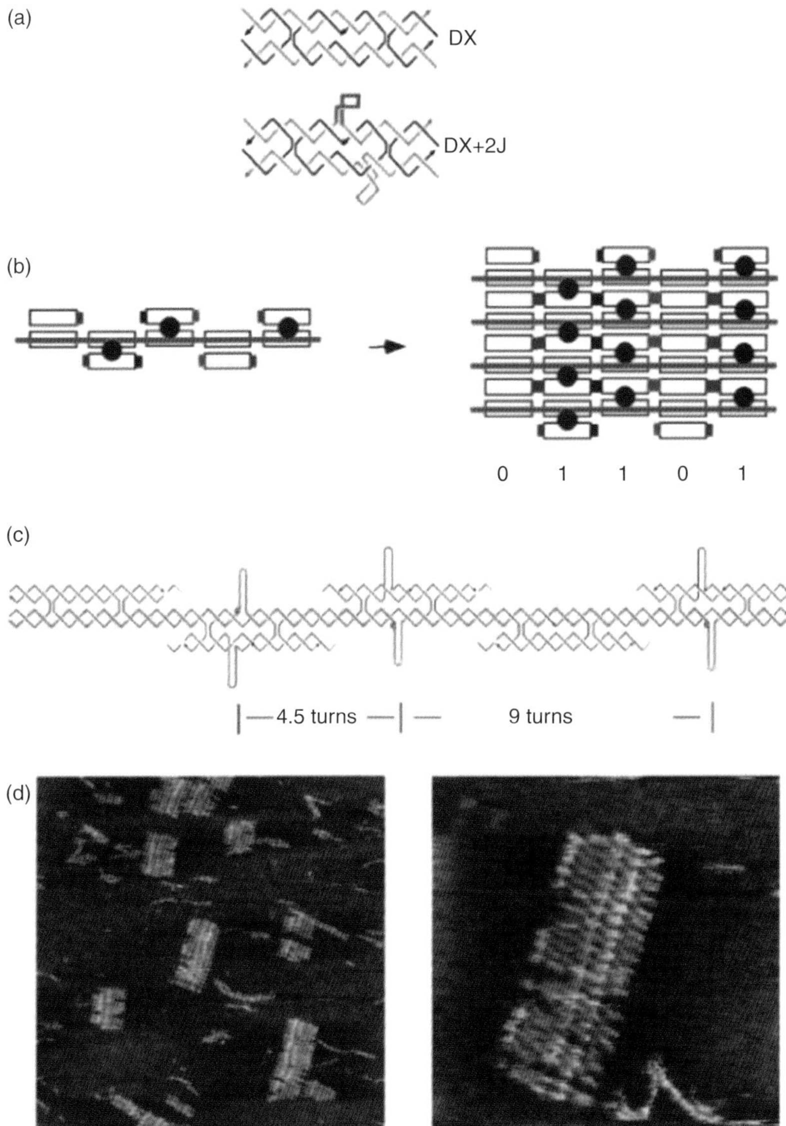

Figure 9-4 *A bar-code predecessor to DNA origami.* (a) *DX and DX+2J motifs.* (b) *A schematic version of the motifs.* (c) *A 1D pattern of these molecules in helical representation.* (d) *AFM images of the bar-code system.* Copyright (2003) National Academy of Sciences, USA.[9.4]

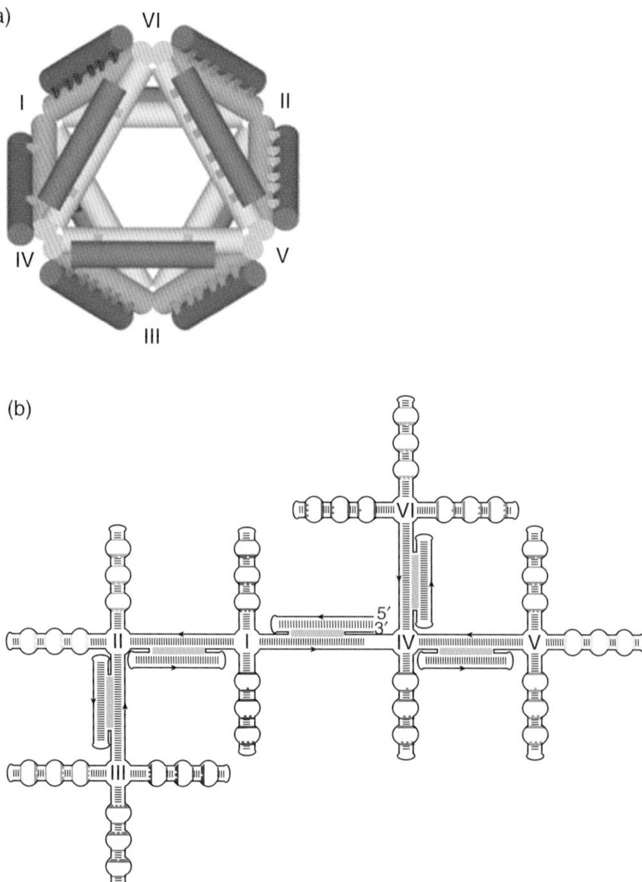

Figure 9-5 *An octahedral predecessor to DNA origami.* (a) *A schematic of an octahedron wherein each edge is a DX or a PX structure.* (b) *An exploded view of the strands.* The long strand is color-coded to indicate the strands for the seven PX motifs. The other five edges show the central strands of the DX structure. Reprinted by permission from Macmillan Publishers Ltd.[3.19]

The extension of origami to 3D was first based largely on the 6HB notion of the 120° angle resulting from using 7-nucleotide separations to orient a collection of helices in 6HB or extended 6HB honeycomb motifs that produce multilayer structures.[9.10] This notion is shown in Figure 9-7, where a series of helices are seen to be organized into a honeycomb layer. This approach is quite broad, and Figure 9-8 illustrates a variety of structures that have been formed in this fashion. The top rows show the designs in oblique and perpendicular views, and the lower rows show TEM images of the actual structures.

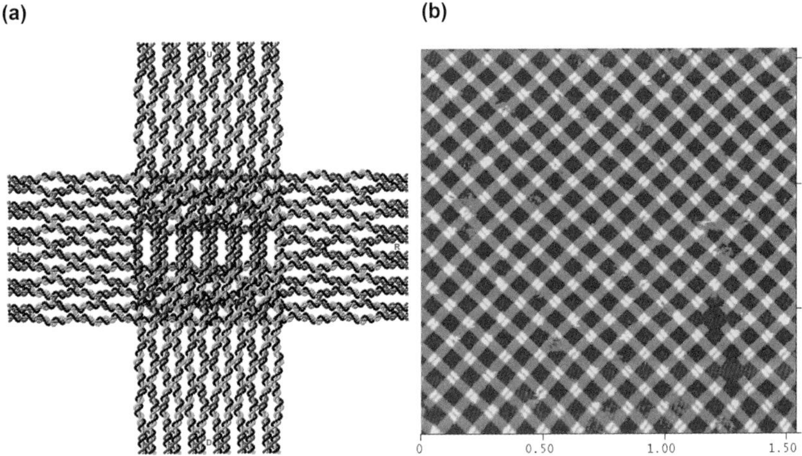

Figure 9-6 *A 2D origami array.* (a) *A DNA origami motif with helix axes pointing at right angles to each other.* (b) *A 2D origami array self-assembled from the motif in (a).* Reprinted by permission from John Wiley & Sons.[9.7]

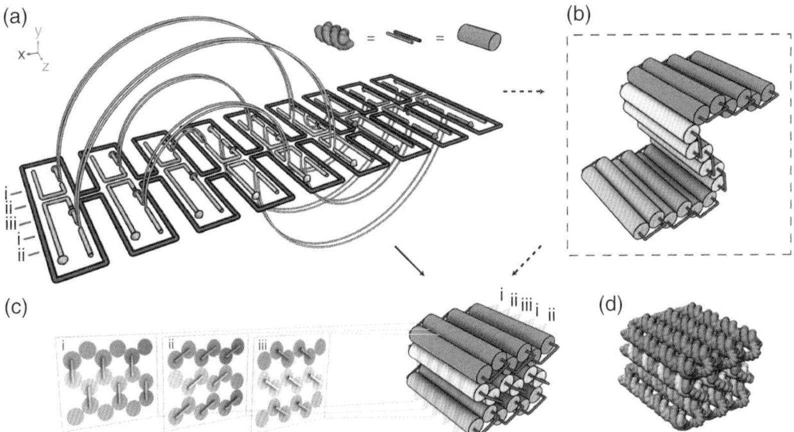

Figure 9-7 *Forming 3D origami motifs with the 6-helix bundle motif.* (a) *The connections between helical domains.* (b) *Schematic of leaving the planar conformation.* (c) *Connections between helices.* (d) *Schematic showing the final structure in both cylindrical and helical representations.* Reprinted by permission from Macmillan Publishers Ltd.[5.48]

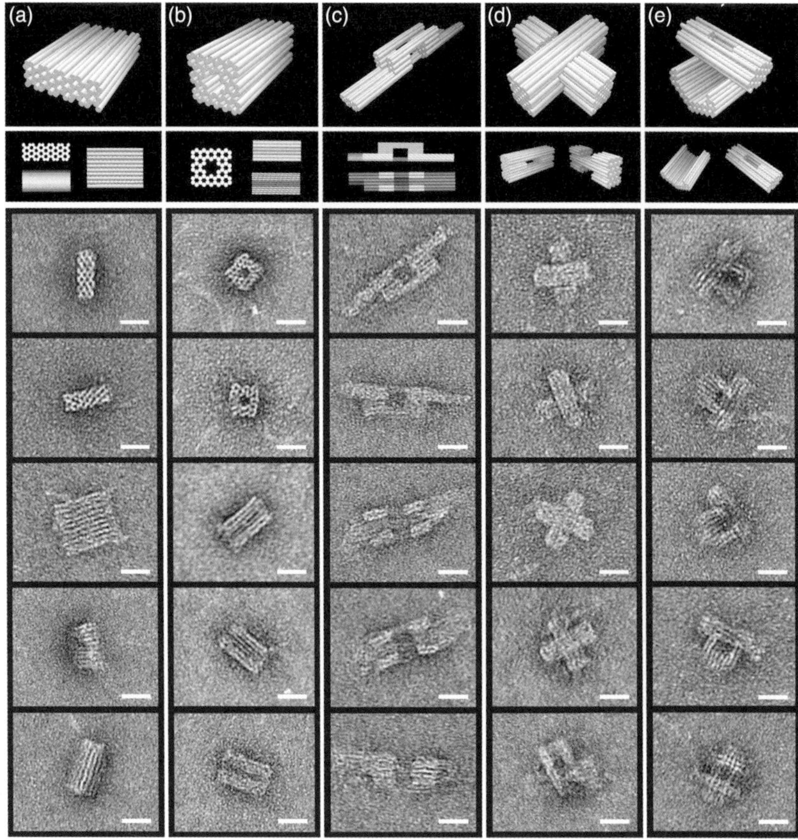

Figure 9-8 *Five 3D origami images.* Two different views of schematics are shown at top, and a series of cryo-EM images are shown for each one below. Reprinted by permission from Macmillan Publishers Ltd.[5.48]

The hierarchical assembly of the image seen in Figure 9-8e is shown in Figure 9-9a. The assembly and electron microscope examination of an icosahedron whose edges are 6HB motifs is shown in Figures 9-9b through 9-9d. The icosahedron is made from three different origami units that have been joined together successfully to produce the target.

The next notable foray into 3D origami came from the Aarhus group.[9.11] They have made an apparently cubic box. It consists of six segments that fold together to make a cube-shaped object, as illustrated in Figure 9-10. Six segments of the M13 viral strand are color-coded, and these colors correspond to the faces in panel b. Panel c shows how these six sides fold to produce the box-like figure. This box has generated quite a bit of excitement,

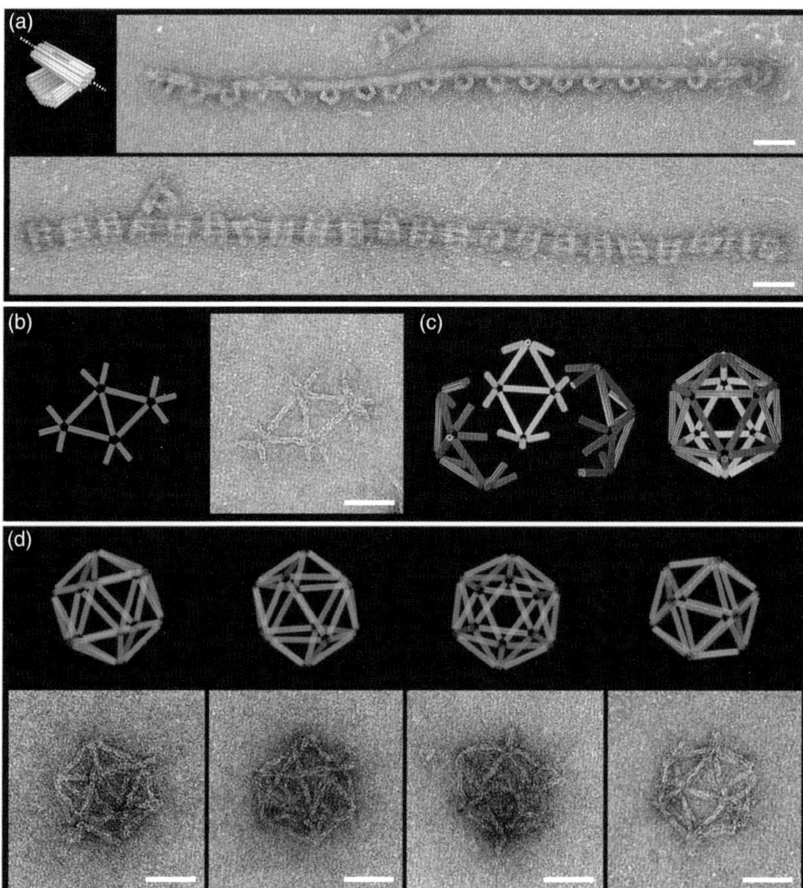

Figure 9-9 *3D origami constructs with multiple components.* (a) *A 1D array of a right-angle motif.* (b) *The basic unit of an icosahedron.* Each edge is a 6-helix bundle. (c) *Three of the units in (b) assemble to form the icosahedron.* (d) *Cryo-EM images of the icosahedron.* Reprinted by permission from Macmillan Publishers Ltd.[5.48]

because it is of an appropriate dimension (36 × 36 × 42 nm) to hold a macromolecule, and thus is implicated for potential utility in macromolecular drug delivery. Figure 9-11 shows that the lid can fold off the box, leaving one edge intact, much like the lid of a jack-in-the-box. Furthermore, the same drawing shows that the lid can be opened selectively. This is because two locks have been installed on the lid, locks that pair with each other but can be removed using the basic Yurke *et al.* toehold techniques described in Chapter 8. The top portion of Figure 9-11 shows how the locks can be

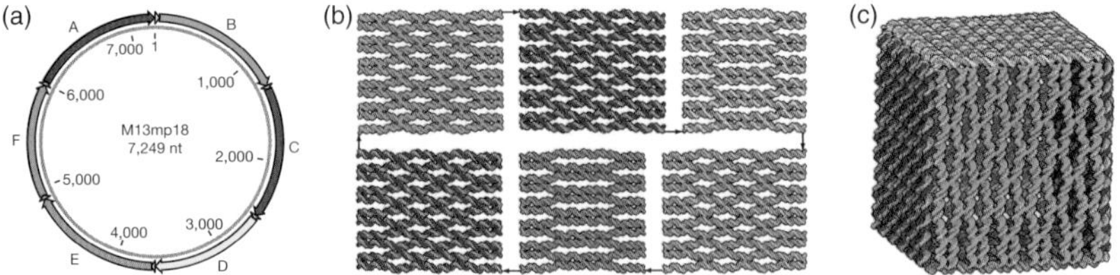

Figure 9-10 *Design of a DNA origami box.* The faces are color-coded on the scaffold in (a), and then on an unfolded surface in (b). (c) shows the box assembled. Reprinted by permission from Macmillan Publishers Ltd.[9.11]

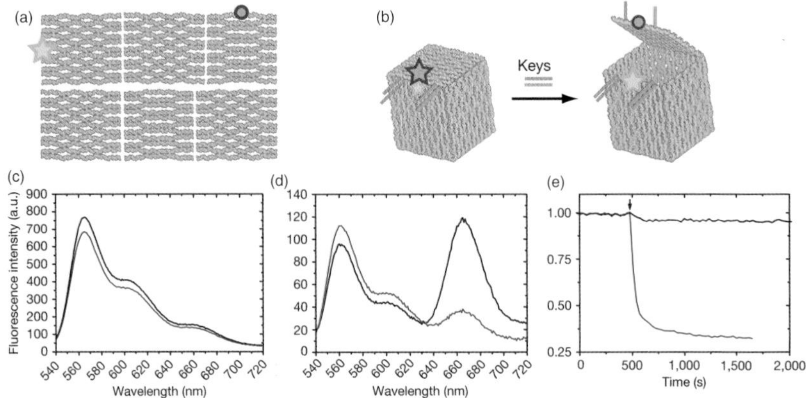

Figure 9-11 *FRET signals resulting from closure of the box in Figure 9-10.* Reprinted by permission from Macmillan Publishers Ltd.[9.11]

released, and the bottom portion shows how a FRET signal can be used to monitor the opening of the box.

One of the most exciting uses of DNA origami has been the work by Liedl *et al.* to build molecules that exhibit the actual properties of tensegrity.[9.12] Tensegrity structures (often attributed to Buckminster Fuller) were initially suggested by the artist Kenneth Snelson, to combine both tension and compression in the same motif; Snelson has built structures such as the statue shown in Figure 9-12a. It is clear that rigid metallic rods resist compression in this structure, while thinner metallic cables provide the tension. The combination of these two elements provides a stable sculpture that can be seen at the Storm King Art Center in New York state. The DNA analog of this structure is shown in Figure 9-12b. This image shows electron micrographs of the same types of elements. The rods are 13-helix bundles of DNA, related to the 6-helix bundles described above, and the roles of

Figure 9-12 *Tensegrity.* (a) *A sculpture by Kenneth Snelson at Storm King Art Center.* (b) *A tensegrity structure built from DNA origami.* Both schematic and cryo-EM images are shown. Reprinted by permission from Macmillan Publishers Ltd.[9.12]

the cables are taken by single strands of DNA whose lengths are set appropriately.

The notions of tensegrity and applying forces to DNA structure lead us to the concept of distorting DNA molecules by design in the context of DNA origami. Figure 9-13 shows how Dietz *et al.*[9.13] have achieved this. The context of the 3D origami 7-nucleotide crossover separation arrangement leads to relaxed bundles, as shown in Figures 9-13a and 9-13b. However, if one alters the relaxed 7-nucleotide crossover separation, other structures will result:

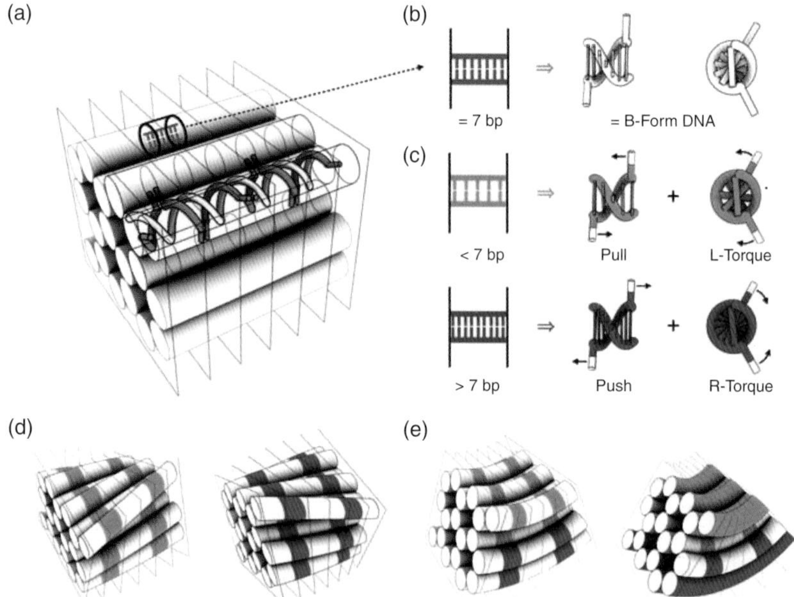

Figure 9-13 *Distortion of DNA by playing with the twist.* The upper panels show that the relaxed conformation for a 6-helix bundle is 7 nucleotides between crossovers. The two torque directions result from fewer or greater than 7 nucleotides between crossover. Uniform and mixed distortions are shown schematically at the bottom left and right, respectively. Reprinted by permission from the AAAS.[9.13]

Figure 9-13c (upper) shows the left-handed torque that results when fewer than seven nucleotides are used for the separation, and Figure 9-13c (lower) shows how the opposite, right-handed torque results if more than seven nucleotides are used to separate the crossovers. Figure 9-13d shows that the whole arrangement will be twisted if all the helices have one or the other torque. Figure 9-13e shows how the combination of deletions (orange) and insertions (blue) can result in bent molecules.

Figure 9-14 illustrates the principles shown in Figures 9-13c and 9-13d: panel a shows a relaxed block of DNA helices, panel b shows a block with a left-handed torque, and panel c shows a block with a right-handed torque. Similarly, Figure 9-15 shows a series of increasingly bent molecules, using the principles illustrated in Figure 9-13e. Figure 9-16 shows a series of intricate DNA molecules that have been formed using these principles, such as gears (panels a and b), a beachball (c), a concave triangle (d), a rounded triangle (e), and a spiral (f).

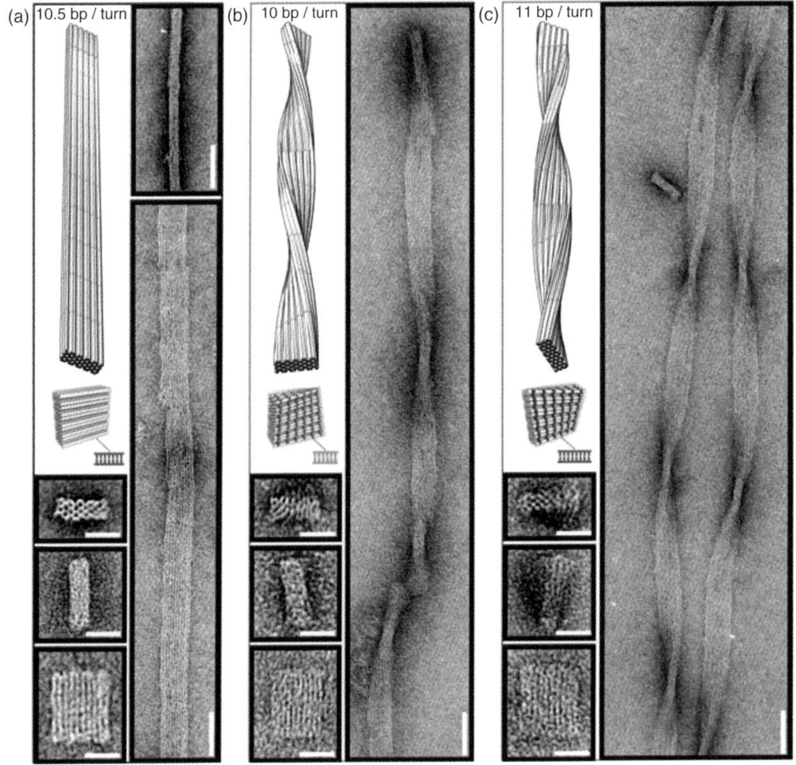

Figure 9-14 *The impact of changing the helicity of origami 3D motifs.* Relaxed, fewer, and greater than 10.5 nucleotide pairs per turn are shown for a DNA origami lath. Reprinted by permission from the AAAS.[9.13]

Recently, Yan and his colleagues have gone beyond hexagonal lattices associated with 120° angles and their variations.[9.14] This group has built DNA structures representing concentric circles and similar notions in 3D. They do this by changing the separation of crossover strands in a circular origami arrangement, as illustrated in Figure 9-17. Figure 9-17a shows a straight arrangement of DNA helices and Figure 9-17b shows the bending that results when the separations differ from full turns but are long. Figures 9-17c and 9-17d show differing views of a three-ring concentric arrangement. Figure 9-17e shows that this notion can be extended to out-of-plane systems, and Figure 9-17f shows various views of the system shown in panel e. Figure 9-18a shows a series of 9-ring sets of concentric rings, and Figure 9-18b shows 11-layer concentric rings with straight edges and rounded corners. The number of base pairs per turn in these structures has been taken from 9 to nearly 12, and the

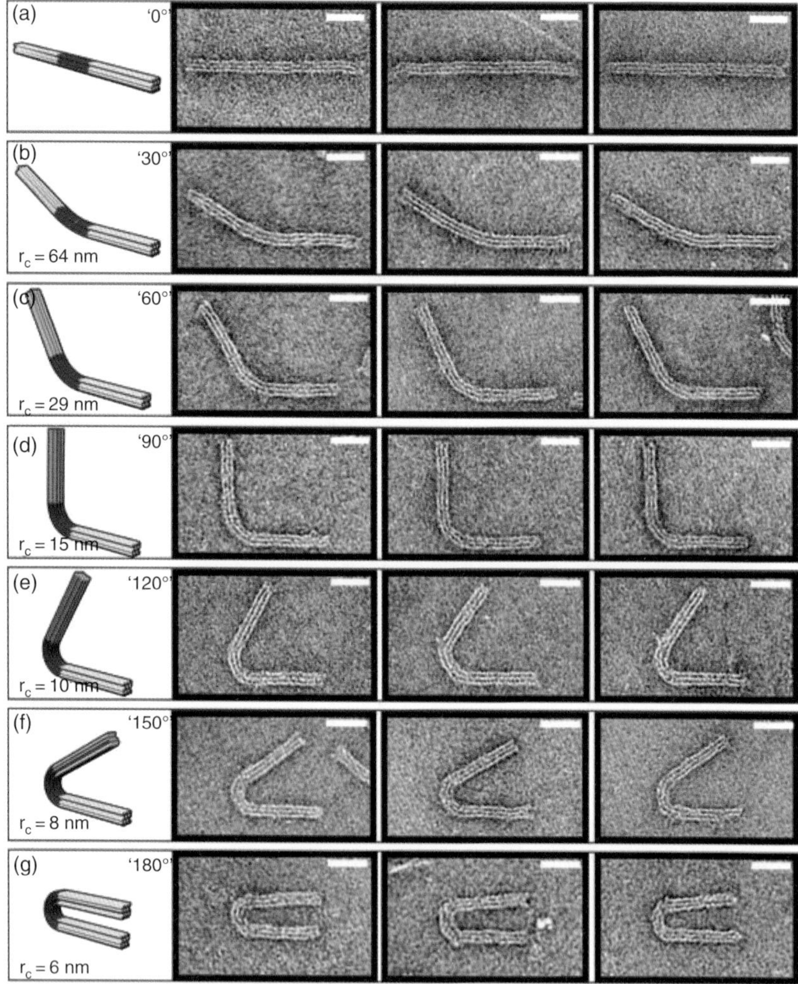

Figure 9-15 *Bending a DNA structure.* The distortions for Figure 9-13e, which show the bending of the motif, have been used to greater and greater extents to bend the structure all the way to a hairpin. Reprinted by permission from the AAAS.[9.13]

structures appear to be robust enough to tolerate these variations. Figure 9-19 illustrates the extension of the thinking here to three dimensions. Panels a, b, and c show schematics of a hemisphere, a sphere, and an ellipsoid, while panels d, e, and f, respectively, show TEM images of these structures. Figure 9-19g shows the schematic design of a flask-like object, whose TEM images are shown in panel h.

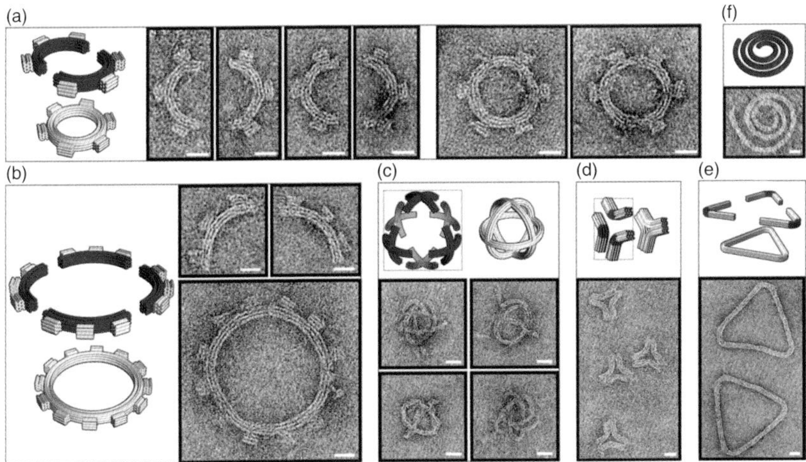

Figure 9-16 *Forming 3D DNA structures into curved motifs.* A variety of curved motifs, gear-like structures, and 3D curves are shown. Reprinted by permission from the AAAS.[9.13]

Overall, the impact of the DNA origami technique has been enormous. Indeed, some people refer to structural DNA nanotechnology as "DNA origami," although, as we have seen and will see later on, there are many other aspects to structural DNA nanotechnology. I have only highlighted a few of the exciting applications of DNA origami here; it would take a whole book by itself to discuss all the experiments and constructions that have been done. The success of DNA origami calls into question, to some extent, the sequence design principles discussed in Chapter 2, since the sequences are already set by the M13 single-stranded molecule. However, it should also be remembered that DNA origami has different success criteria from some of the other aspects of structural DNA nanotechnology. As we noted at the beginning of this chapter, its precision is inherently somewhat lower, and, at this writing, it has not been subjected to some of the more critical tests that other aspects of the system have received.

DNA bricks. A recent development in this area is the advent of DNA bricks in both 2D and 3D. Like DNA origami, this development, pioneered by Peng Yin's laboratory, is a consequence of the falling price of DNA (see Chapter 12 for a fuller description of this phenomenon), combined with the relatively low quantities that are needed for many modern experiments. The first implementation of this work was reported in 2012.[9.15] The approach is shown in Figure 9-20. The idea is that a large number of different individual strands can be assembled into a "molecular canvas." The complex shapes,

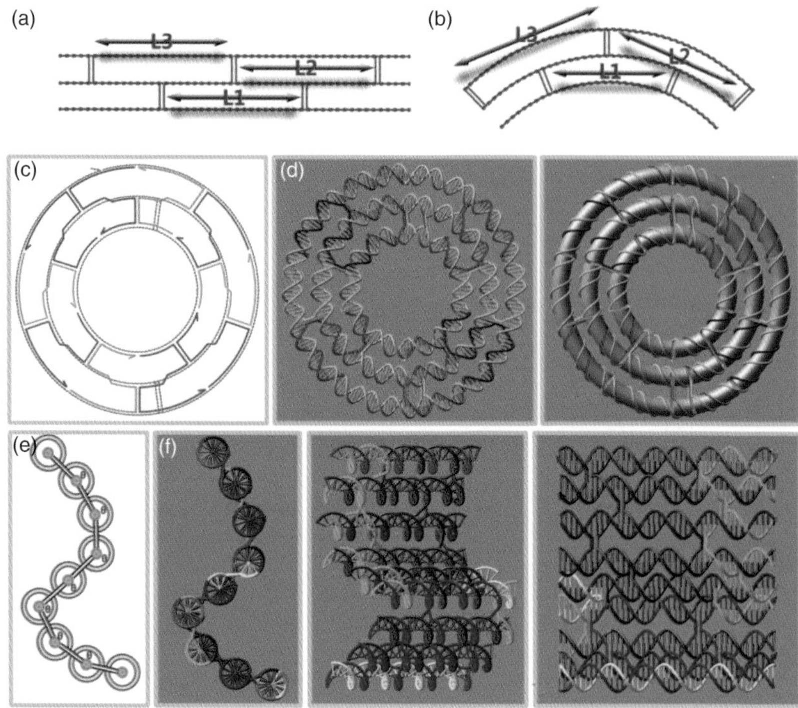

Figure 9-17 *Building curved origami structures in 3D.* The top two rows show how to make concentric DNA rings in a planar structure. The bottom row shows that this system is not confined to 2D. Reprinted by permission from the AAAS.[9.14]

such as the eagle head, shown schematically in the drawing are produced by selecting the desired strands from the canvas, and then including only those strands in the final construct. This is done robotically, to simplify the work that needs to be done. Note that if a strand is on a border, then a second version of it, designed not to interact with other strands, will be used. Figure 9-21 shows some of the 2D images that result from this approach.

In further work, the Yin and Shih laboratories have advanced the system to 3D.[9.16] Rather than using strands of about one turn (4 units in 42 nucleotides, as seen in Figure 9-20), they shorten each segment to eight nucleotides, so that it represents about 3/4 of a turn. This leads to the roughly orthogonal array shown schematically in Figure 9-22. Thus, for the investment in a large number of different DNA strands (see the discussion about DNA costs in Chapter 12), 3D objects are available to the person who wants to make them. Figure 9-23 shows both the programming of a given object and a sampling of the 3D objects that

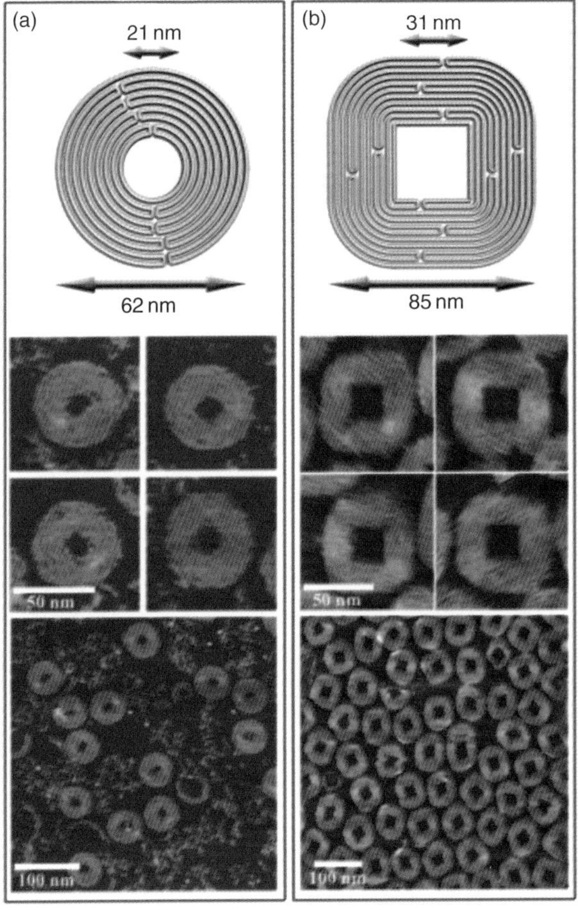

Figure 9-18 *Concentric DNA origami structures.* Both round structures and squares with curved corners are shown. Reprinted by permission from the AAAS.[9.14]

have been made in this fashion. By changing the design slightly, 3D objects based on hexagonal, rather than orthogonal, lattices have also been designed. The key points about DNA origami and DNA bricks are that large constructs are entailed, and the precision is also high. In these systems, we have left the realm where every nucleotide pair can have an impact on the final results. This is important for many investigators who care about what they can make on the multi-nanometer scale, and are less concerned with the fine molecular details of the construct.

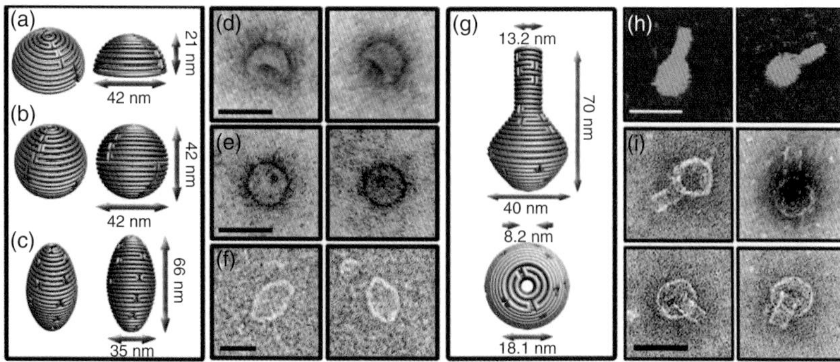

Figure 9-19 *Extending curved DNA origami into three dimensions.* Hemispherical, spherical, and ellipsoidal images are shown at left. A vase-shaped origami is shown at the right. Reprinted by permission from the AAAS.[9.14]

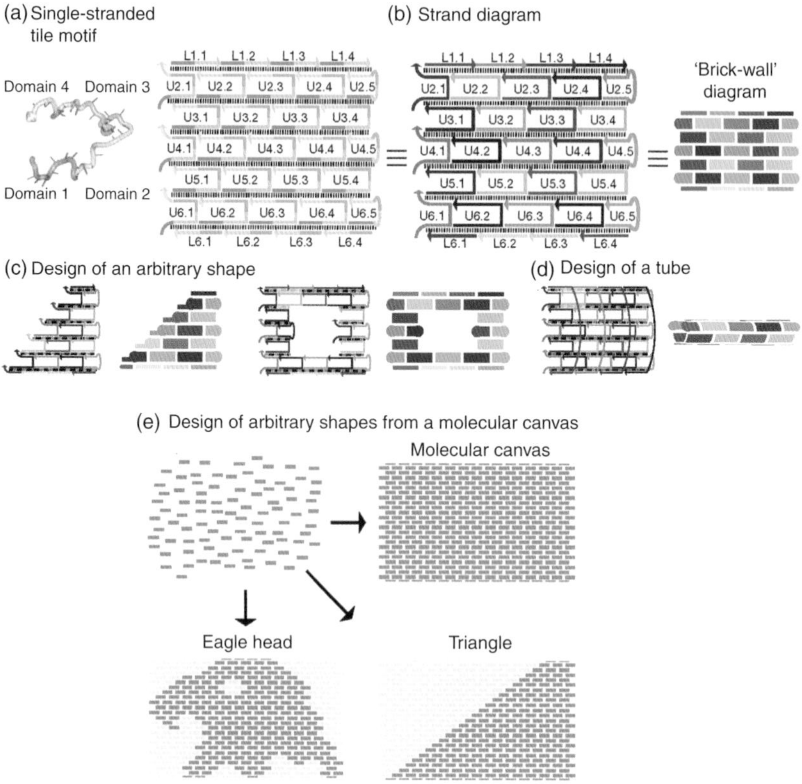

(a) Single-stranded tile motif

Domain 4 Domain 3

Domain 1 Domain 2

(b) Strand diagram

'Brick-wall' diagram

(c) Design of an arbitrary shape

(d) Design of a tube

(e) Design of arbitrary shapes from a molecular canvas

Molecular canvas

Eagle head

Triangle

Figure 9-20 *DNA bricks.* (a) *A single-stranded motif.* (b), (c), and (d) *"Brick-wall" diagrams show how this can tile a plane or form a tube.* (e) *A series of single-stranded "brick" motifs form a canvas.* By eliminating some of them from the canvas, patterns can be formed.[9.15]

Figure 9-21 *Collections of 2D patterns formed by DNA bricks.*

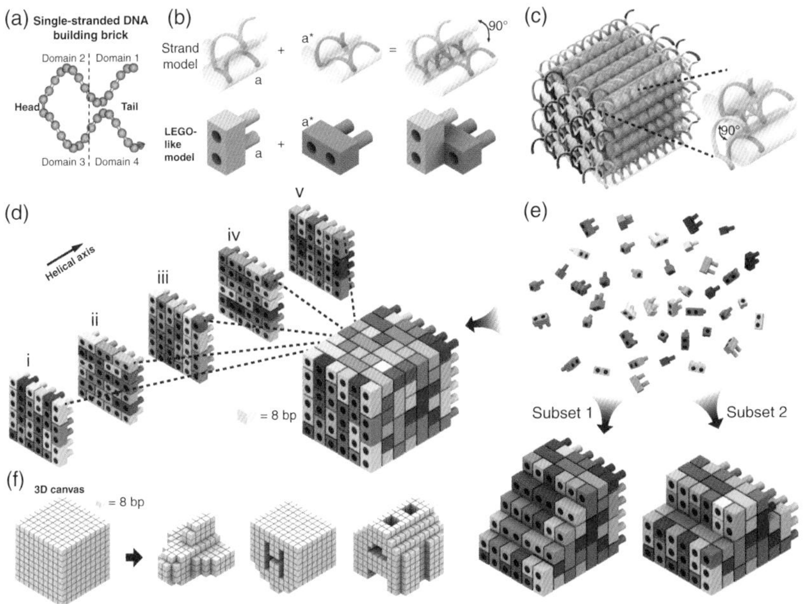

Figure 9-22 *Extending bricks to 3D.* Eight nucleotides are an approximation to 90° angles, and the same "canvas" strategy of Figure 9-20 is extended to a 3D block canvas here. Reprinted by permission from the AAAS.[9.16]

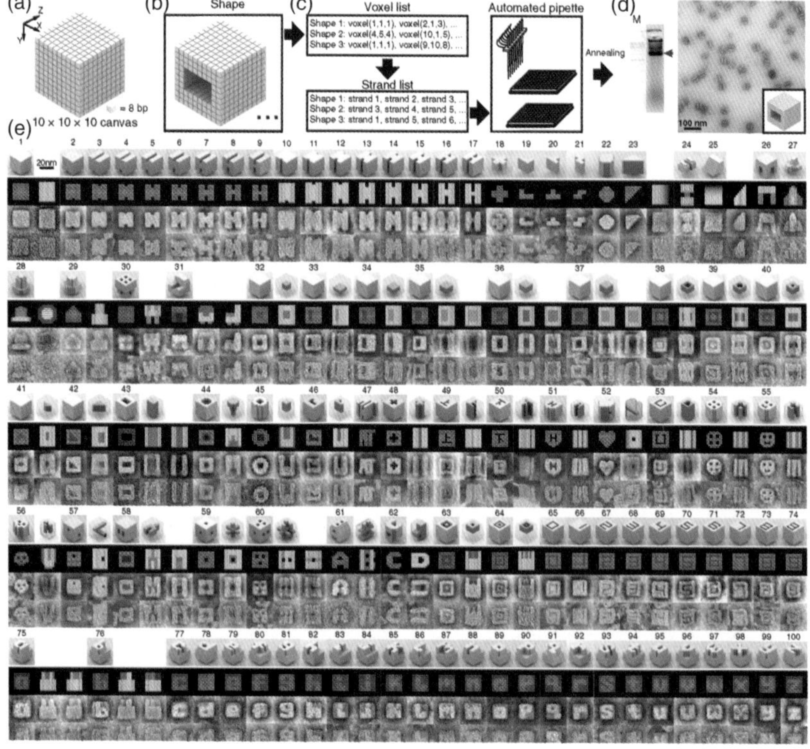

Figure 9-23 *Collections of 3D patterns formed by DNA bricks.* Reprinted by permission from the AAAS.[9.16]

References

9.1 P.W.K. Rothemund, Scaffolded DNA Origami for Nanoscale Shapes and Patterns. *Nature* **440**, 297–302 (2006).

9.2 S.M. Douglas, A.H. Marblestone, S. Teerapittayanon, A. Vasquez, G.M. Church, W.M. Shih, Rapid Prototyping of 3D DNA-Origami Shapes with caDNAno. *Nucl. Acids Res.* **37**, 5001–5006, (2009).

9.3 X.C. Bai, T.G. Martin, S.H.W. Scheres, H. Dietz, Cryo-EM Structure of a 3D DNA-Origami Object. *Proc. Nat. Acad. Sci. (USA)* **109**, 20012–20017 (2012).

9.4 H. Yan, T.H. LaBean, L.P. Fang, J.H. Reif, Directed Nucleation Assembly of DNA Tile Complexes for Barcode-Patterned Lattices. *Proc. Nat. Acad. Sci. (USA)* **100**, 8103–8108 (2003).

9.5 Z. Shen, H. Yan, T. Wang, N.C. Seeman, Paranemic Crossover DNA: A Generalized Holliday Structure with Applications in Nanotechnology. *J. Am. Chem. Soc.* **126**, 1666–1674 (2004).

9.6 P.K. Maiti, T.A. Pascal, N. Vaidehi, J. Heo, W.A. Goddard, III, Atomic Level Simulations of Seeman Nanostrutures: The Paranemic Crossover in Salt Solution. *Biophys. J.* **90**, 1463–1479 (2006).
9.7 W. Liu, H. Zhong, R. Wang, N.C. Seeman, Crystalline Two-Dimensional DNA Origami Arrays. *Angew. Chemie* **50**, 264–267 (2011).
9.8 F. Mathieu, S. Liao, C. Mao, J. Kopatsch, T. Wang, N.C. Seeman, Six-Helix Bundles Designed from DNA. *Nano Letters* **5**, 661–665 (2005).
9.9 S.M. Douglas, J.J. Chou, W.M. Shih, DNA-Nanotube-Induced Alignment of Membrane Proteins for NMR Structure Determination. *Proc. Nat. Acad. Sci. (USA)* **104**, 6644–6648 (2007).
9.10 S.M. Douglas, H. Dietz, T. Liedl, B. Högberg, F. Graf, W.M. Shih, Self-Assembly of DNA into Nanoscale Three-Dimensional Shapes. *Nature* **459**, 414–418 (2009).
9.11 E.S. Andersen, M. Dong, M.M. Nielsen, K. Jahn, R. Subramani, W. Mamdouh, M.M. Golas, B. Sander, H. Stark, C.L.P. Oliviera, J.S. Pedersen, V. Birkedal, F. Besenbacher, K.V. Gothelf, J. Kjems, Self-Assembly of a Nanoscale DNA Box with a Controllable Lid. *Nature* **459**, 73–77 (2009).
9.12 T. Liedl, B. Högberg, J. Tytell, D.E. Ingber, W.M. Shih, Self-Assembly of Three-Dimensional Pre-Stressed Structures from DNA. *Nature Nanotech.* **5**, 520–524 (2010).
9.13 H. Dietz, S.M. Douglas, W.M. Shih, Folding DNA into Twisted and Curved Nanoscale Shapes. *Science* **325**, 725–730 (2009).
9.14 D.R. Han, S. Pal, J. Nangreave, Z.T. Deng, Y. Liu, H. Yan, DNA Origami with Complex Curvatures in Three-Dimensional Space. *Science* **332**, 342–346 (2011).
9.15 B. Wei, M.J. Dai, P. Yin, Complex Shapes Self-Assembled from Single-Stranded DNA Tiles. *Nature* **485**, 623–627 (2012).
9.16 Y.G. Ke, L.L. Ong, W.M. Shih, P. Yin, Three-Dimensional Structures Self-Assembled from DNA Bricks. *Science* **338**, 1177–1183 (2012).

10
Combining structure and motion

Most of this book has been devoted to structure, although in Chapter 8 we talked about DNA-based nanomechanical devices. It is a natural thing to wish to combine devices with structures, so that one can place the devices and get them to work in a specific structural context. To be sure that the motion is occurring in the device whose state is being varied, as opposed to some other devices that might be in solution, direct structural observation of the device, or an array of devices, is desirable. Therefore, it is necessary is to have a structure large enough to accommodate both a machine and the attachments necessary to demonstrate the motion.

Combinations before origami. The first attempt at doing this was in the era before DNA origami. Thus, the way to provide a structural context for a nanomechanical device was a 2D array. We saw in Chapter 7 that a TX array can be connected 1–3 to its neighbors (see Figure 7-8), thereby generating a gap that provides a little space for accommodating the attachments. This space provides an attachment point where a nanomechanical device can be placed. The PX-JX_2 device was used in the first example below.[10.1] This device was converted to a cassette that could be fitted into the slot by the addition of an extra helix. The cassette consists of the PX-JX_2 device with another domain added to it on one end, as shown in Figure 10-1. Panel a shows the cassette in the PX state in a frontal view. A third helical domain is visible on the lower left. Panel c shows the same cassette in the JX_2 state. You should note that the PX device is formed with crossovers between strands of the same polarity, but the cassette has been added by fusing strands of the opposite polarity. So as to visualize the motion of the device in the context of the array, it was necessary to add a hairpin marker, which appears as a circular magenta helix with yellow base pairs viewed down its axis. It is in front of the cassette device in panel a, and behind it in panel c. The cassette is shown obliquely in panels b and d, where the hairpin is more readily visible.

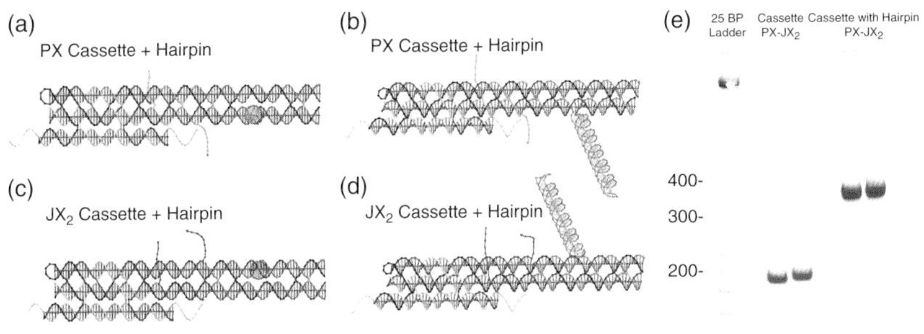

Figure 10-1 *A PX-JX$_2$ device incorporated into a cassette.* (a) *The PX state of the PX-JX$_2$ cassette shown in a perpendicular view.* (b) *The PX state shown in an oblique view.* (c) *The JX$_2$ state of the PX-JX$_2$ cassette shown in a perpendicular view.* (d) *The JX$_2$ state shown in an oblique view.* The top two domains are the device itself, while the bottom domain anchors the device into a 2D DNA array. The magenta double helical hairpin "robot arm" enables visualization of the operation of the device by AFM. (e) *A gel showing the mobility of the cassette and of the cassette containing the hairpin.* The PX state always migrates slightly more rapidly than the JX$_2$ state. Reprinted by permission from the AAAS.[5.14]

The design of the 2D array is illustrated in Figure 10-2, which shows the eight different TX tiles that constitute the repeating unit. One extra tile is shown at the bottom, and a full second repeat is drawn on the right. The cassette, inserted vertically into the array, is drawn in red, and one tile away another vertical tile, drawn black, is labeled M. This is a "marker tile," inserted so that "up–down" directionality in this view can be established, so we know which way the indicator helix is pointing. A full-on view of the marker tile is shown below the array. A schematic of the same array is shown in Figure 10-3, with two vertical repeats and four horizontal repeats. The directions of the indicator helices are shown for both states of the array, the PX state on the left and the JX$_2$ state on the right. It should be clear that in the PX state the indicator helix is pointing in towards the marker tile, while it is pointing away from the marker tile in the JX$_2$ state. AFM images of the operation of the device are illustrated in Figures 10-4 and 10-5. Figures 10-4a and 10-4b show the entire array undergoing the PX $\rightarrow$ JX$_2$ and the JX$_2$ $\rightarrow$ PX transitions, respectively. There are visual guides to indicate a few unit cells, as well as the position of the indicator arm. Figure 10-5 is a zoom of an individual unit cell undergoing the two transitions.

A capture system. The advent of DNA origami (Chapter 9) simplified the combination of devices and structures immensely. Rather than combining a large number of tiles to create an area of adequate extent to demonstrate the motion of a nanomechanical device by AFM, it is only necessary now to build a single origami to act as the ~10 000 nm^2 superstructure on whose surface the

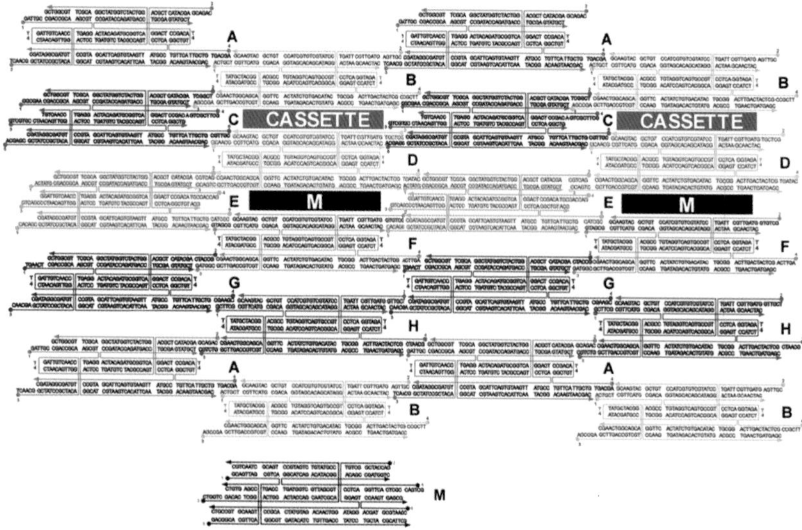

Figure 10-2 *The 2D array into which the PX-JX$_2$ cassette is bound.* There are eight unique TX motifs forming the array, as indicated by the color-coding. The cassette is indicated as being bound perpendicular to the array. The M tile is a marker tile also bound perpendicular to the array. Its TX structure is indicated at the bottom of the drawing. Reprinted by permission from the AAAS.[5.14]

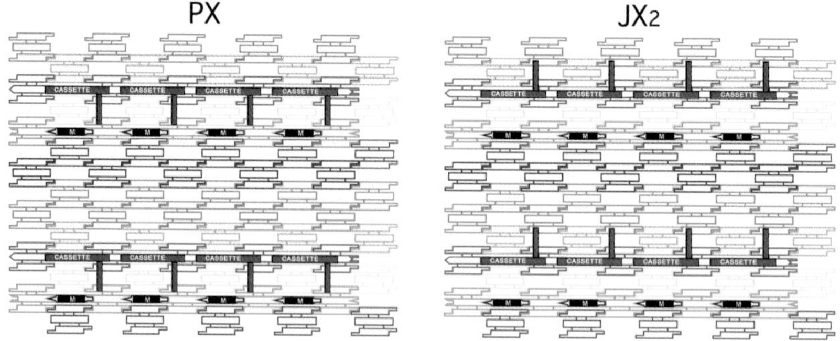

Figure 10-3 *Schematic depiction of the operation of the device in the 2D array.* At left, the cassette is shown in the PX state, and the robot arm points towards the marker tile. At right, the cassette is shown in the JX$_2$ state, with the robot arm pointing away from the marker tile. Reprinted by permission from the AAAS.[5.14]

action of the device needs to be shown. An early example of utilizing DNA origami tiles as the support for molecular motion entailed the insertion of two different PX-JX$_2$ cassettes into a single origami tile to create a "capture" system for a variety of shapes.[10.2]

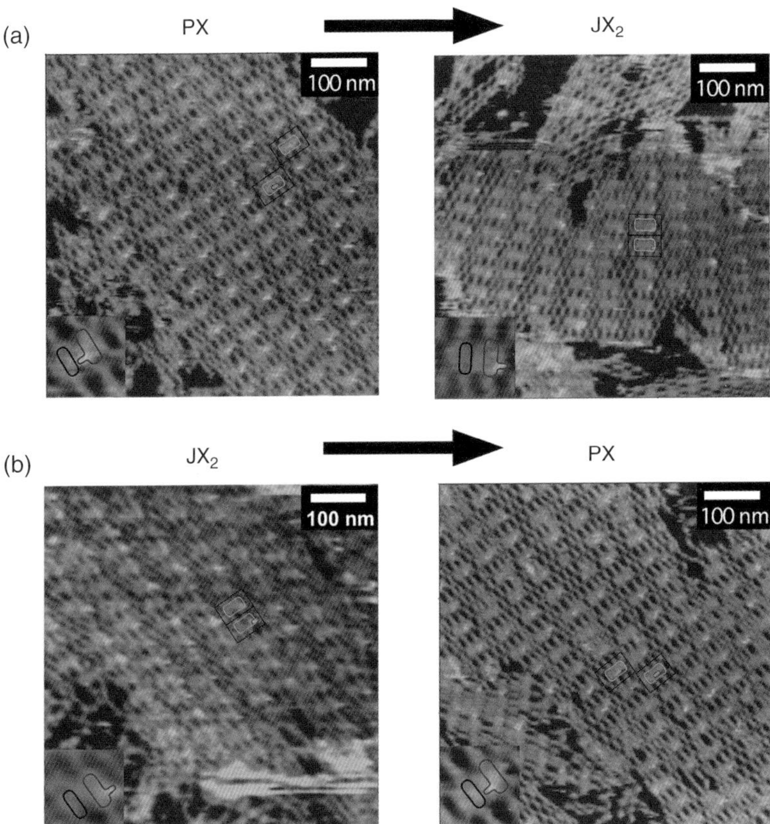

Figure 10-4 *Long views of the operation of the robot arm.* (a) *The shift from the PX state to the JX_2 state.* The left side shows the robot arm in the PX state. The right side shows the robot arm in the JX_2 state. (b) *The shift from the JX_2 state to the PX state.* The left side shows the robot arm in the JX_2 state. The right side shows the robot arm in the PX state. Reprinted by permission from the AAAS.[5.14]

The idea is shown in Figure 10-6. There are four panels to this drawing, each showing the four states available to this system in a side view. At the bottom of each panel is a long blue rectangle labeled "Array"; this is the DNA origami tile that acts as the superstructure for the system. A PX-JX_2 device is anchored into the origami tile by a domain of DNA duplex, drawn as a green box with sticky ends to the left and the right of the box. There are two such anchors, one on the left and one on the right. The PX-JX_2 device itself is attached to each of the anchors. The tips of the devices contain sticky ends that point towards the center of the system; the sticky ends on the left are labeled A and B and those on the right are labeled C and D. In the top panel,

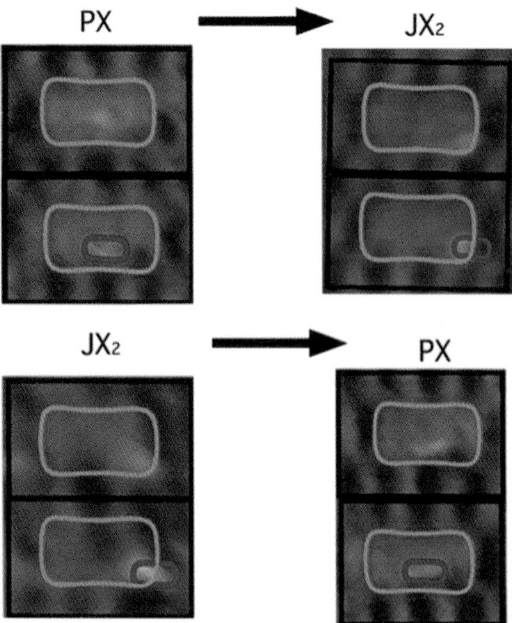

Figure 10-5 *Zooms of the operation of the robot arm.* Each panel contains two unit cells, which are outlined in black. The device is circled in blue in each unit cell, while the robot arm is circled in red in each lower unit cell, as an aid to interpretation. The top two images show the transition from the PX state to the JX_2 state, while the bottom two images show the transition from the JX_2 state to the PX state. Reprinted by permission from the AAAS.[5.14]

both the left device and the right device are in the PX state; thus, sticky end A is above sticky end B on the left, and likewise sticky end C is above sticky end D on the right. Because they are organized in this fashion, they are able to bind the species labeled "Capture-1" and drawn in magenta at the center of the panel. This happens because the sticky ends on the bottom two layers of Capture-1 are complementary to this arrangement of sticky ends: A′ above B′ and C′ above D′. Note that there are three layers to Capture-1, but the sticky ends are only on the bottom two layers.

In the panel second from the top, the left device is still in the PX state, but the right device is in the JX_2 state, so sticky end D is now above sticky end C. Consequently, the arrangement of sticky ends is able to capture the species labeled "Capture-2" (drawn in blue) from solution, rather than Capture-1. The third panel presents a third arrangement of sticky ends, with B above A on the left, reflecting the JX_2 state of the left device, and

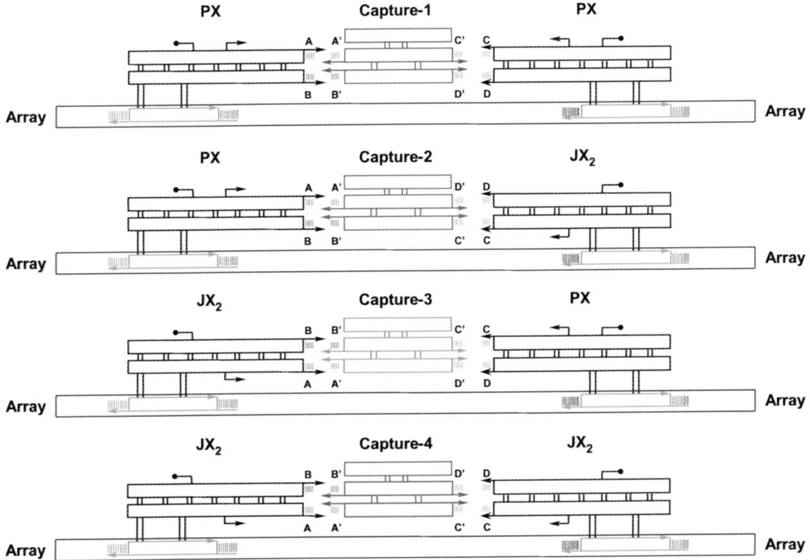

Figure 10-6 *The capture system.* Two different PX-JX_2 cassettes have been sunk into a DNA origami structure, which is labeled "Array." The third (bottom) domain of the cassette binds to the origami structure. The four different panels show, from top to bottom: both cassettes in the PX state, the left cassette in the PX state and the right cassette in the JX_2 state, the left cassette in the JX_2 state and the right cassette in the PX state, and, at bottom, both cassettes in the JX_2 state. The sticky ends on the two devices are labeled A and B on the left, and C and D on the right. A is above B and C is above D in the PX state; B is above A and D is above C in the JX_2 state. Consequently, the system can capture four different species, whose sticky ends match those of the cassettes, indicated as "Capture-1," "Capture-2," "Capture-3," and "Capture-4" in going from top to bottom of the drawing.[10.2]

C above D on the right because this device is in the PX state. Hence, the species labeled "Capture-3" (drawn in green) is captured from solution. In the bottom panel, both devices are in the JX_2 state, so B is above A and D is above C, leading to the capture of the species labeled "Capture-4" and drawn in red.

There are three layers to each of the Capture species, but so far we have only discussed the recognition of the sticky ends on the bottom two layers. The top layer of each of these molecules contains a geometrical feature that can be distinguished in the AFM. Figure 10-7 shows schematics (panel a) and AFM images (panel b) of this system. Section i shows just the naked origami tile. Note that it is notched asymmetrically, so that directionality is established. The slots in the origami tile to accommodate the anchors are visible in the

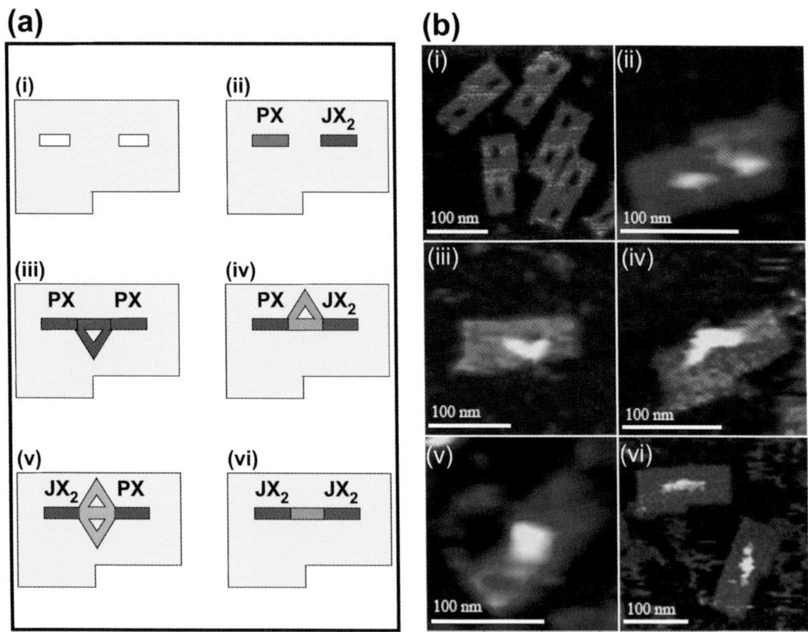

Figure 10-7 *The capture system viewed from the top.* (a) *Schematics of the four species.* (i) The origami structure without the cassettes; note the notch at the bottom. (ii) The origami with the cassettes, indicating the color-coding, green for PX and purple for JX_2. (iii) Capture-1, a triangle pointing towards the notch of the origami structure. (iv) Capture-2, a triangle pointing away from the notch. (v) Capture-3, a diamond-shaped structure. (vi) Capture-4, a linear structure. (b) *AFM images of the four species.* The six panels show the same species indicated in the panels of (a).[10.2]

corresponding AFM image. Section ii shows the origami tile containing the devices, but without the Capture molecules present. The top layer of Capture-1 is a triangular feature that is designed to point towards the notch (section iii), and the top layer of Capture-2 is a similar triangular feature that is designed to point away from the notch (section iv). These are visible in AFM images iii and iv, respectively. Capture-3 is a diamond-shaped double triangle, and Capture-4 is a linear DNA feature. These structures are visible in AFM images v and vi, respectively.

Thus, the system appears to work. However, evaluating such a system is tricky. We might have been cherry-picking the correct results. As we will also note in Chapter 12, systems such as this one are subject to inherent chemical errors. The problem arises from what might be called "kinetic traps." Thus, if the target capture molecule is Capture-1, the sticky ends on the system are A above B and C above D. However, the left side of Capture-2 (A above B and D above C) can bind to this configuration, and Capture-3 (B above A and C above D) can bind to the right side of the configuration. Experience with this system shows that these problems do occur, and require an unbiased solution. To state the problem in other words, we do not experience many problems of incorrect tile placement if correct tiles (e.g., the complex 10-tile 2D array in Figure 10-2) are competing with incorrect tiles, but we can expect problems when half-correct tiles are competing with the correct tiles. The completely incorrect tile is never bound, but the half-correct tile is often found to bind.

A blind error-correction protocol was developed that seems to work, at least in this case. The tiles are heated to a slightly non-permissive temperature. When this is done, half-correct tiles replace incorrect tiles but correct tiles replace half-correct tiles, and not the other way around. Figure 10-8 shows how this works. The identity of the captured molecules is color-coded by the arrows pointing at the origami tiles: diamond, black; line, red; triangle pointing away from the notch, blue (none in these images); triangle pointing towards the notch, magenta; damaged unit, white. In panel a, a mixture of the four capture molecules has been applied to the origami. This is not necessary, because all four tiles need not be added at once, only in successive order. In panel b, the linear molecule has been applied, using the binding correction protocol described in the text. In panel c, the triangle pointing to the notch has been applied to the material in panel b and the correction protocol has been applied. In panel d, the diamond has been applied to the material in panel c and the correction protocol has been applied. In panel e, the triangle pointing away from the notch has been applied to the material in panel d and the correction protocol has been applied. Only diamonds are visible in panels d and e. Panels f–i show the same procedure, but in a different order. The triangle pointing to the notch, the triangle pointing away from the notch, the linear element, and the diamond have been applied, respectively. Again, only diamonds are visible in panel i.

A molecular assembly line. One of the holy grails of nanotechnology has been the molecular assembler to build new molecular species from the bottom up. This target has been achieved on the nanometer scale, although not yet on the atomic scale, by using DNA nanotechnology.[10.3] There are three essential elements to a nanofactory: (1) a programmable assembler – on the macroscopic scale this is either a worker or a robot that is given instructions and which then carries them out; (2) an interstation conveyor to move the growing product to different places along the assembly line; and (3) a factory or framework to organize the entire process. In the first DNA-based nanofactory, the

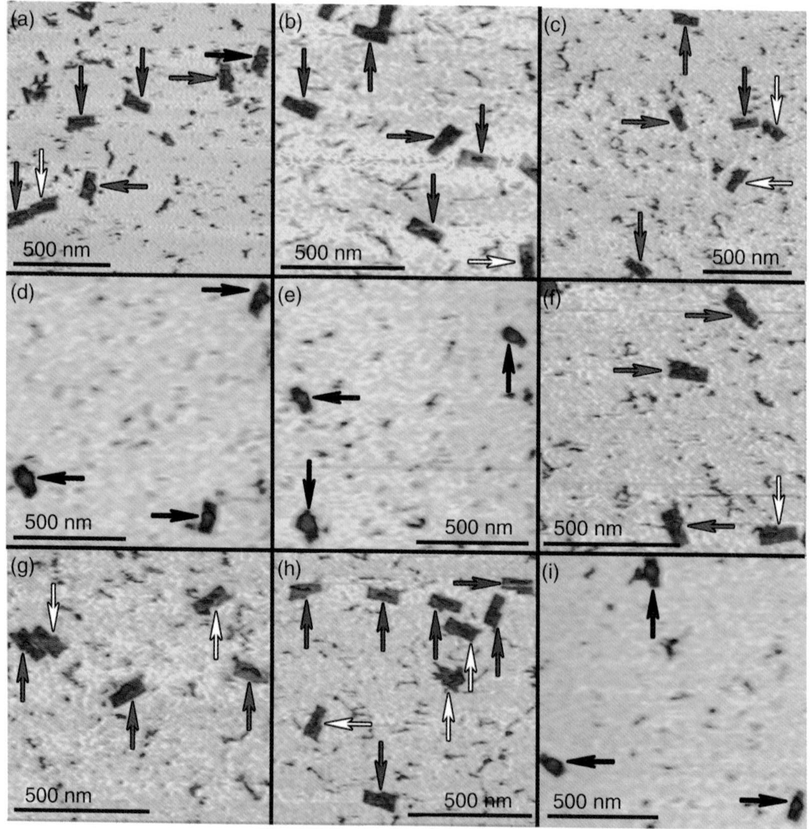

Figure 10-8 *Application of the correction procedure.* This is described fully in the text.[10.2]

programmable assembler is the same PX-JX_2 cassette that was used in the preceding section as part of the capture system, although the cassette is connected by two helices (double cohesion), rather than by a single sticky end pair. Three of them are lined up, and can be programmed to add a component to the growing product or not. The conveyor and the product have been combined into a walker that somersaults along a pathway, and grows as items are added to it. The walker is just the tensegrity triangle that was described in Chapter 7. The walker is analogous to the chassis of a car that is moving along an assembly line in an automobile factory. The factory/framework is provided by a DNA origami construct.

The whole arrangement is illustrated in Figure 10-9. This drawing shows the attachment of all three cargoes to the moving tensegrity triangle chassis.

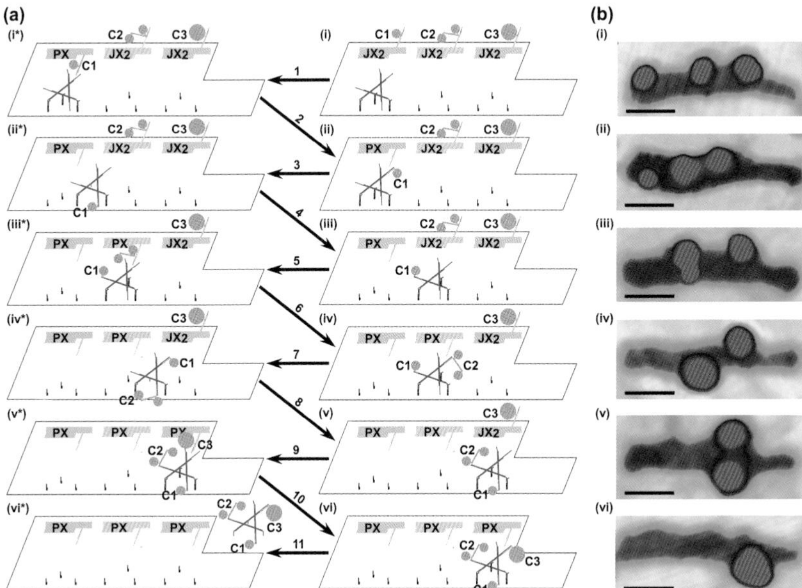

Figure 10-9 *The operation of a programmable nanoscale assembly line.* Twelve images (11 steps) are shown in the addition of three components to a "chassis"/walker which is a tensegrity triangle. The right-hand column of the schematics corresponds to the faux-color AFM images at the right. In the first step (i), the cassettes are all in the "off" state, with the three cargoes pointing away from the path of the walker. In the first step, the left-most cassette is placed in the "on" state (i*), after which it attaches to the chassis (ii). In the next stage, the walker takes a half-step (ii*) and then another half-step (iii) to place it below the second loading station. These steps are then repeated to add the second cargo at (iv). Further repetition leads to the walker being below the third station (v), after which the third cargo is added (vi). The final step is release of the completed complex.[5.55]

Panel a is a schematic that shows all the steps in the process. Panel b shows AFM images of the right-hand column of the schematics in panel a. The system starts with all three loading stations in the "off" position, which corresponds to the JX_2 state of the PX-JX_2 device. Cargo 1 is a 5 nm gold nanoparticle, cargo 2 is a pair of coupled 5 nm gold nanoparticles, and cargo 3 is a 10 nm gold nanoparticle. These are all shown in a(i) as pointing away from the tensegrity triangle. In a(i*), the first position is converted to the PX state, which corresponds to the "on" position. When this happens, cargo 1 is added to the walker, as seen in a(ii). The walker is then programmed to take a half-step a(ii*) and then a second half-step, which places it below the second loading station, as seen in a(iii). Note that the gold cargoes show up in faux color in the positions where they are supposed to be in the AFM images in panels b(i), b(ii), and b(iii). The process is repeated for the second cargo, as shown in a(iii*) and a(iv), adding the pair of gold nanoparticles to the moving walker. A third repetition of the process is shown in the remaining steps.

Figure 10-10 shows the details of the operation of the assembly line. Panel a shows that the three domains of the tensegrity triangle have hands H1, H2, and H3 at one end for grabbing the cargoes. Likewise it walks on feet F1, F2, and F3, which are at the other ends of the domains. Foot F4 is used to orient the walker towards the cargo loading stations. Panel a(i) shows the triangle in a form the shows the helices as ladders, while a(ii) illustrates the helical nature of the motif. Panel b shows the way in which the triangle walks. It is held on position by strands A-1 and A-2, as well as by orienting-strand A-4. Fuel (unset) strands are then added to remove A-1 and A-4, so it is only attached by A-2. Then a new positioning strand, A-3, is added so the walker is now attached by strands A-2 and A-3, having advanced 120°. Panel c shows how proximity of the C1-bearing arm leads to its attachment to the walker. First there is cohesion between the unpaired segment attached to C1 and an unpaired part of H1. Following that, the second step entails branch migration so that more base pairs are formed between C1 and H1 than between C1 and the strand attached to the cargo station. Note that there are more nucleotide pairs formed at each of the successive steps of panel c; this is the thermodynamic driving force for the cargo transfer.

Figure 10-11 shows that this is a programmable system. Panel a shows that there are eight different pathways through the construction cycle, depending on how the cargo stations are programmed. The eight possible products are shown at the right, including the addition of no cargo, each of the individual cargoes, any of the three pairs of cargoes, or the triple addition shown in detail in Figure 10-9. Panel b shows each of those products at the end of the passage of the chassis through the system, and panel c contains transmission electron micrographs of all of the different products. Since this system was developed, numerous other origami-based systems have been produced that combine a series of steps and the operation of either logic gates[10.4,10.5] or systems moving autonomously on a directed pathway.[10.6] The future of this type of device is limited only by the talent and imagination of the investigators. What is described here is a "factory" built to produce a single product. The future obviously lies with those who can do the same thing to produce multiple products through a recycling procedure.

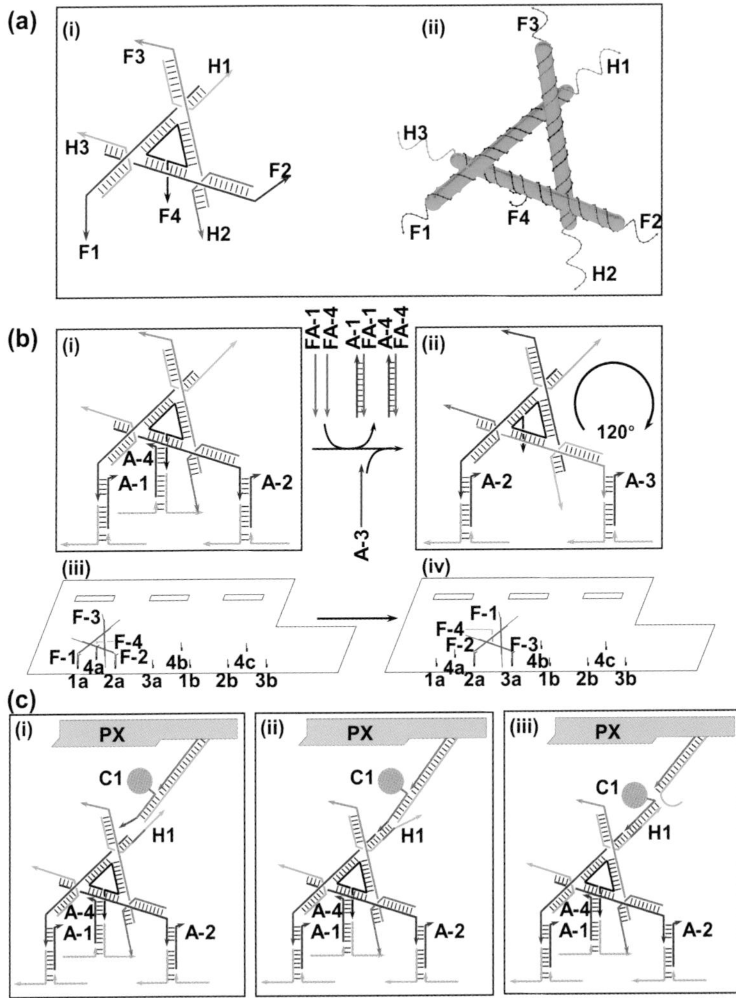

Figure 10-10 *The components of the system.* (a) *The "chassis"/walker.* (i) The walker has three hands (H1, H2, and H3) for grabbing cargo, and three feet (F1, F2, and F3) for walking. It has an additional foot (F4) to orient it towards the loading station. (ii) The walker is shown with its DNA strand structure visible. (b) *The walker takes a half-step.* (i) Strands A-1 and A-2 hold the walker in place, while its orientation is established by A-4. Fuel strands release A-1 and A-4 and another strand A-3 is added so that the walker is re-established a half-step to the right (ii). The walking takes place on the origami surface, shown in (iii) and (iv). (c) *Addition of cargo 1 to the walker.* (i) There is a single-stranded region on both H1(red) and the strand extended to it bearing the cargo. (ii) Those single-strands pair, and then the blue single-stranded portion on H1 displaces the strand holding cargo 1, leading to (iii) transfer to the walker. Increased base pairing drives the reaction at all stages.[5.55]

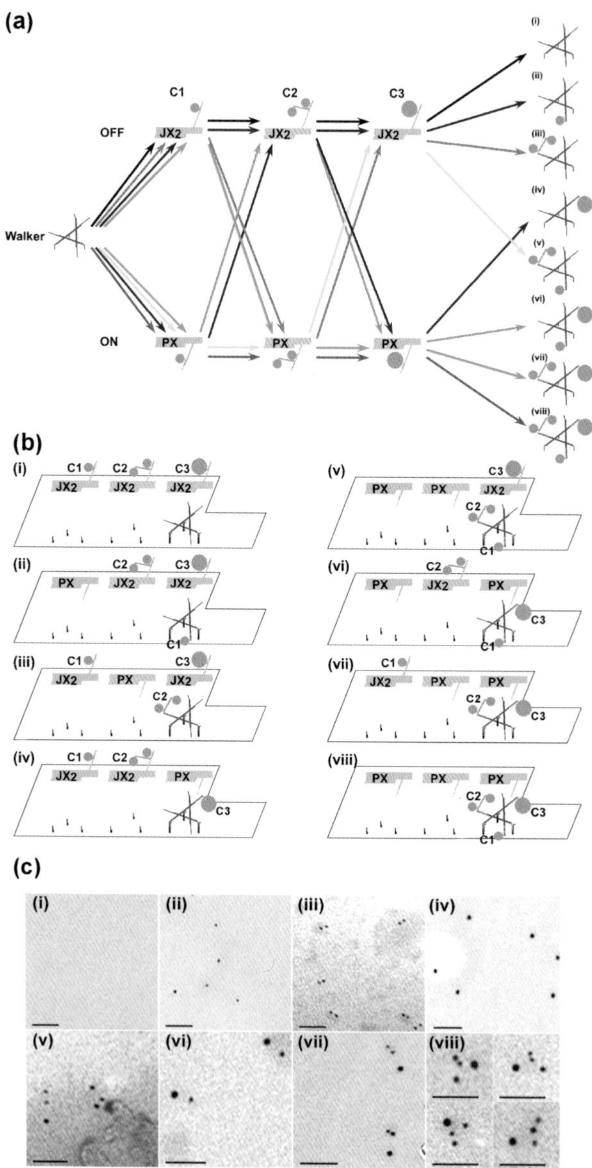

Figure 10-11 *Control over the products of assembly.* (a) *The eight pathways through the system.* Each of the loading stations can be in the ON state or in the OFF state, for a total of eight products, shown at the right. (b) *The conclusion of the assembly process for each of the eight pathways.* Pathway (viii) is the one treated in detail in Figure 10-9. (c) *TEM images of the products of each of the eight pathways.* The images correspond to the products shown schematically in (b).[5.55]

References

10.1 B. Ding, N.C. Seeman, Operation of a DNA Robot Arm Inserted into a 2D DNA Crystalline Substrate. *Science* **314**, 1583–1585 (2006).

10.2 H. Gu, J. Chao, S.J. Xiao, N.C. Seeman, Dynamic Patterning Programmed by DNA Tiles Captured on a DNA Origami Substrate. *Nature Nanotech*. **4**, 245–249 (2009).

10.3 H. Gu, J. Chao, S.J. Xiao, N.C. Seeman, A Proximity-Based Programmable DNA Nanoscale Assembly Line. *Nature* **465**, 202–205 (2010).

10.4 E.S. Andersen, M. Dong, M.M. Nielsen, K. Jahn, R. Subramani, W. Mamdouh, M.M. Golas, B. Sander, H. Stark, C.L.P. Oliviera, J.S. Pedersen, V. Birkedal, F. Besenbacher, K.V. Gothelf, J. Kjems, Self-Assembly of a Nanoscale DNA Box with a Controllable Lid. *Nature* **459**, 73–77 (2009).

10.5 S.M. Douglas, I. Bachelet, G.M. Church, A Logic-Gated Nanorobot for Targeted Transport of Molecular Payloads. *Science* **335**, 831–834 (2012).

10.6 K. Lund, A.J. Manzo, N. Dabby, N. Michelotti, A. Johnson-Buck, J. Nangreave, S. Taylor, R.J. Pei, M.N. Stojanovic, N.G. Walter, E. Winfree, H. Yan, Molecular Robots Guided by Prescriptive Landscapes. *Nature* **465**, 206–210 (2010).

11
Self-replicating systems

This book is not about biology, but it is hard to separate DNA from biology. Nature uses DNA to store and replicate the information for living organisms, so it is natural to ask whether DNA motifs can be used to replicate information and to select materials that are the most propitious for particular environments. Replication can be used to provide exponential growth in molecular or cellular populations. However, one can ask whether there is any merit to doing that, rather than just making a large batch of whatever you want. Certainly chemicals are not produced by replication; when a large amount of material is wanted, the synthesis methods are just scaled up, from, say, the millimole scale to the mole scale.

Evolution and selection. The key element here is the notion of selection and evolution. If there are many different species produced, then it is useful to be able to select for the best one, given particular selection criteria. If there is a given circumstance existent in the medium, then it is useful to be able to select those individuals, be they molecules, cells, or larger organisms, that are best suited to survive under the selection criteria. If only those particular individuals survive the selection criteria, or if they are better suited to replicate, then self-replication is useful: the best-fitted individuals can then dominate the succeeding populations within that environment. This is hardly new wisdom. The nature of natural selection was pointed out by Darwin[11.1] and Wallace[11.2] in the nineteenth century. The idea of applying this notion to molecules (viral genomes) was first suggested by Spiegelman in the last third of the twentieth century,[11.3] and then was taken up by Ellington and Szostak[11.4] and by Tuerk and Gold[11.5] some years later. In the latter cases, the investigators made partially random linear RNA molecules thought likely to fold to give a specific phenotype, either enzymatic or binding activity. Those molecules best suited to have this phenotype were selected and amplified repeatedly until a few molecular species dominated the population and their sequences could be identified.

Can the same thing be done with unusual DNA motifs? It is certainly an open challenge to optimize detailed structural properties in the 1–4 Å range based on selection procedures involving conventional phosphoramidites. For example, to optimize the lengths of the segments in the tensegrity-triangle crystals described in Chapter 7, mixing various bases in the synthesizer would not really do the trick. Even contaminating the phosphoramidites with dinucleotide phosphoramidites would be of marginal value if crystals were being sought. Some of the key structural features described in the foregoing chapters would certainly be extremely hard to optimize through selection, because competition between molecular species, be they crystalline components or machine components, would be very hard to set up. Nevertheless, some things could be optimized, for example the ability to survive in a particular environment.

1D emulation of Watson–Crick replication using large motifs. First, let us describe a system in which large DNA motifs emulate the bases in a Watson–Crick double helix. The system we will describe[11.6] is notable for its ability to replicate the information content of a linear array, just like strands of DNA in biology. However, the components used are more complex. Rather than an individual nucleotide pair, a bent version of the TX motif, called BTX, was used. The angle between the plane of the first DX of the TX and the second DX of the TX is 120°, just as in the 6-helix bundle shown in Figure 7-17. Figure 11-1a shows two of these molecules, one on top and one on the bottom. The left panel of Figure 11-1a shows a longitudinal view, and the right side shows a cross section. They are joined by four different sticky ends, two on each of the first and third helices of the BTX motif. A view of the sequence is shown in Figure 11-1b; this image also shows a variant of the molecule with four hairpins attached to the covalent structure of a BTX. These hairpins are visible in the AFM, and so can be used to distinguish a species of BTX that contains them from a species that does not. The reason for using the hairpins is that biotin–streptavidin, which could also distinguish between two species, is not adequately reliable for use in all cases. Sometimes, a streptavidin simply does not bind to a biotin. Even though the probability of binding is greater than 90%, that's not enough if four different sites need to be bound in every molecule.

Self-replication requires a seed to template the pattern to be replicated. Figure 11-2 shows the formation of the seed. In this case, there are green tiles and red tiles, and the red tiles contain a biotin. The tiles are mixed together with specific long sticky ends that make a pattern of ABBABAB, or, as shown in the figure, red, green, green, red, green, red, green. In this case, the seed tiles were imaged with streptavidin binding to the red tiles. The AFM image of this

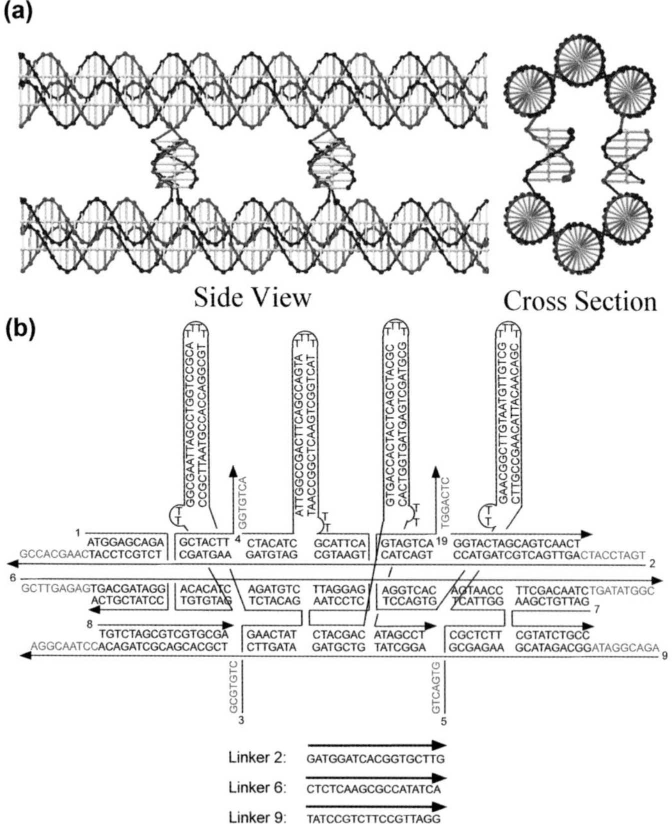

Figure 11-1 (a) *The P6HB motif.* This representation shows the way in which two BTX domains are connected by four lateral cohesive to form the P6HB motif. The cross section view shows two of the four helices that are formed by the lateral cohesive interactions. The interactions at the rear are eclipsed in this projection. (b) *The sequence and structure of the B' BTX tile.* Four helical domains are shown attached perpendicular to the BTX motif so that they will create a topographic feature that can be detected in the atomic force microscope (AFM).[11.6]

structure is shown in Figure 11-3. It is clear from the zoomed lower images that the desired pattern has been formed. The biotin-containing tiles are fatter than those that lack the biotin; this is because streptavidin binds in the examples shown. The ABBABAB pattern is clear in these images.

In this case, the seed or mother pattern is attached to a solid support, called a dynabead, as shown in Figure 11-4. There are two directions of cohesion needed in this system. First, the tiles that will become the daughter array must individually recognize the seed. This is the vertical direction in Figure 11-4.

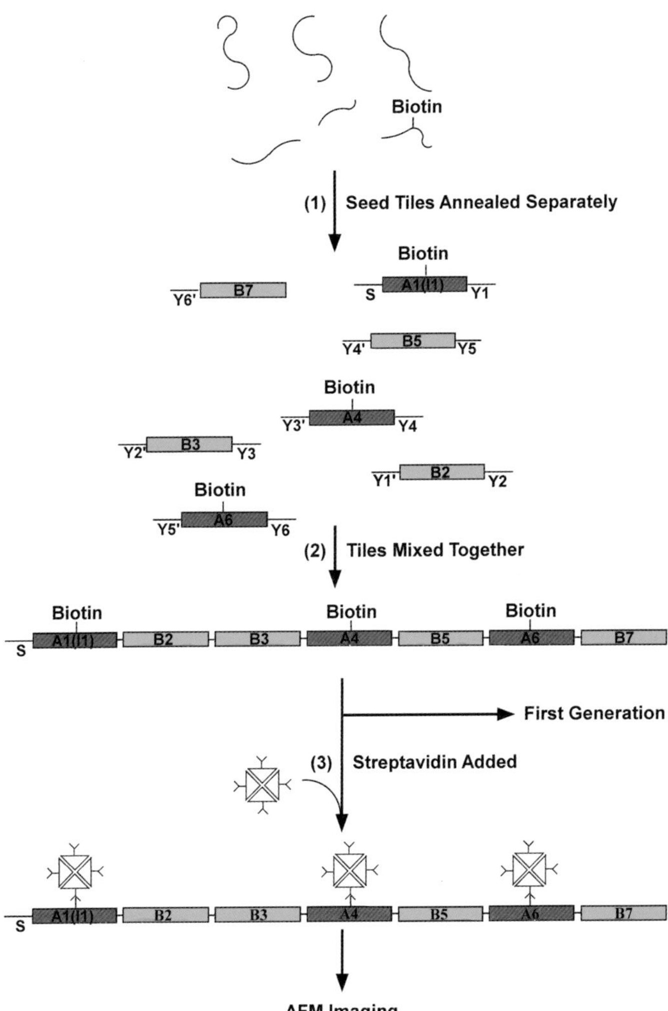

Figure 11-2 *Seed formation.* This drawing shows, in step 1, how the individual strands of the seed tiles are self-assembled in separate vessels to produce seven different BTX tiles flanked by three sets of unique sticky ends, labeled with Y and a number. Primed numbers are complementary to unprimed numbers. The red tiles are the A tiles and the green tiles are the B tiles. The A tiles contain a biotin group to enable their decoration by streptavidin. The tile labeled A1(I1) is the initiator tile. The strand labeled S on its left can bind to a dynabead during the replication process. Step 2 shows that these tiles produce 7-unit seeds when they are mixed together. The tiles are prepared for AFM imaging by the addition of streptavidin (step 3).[11.6]

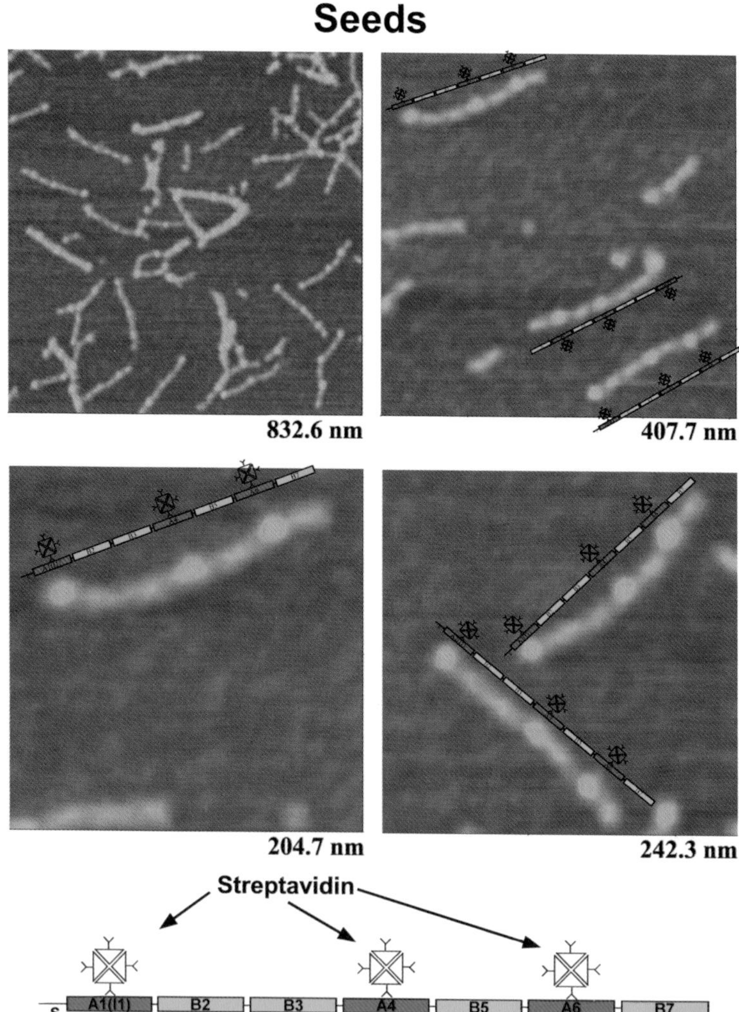

Figure 11-3 *AFM images of seeds.* The image at the upper left shows a typical field of slightly less than a square micron. A large number of seeds are present, along with some multimeric complexes. The other three panels are zoomed images. A schematic image of each seed is shown next to the seed.[11.6]

Then they must be linked together in some fashion, in the longitudinal direction. This is the horizontal direction in Figure 11-4. Reliability in the longitudinal direction is essential. With this design, chemical crosslinking has only two opportunities for each position to do this, and its reliability is not high enough to guarantee a link. Consequently, non-covalent linking was done here. Two

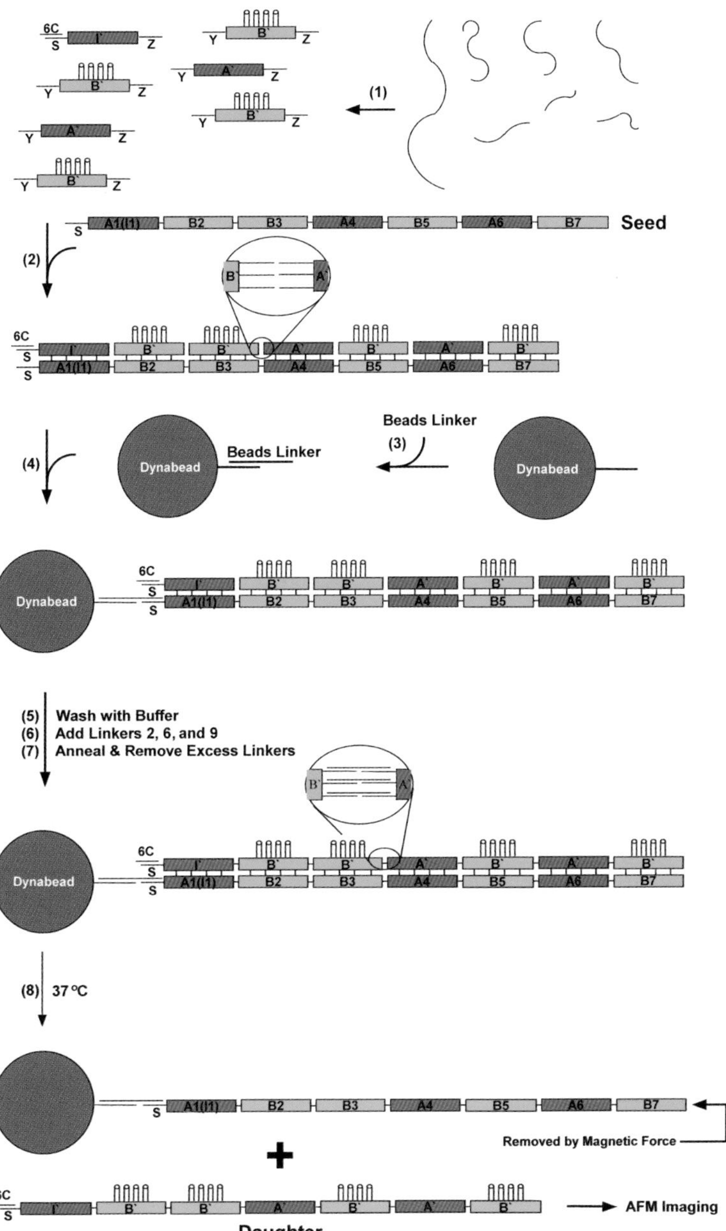

Figure 11-4 *Replication of the seed pattern in the first generation.* Strands are annealed in step 1, where the tiles are all flanked by the same connectors, designated Y and Z. The initiator tile contains a protected S-strand, paired with a cover strand, 6C. The B′ tiles contain the 4-hairpin markers for AFM imaging. In the presence of the seed tile (step 2), the strands assemble into a pattern mimicking the seed pattern. The magnetic dynabead is prepared in step 3, and attached to the seed (step 4). This is followed by a wash step, the addition of linkers, and their annealing (steps 5–7). Heating the system to 37 °C results in the separation of the daughter 7-tile complex and the seed (removed magnetically).[11.6]

nonamers of DNA extended towards each other between each tile, and they were linked non-covalently by an octadecamer. This linkage was strong enough to withstand the heating (37 °C) needed to separate the mother from the daughter on successive cycles. Because the seed was connected to a solid support (Figure 11-4), it was possible to separate the two generations in the pot and examine them. However, the lack of reuse of the solid support prevented exponential growth of succeeding generations. Note that, in this system, either the green tiles or the red tiles could carry the marker in any generation. In the example shown in Figure 11-4, the green tiles carry the marker. The successful copying of the pattern by this process is seen in Figure 11-5, where a number of examples are shown in AFM images. Figure 11-6 shows that this process can be carried on for two generations. In the second generation, it is the red tiles that contain the marker.

Replication of origami tiles. Although the experiments described above demonstrate the ability to replicate a 1D informational arrangement with DNA motifs, they do not demonstrate exponential growth. Nevertheless, such experiments have been performed, in this case with DNA origami. The difficulties of working with extensive arrays of origami tiles meant that the most successful experiments were conducted with dimers of origami. In an emulation of Watson–Crick base pairing, A and T symbols were written on origami tiles, as shown in Figure 11-7. In further emulation of ordinary DNA, the A tile was made complementary to the T tile, so that a T–T origami dimer would yield an A–A origami dimer in the next generation. The process was initiated, as in the previous experiment, with a seed tile, one reading T–T. This tile was linked together successfully by numerous long sticky ends. Figure 11-8 shows how the T–T dimer seed (panel i) complements A-tiles (panel ii), so that a dimer of A–A is produced from the seeds in the first generation.

The replication cycle is shown in Figure 11-9. Starting with T–T seeds at the upper left, A–A daughter tiles hybridize with them to produce A–A dimers in the recognition complex. The A–A dimers are then crosslinked together covalently by the application of ultraviolet (UV) light. The recognition complex is then dissociated into A–A dimers and T–T seeds. The T–T seeds continue to cycle, but the A–A dimers, now covalently crosslinked, can act as seeds themselves, as seen on the right part of the figure. T daughter tiles can pair with the A–A dimer to form another T–T/A–A recognition complex. The daughter tiles in this complex are also crosslinked by UV light, and then the recognition complex is dissociated by heat, thereby generating two flavors of seeds, A–A dimers and T–T dimers. Note that the A–A seed dimers cycle again (top of right panel), while the T–T seed dimers also can cycle (top of left panel).

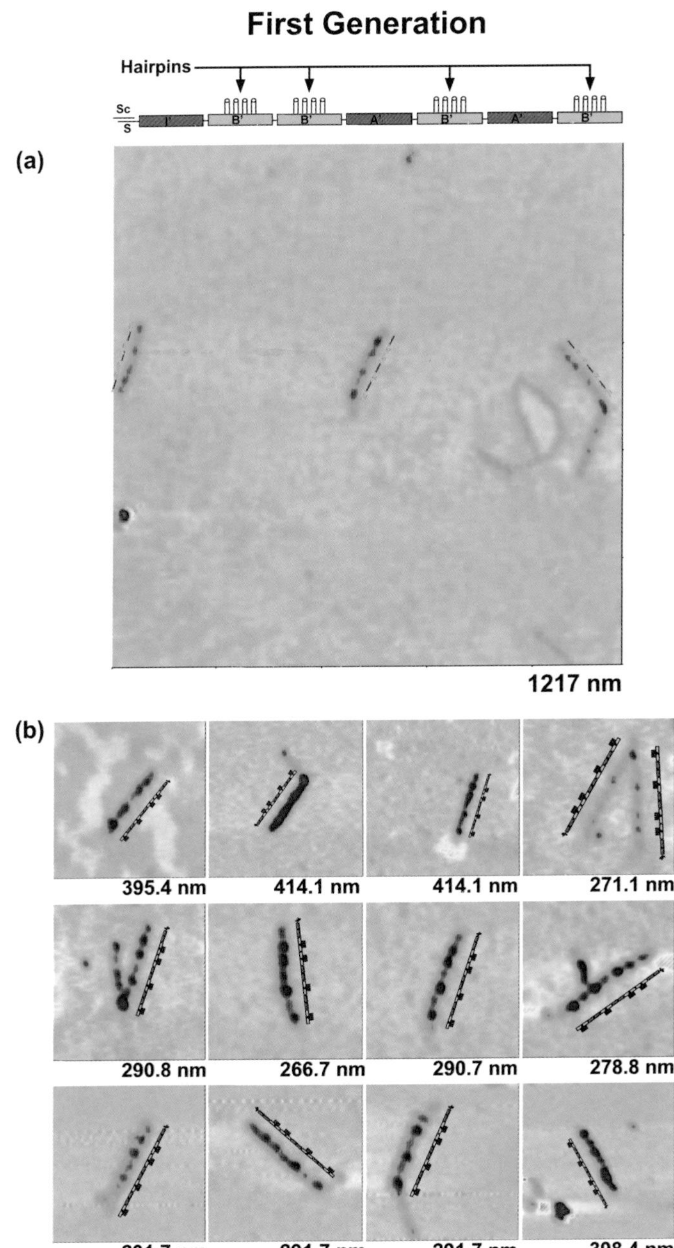

Figure 11-5 (a) *AFM microscopy of first-generation constructs showing a typical field slightly larger than a square micron.* (b) *Zoomed images of heptameric daughter complexes.* These images, flanked by explanatory schematic images, demonstrate that the ABBABAB pattern has been replicated successfully.[11.6]

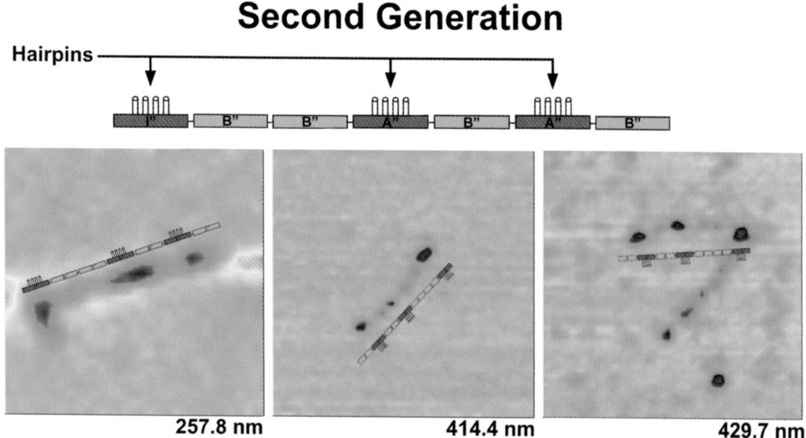

Figure 11-6 *AFM images of second-generation molecules.* These zoomed images show the pattern that was programmed in the original seed tile.[11.6]

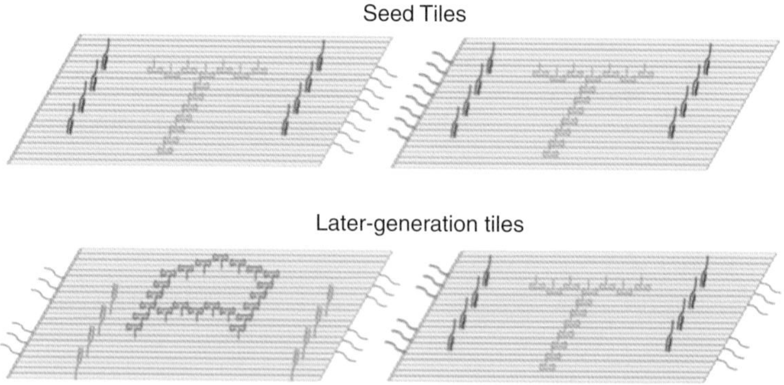

Figure 11-7 *DNA origami tile design and self-replication cycling.* Seed- and later-generation tile design (seed- and first-generation tile (A-tile) design) comprising three domains: (1) DNA hairpin structures to create a topographic feature that can be detected by AFM (seed- and second-generation tiles labeled with T and first-generation tiles labeled with A); (2) vertical sticky ends (red), which pair with the successive-generation tiles; (3) both T and A have the same horizontal sticky ends as linkers (pink with ^{CNV}K and turquoise without it) to connect other tiles in the same generation, with sticky ends on the right side of each tile complementary to those on the left side.

Thus, the system can continue to cycle over and over again, leading to exponential growth.

The exponential growth of the system is visible in Figure 11-10. The ordinate is in units of log 2, so it can be seen that every cycle produces twice as much

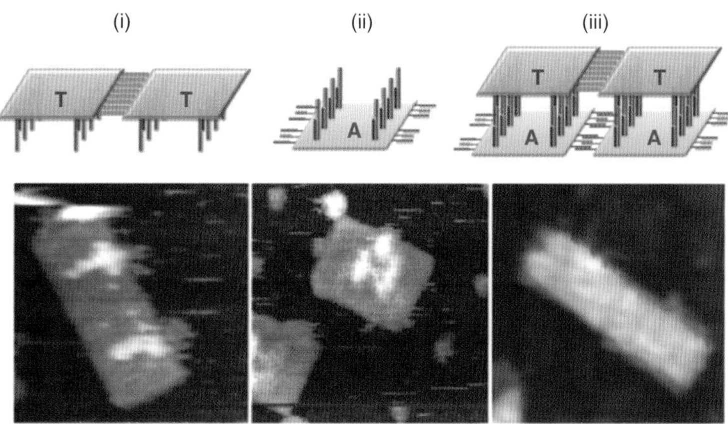

Figure 11-8 *Self-replication of DNA seeds.* Schematics and AFM images of a DNA dimer seed, T–T, a first-generation tile (A-tile), and a double-layer complex of seed and two first-generation tiles.

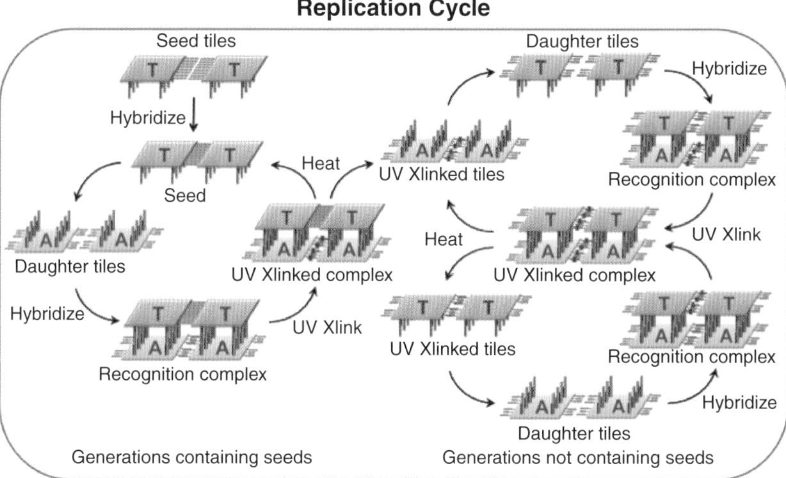

Figure 11-9 *The self-replication cycle.* Self-replication cycling of the dimer seed system, including T–T seed formation with special horizontal complementary sticky ends, recognition and hybridization of daughter tiles to seeds with vertical bonds, formation of new-generation dimers using horizontal bonds, and the ^{CNV}K photo-crosslinking reaction and separation of the two successive generations by heating of the system to ~46 °C. The left-side cycles include seed tiles and the right-side cycles do not.

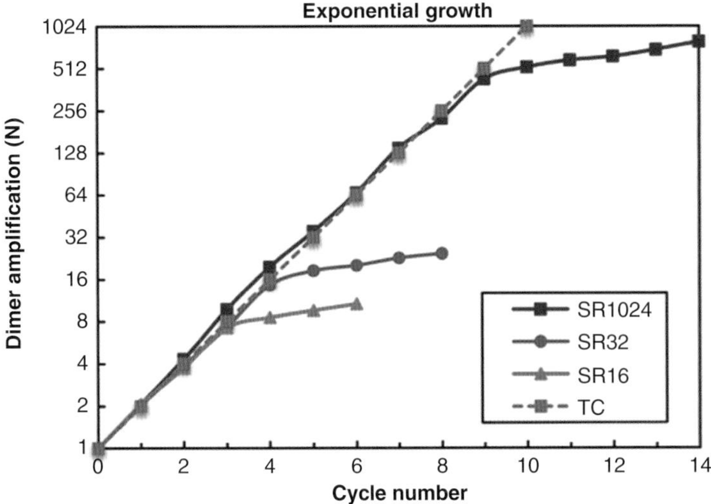

Figure 11-10 *Dimer amplification.* The amplification of dimers (including seeds and later generations) increased exponentially before leveling off as the supply of substrates was exhausted. Red, blue, and green curves represent self-replication cycling containing seeds, first-generation, and second-generation tiles with ratios of 1:1024:1022, 1:32:30, and 1:16:14, respectively. The purple curve is the theoretical curve for exponential growth. The dimer amplification factor was calculated at the end of each cycle.

material as the previous cycle. If one has only enough potential daughter origami tiles to make N cycles, the exponential doubling continues through cycle N−1. After this the system peters out because there are not enough daughter tiles present for another doubling. This is observed whether the target is N = 4, N = 5, or N = 10. Exponential growth takes place through the (N−1)th cycle. At a certain point, to save tiles, it is useful to resort to serial growth, whereby a certain amount of material from a previous cycle inoculates a fresh medium. Figure 11-11 shows an example of this, where a total of 24 cycles are performed, with about 14 million (instead of 16 million) progeny produced in principle (X. He, R. Sha, Y. Mi, P.M. Chaikin, N.C. Seeman, Self-Replication, Exponential Growth, Selection and Competition in Systems of DNA Origami Tiles. *Science*, submitted 2014).

We introduced this chapter with the notion of selection being our goal. This remains our goal and that of the other workers in this area. In the system described above, control of local features would facilitate the achievement of this target. When achieved, DNA nanotechnology will join those biological systems that demonstrate evolution. This will enable the selection of DNA-based materials for properties that prove optimal for particular environments.

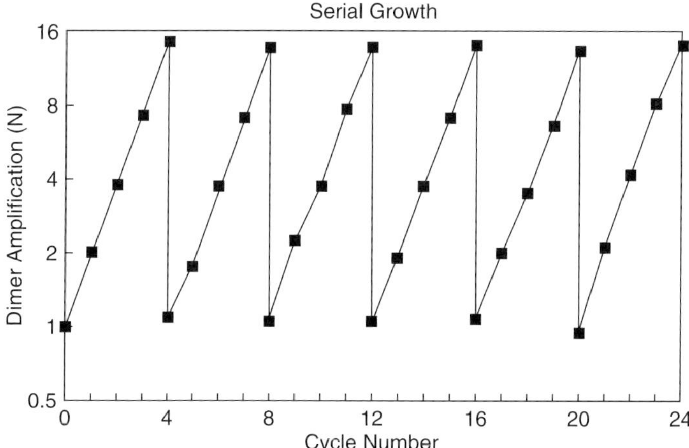

Figure 11-11 *Amplification of a serial transfer experiment.* A self-replication amplification curve that was obtained by six successive replications, allowing approximately 14-fold amplification before transfer of 8% of the mixture to the next replication tube, which contained a fresh supply of later-generation monomers.

References

11.1 C. Darwin, *The Origin of Species*, London, John Murray (1859).

11.2 A.R. Wallace, *On the Tendency of Varieties to Depart Indefinitely from the Original Type*, London, Linnean Society of London (1858).

11.3 D.R. Mills, R.L. Peterson, S. Spiegelman, An Extracellular Darwinian Experiment with a Self-Duplicating Nucleic Acid Molecule. *Proc. Nat. Acad. Sci. (USA)* **58**, 217–224 (1967).

11.4 A.D. Ellington, J.W. Szostak, *In Vitro* Selection of RNA Molecules that Bind Specific Ligands. *Nature* **346**, 818–822 (1990).

11.5 C. Tuerk, L. Gold, Systematic Evolution of Ligands by Exponential Enrichment: RNA Ligands to Bacteriophage T4 DNA Polymerase. *Nature* **249**, 505–510 (1990).

11.6 T. Wang, R. Sha, R. Dreyfus, M.E. Leunissen, C. Maass, D. Pine, P.M. Chaikin, N.C. Seeman, Self-Replication of Information-Bearing Nanoscale Patterns. *Nature* **478**, 225–228 (2011).

12
Computing with DNA

Anybody who has reached this point in the book realizes that DNA and its congeners are special molecules because of their ability to encrypt information. We have seen a variety of examples here of how that information can be used to direct the folding of molecules to produce specifically shaped and organized molecular species. Likewise, everybody alive during the twenty-first century is aware that we live in an age of information. The information that we use is usually encrypted in electronic bits, rather than in DNA; nevertheless, some information has already[12.1,12.2] been stored specifically in DNA molecules, as an alternative to electronic or print media. Of course, the key way in which we are exposed to information in our daily lives is through our computers and their variations: pads and smartphones. These circumstances lead one to ask if it might be useful to try doing computation with DNA. There is a large community of investigators who work in the field of molecular computation, and, more than any other, this community has contributed valuable ideas and workers to the field of structural DNA nanotechnology. The treatment of this topic here is only a simplified introduction to a few topics in an elementary form.

DNA in logical computations: the Adleman experiment. The first use of DNA in computation was done by Leonard Adleman.[12.3] He solved a Hamiltonian path problem using DNA molecules. The problem he solved, as has been the case for much of the computation performed with DNA, was a toy problem, one that could be solved simply in one's head, but which also represented a class of problems that are potentially challenging to traditional computational techniques. The Hamiltonian path problem is related closely to the "traveling salesman" problem, the optimization of a route through a number of cities in a territory. If there are only a few cities, the problem is easy to work out, but the number of solutions becomes enormous if there are many cities. Thus, if there are 10 cities, the number of possible answers is proportional to 10!, or about 11 million, a large but tractable number. However, if

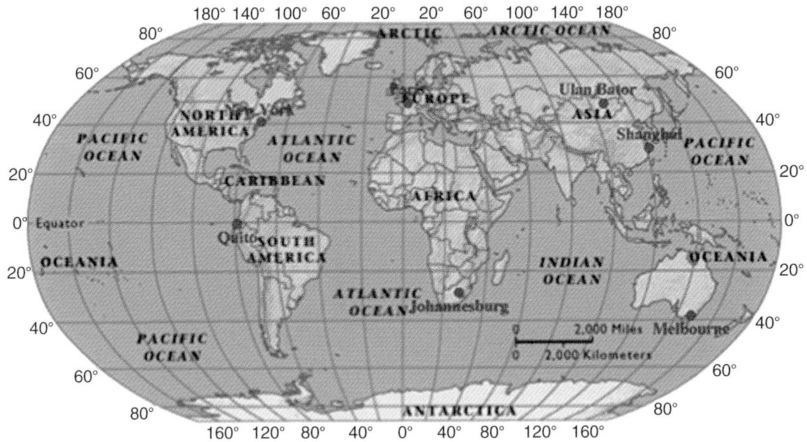

Figure 12-1 *Seven cities for the Adleman experiment.* Seven cities are indicated: Quito, New York, Paris, Johannesburg, Ulan Bator, Melbourne, and Shanghai.

there are 100 cities, the number of possible answers is proportional to 100!, which is not a number readily handled by a computer that examines each possibility in turn. The notion behind molecular computing is that the huge multiplicity of molecules in a pot can be treated in parallel, rather than sequentially, thereby providing a vast acceleration in calculation speed. For example, there are $\sim 10^{12}$ molecules in a picomole of strands, 10^{15} if each one is present 1000 times. At typical micromolar concentrations, these molecules would only take up a milliliter, one cubic centimeter. If they were operated on in parallel, an answer could be derived relatively quickly, at not a large expense of material or of time. Thus, even if an operation performed in the lab took 1000 seconds, a factor of 10^{12} slower than the nanosecond that a computer cycle takes, if this operation were performed on 10^{15} or 10^{18} molecules, one would still be ahead of the game. The logic behind this notion and the availability of varied nucleic acid sequences derives from the combinatorial chemistry approaches pioneered by Ellington and Szostak[12.4] and by Tuerk and Gold[12.5] in the early 1990s.

Let's look at what Adleman did. Figure 12-1 contains a map of the world showing seven cities: Quito, New York, Paris, Johannesburg, Melbourne, Ulan Bator, and Shanghai. In the Hamiltonian path problem he solved, there is an incomplete set of routes between these cities, and the goal is to find a route from an origin city to a target city that goes through all of the cities once. Figure 12-2 shows the routes that Adleman had available to use. Thus, there are 1-way routes from Quito, for example, to New York, to Shanghai, and to Ulan Bator.

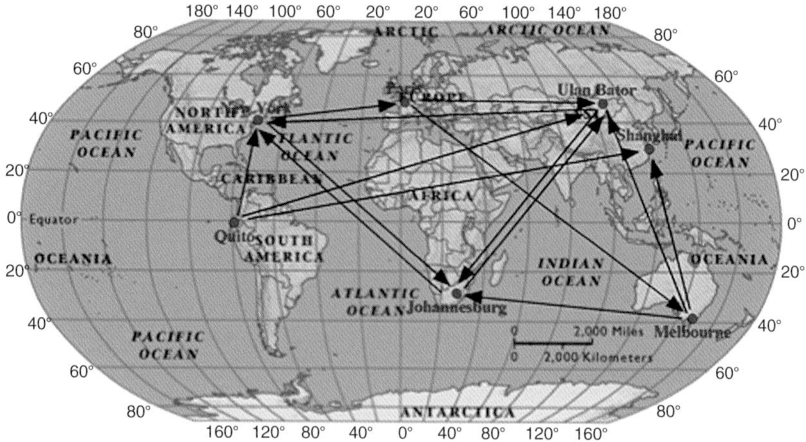

Figure 12-2 *Routes for the Adleman experiment.* An incomplete series of routes is shown by black arrows between the cities indicated in Figure 12-1.

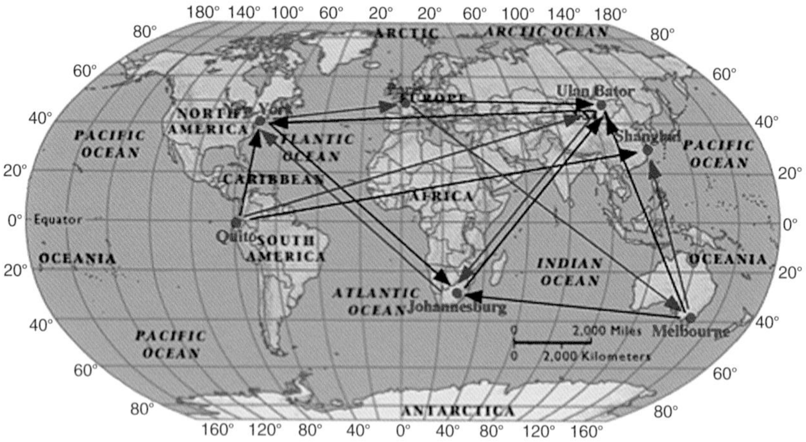

Figure 12-3 *A Hamiltonian path between the seven cities.* The components of the Hamiltonian path are shown in magenta.

Similarly, there are routes both ways between New York and Johannesburg and between Johannesburg and Ulan Bator, but there is no route at all between Paris and Johannesburg or between New York and Shanghai. Figure 12-3 shows a Hamiltonian path between Quito and Shanghai: starting in Quito, one goes to Ulan Bator, then to Johannesburg, then to New York, then Paris, then Melbourne, and finally to Shanghai.

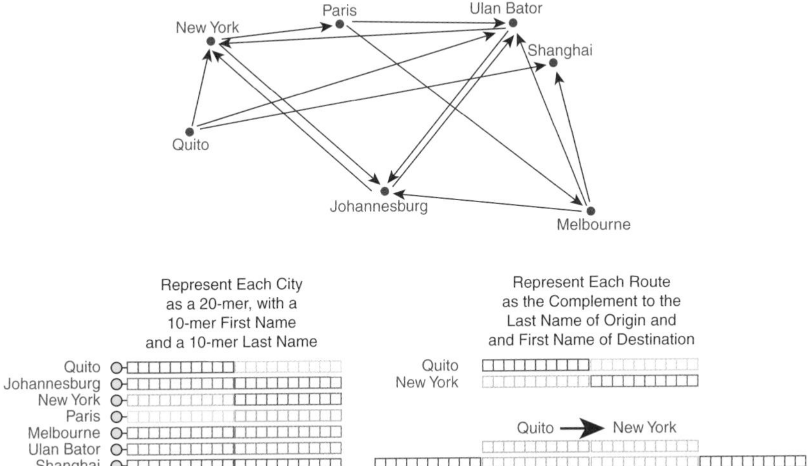

Figure 12-4 *The scheme for elucidating the Hamiltonian path with DNA.* The intercity routes are shown at the top. At lower left are the cities shown, each with a 10-nucleotide first name and a 10-nucleotide last name. Each of the yellow circles on the 5′ end of the 20-mer representing a city represents a phosphate, enabling the strand to be ligated to another strand. At lower right is a representation of the routes, the complement to the last name of the origin city fused to the complement to the first name of the destination city. The routes are unphosphorylated. As an example, the route from Quito to New York is shown.

So let's see how Adleman solved the problem. Figure 12-4 shows that he represented every city as a DNA 20-mer; he gave every city a "first name," the first 10 nucleotides, and a "last name," the 11th to 20th nucleotides. All of his cities were phosphorylated on their 5′ ends, so they could be ligated together to form a longer itinerary. He represented all of the routes as DNA 20-mers complementary to the last name of the origin city and the first name of the destination city. The joining of Quito to New York is shown, to exemplify this strategy. The key point here is that T4 DNA ligase only ligates two molecules separated by a nick, not just any two molecules in the reaction vessel. Thus, the routes are acting as catalysts to promote the ligation of any two 20-mers representing successive cities for which a route exists, but a pair of cities for which a route does not exist will not be ligated together. The notion is that *all* pairs for which routes exist will be joined, but the ligase will not stop there. Every possible route of whatever reasonable length will be formed by the ligation reaction.

So how was the correct answer selected from all the possible product strands? To begin, it was clear that, with seven cities in the answer, the only strands containing the correct answer had to be 140 nucleotides long. These were readily separated from the other strands present just by running a gel. Second, the correct sequence had to begin with Quito and end with Shanghai. The way to select these strands is to perform polymerase chain reaction (PCR) with primers complementary to Quito and to Shanghai. Strands not starting at Quito and ending at Shanghai would fail to be amplified by this procedure. Note that the other strands would not be eliminated, but they would fail to be amplified. Finally, the presence of each of the other five cities (Ulan Bator, Johannesburg, New York, Paris, and Melbourne) had to be ascertained individually, by binding the candidate strands to beads that contained complements to those cities. In the end, sequencing provided the correct sequence for the Hamiltonian path, of which only one existed in this case.

Solving a 3-colorability problem. Another example of DNA-based computing can be found in the following example, where the 3-colorability of a simple graph was established.[12.6] 3-Colorability means that it is possible to color the vertices of a graph using three colors, and no edge of the graph is flanked by the same colors. Figure 12-5 shows the graph that is being examined. Panel a shows the nodes of the graph: vertex 5 is 2-connected, vertex 2 is 4-connected, and the other vertices are all 3-connected. Panel b shows a DNA representation of the graph: vertex 5 is represented by a "2-arm junction," two helices connected by single-stranded spacers. Vertex 2 is represented by a 4-arm junction, and the other vertices are represented by 3-arm junctions. Panel b also shows a successful coloring of the graph using three colors: red, blue, and green; in this representation, vertices 2 and 5 are colored green, vertices 1 and three are colored blue, and vertices 4 and 6 are colored red.

Panel c shows how the graph is assembled for the experiment. Each of the vertices is tailed in a long sticky end. Two vertices can always be set to a particular color, say blue and green. The other vertices must be present in three copies: a red copy, a blue copy, and a green copy. Each is tailed by a 24-nucleotide sticky end that identifies it as being a particular one of the vertices, so the correct connectivity of the resulting graphs is established in the two 8-nucleotide portions on either end of the sticky end. The coloration of a particular graph is carried in the central eight nucleotides at the center of the sticky end. That part of the sticky end contains a sequence that forms a restriction enzyme recognition site. Note that restriction enzymes recognize specific sequences (hexamers here) and cleave DNA only there. When the colored sticky ends are identical (i.e., the vertices flanking the edge are the same color), then the restriction site is complete and the sequence is cleaved. When they are different, then the site is incomplete and the enzyme cannot cleave the edge of the graph; those edges survive, while the others are

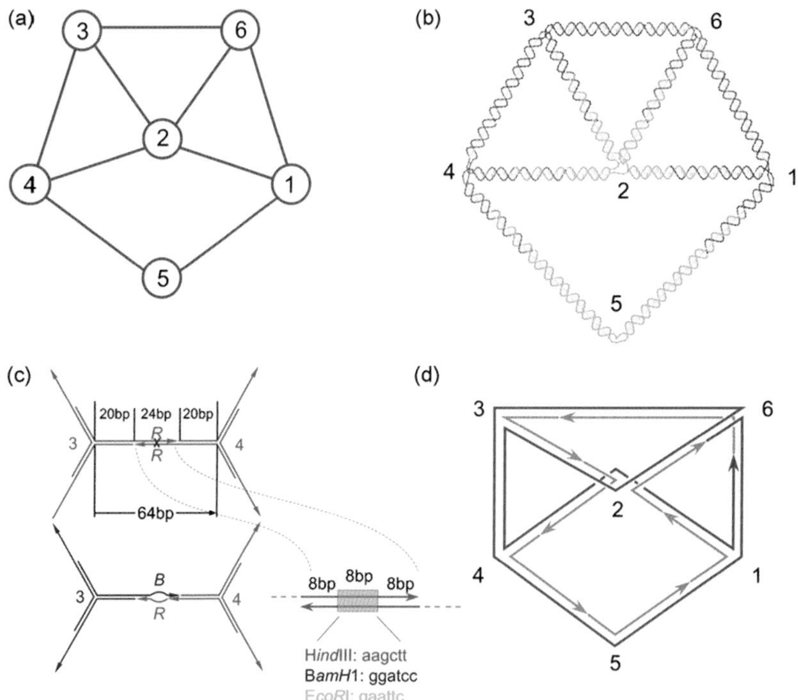

Figure 12-5 *Solution to a 3-colorability problem: schematics describing this computational system and experiments.* (a) *The connectivity of the graph and the numbering of the vertices.* (b) *A molecular representation of the graph.* The color assignments of the vertices and of the branched junctions that flank them are indicated. The molecule shown is a cyclic single strand of DNA. (c) *The structure of the sticky ends.* A homochromic edge between vertices 3 and 4 is shown in red above a blue–red heterochromic edge. The homochromic edge contains a restriction site for Hind III in the middle of the sticky end, whereas the heterochromic edge has two distinct mismatched restriction sites, indicated as a "bubble." The possible enzyme sites are shown color-coded, as is the structure of the sticky ends. The portions of the sticky ends flanking the restriction sites (first eight and last eight nucleotides) contain vertex–edge-specific sequences. (d) *The structure of the reporter strand.* The graph is shown in a representation with the helices unwound. Each edge has an integral number of helical turns. The reporter strand visiting every edge is illustrated in magenta. The strands that are not phosphorylated, and hence not ligated to the reporter strand, are illustrated as individual strands in gray. Arrows indicate the 3′ ends of the strands. Reprinted by permission from Elsevier.[5.3]

destroyed. This is a good example where computer science must accommodate chemistry. Although the scheme above is foolproof in principle, it is not possible to ligate all the strands successfully. Consequently, a reporter strand was used to establish the coloring scheme. The gray strand in panel d goes

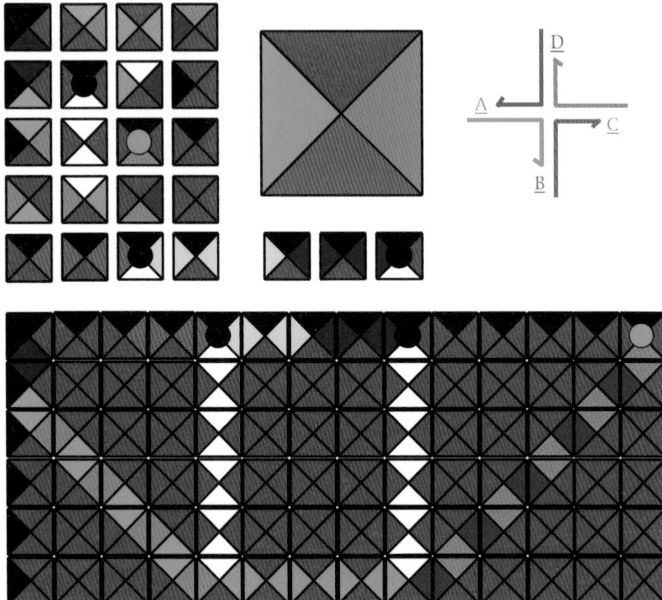

Figure 12-6 *Wang tiles forming a pattern that represents the addition of 5 to 9, summing to 14.* A series of 23 small Wang tiles are shown on the upper left and at the bottom of the upper central portion of the drawing. Their assembly according to the Wang rules (identically colored edges are shared) is shown at the bottom. The relationship between a Wang tile and a 4-arm branched junction with sticky ends is shown by color-coding in the upper central and upper right portions of the drawing.[12.9]

through all edges, and serves to establish the coloring scheme.[12.7] Thus, this scheme is somewhat limited, and would need improved ligation chemistry to be successful in significantly larger systems.

Algorithmic assembly. A stimulating possibility that has so far been fulfilled only partially is algorithmic assembly, first suggested by Erik Winfree.[12.8] The idea is to use DNA motifs (tiles) as logic gates that can be used to produce patterns with fewer tiles than are absolutely necessary. The notion is predicated on the notion of Wang tiles. Figure 12-6 is an example of addition using Wang tiles (adapted from Figure 11-4.1 of reference 12.9). The upper portion shows 23 small tiles with differently colored edges. The arrangement at the bottom of the figure shows how they can be assembled if one requires every tile to assemble by having similarly colored edges bind together. Thus, red edges bind to red edges, green to green, blue to blue, yellow to yellow, etc. It turns out that the assembly shown, which is unique if the tiles are not allowed to rotate, demonstrates an addition, the sum of 5 and 9

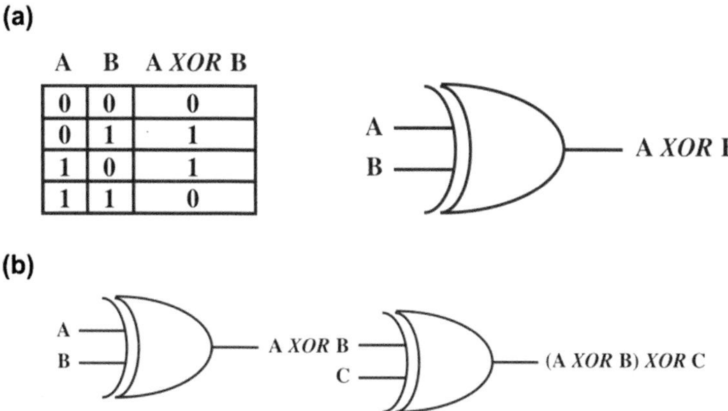

Figure 12-7 *The XOR operation.* (a) *The truth table for XOR, and its symbol.* (b) *Cumulative XOR.* The output of one XOR operation becomes part of the input to another XOR operation.

to yield 14. Starting from the upper left corner, the green–blue diagonal assembles, while a white vertical path starts at the fifth tile in the top row. Where they intersect at the bottom, the blue/red/orange/white tile is inserted, and the pathway changes from a diagonal to a horizontal pathway, as a consequence of the orange–orange interactions. This pathway meets another vertical pathway descending from the ninth tile in the top row. Where they intersect, the orange/red/purple/white tile is inserted, changing the direction of the pathway again. The purple–gray diagonal then intersects the top row at the 14th tile. Tiles that assemble in this fashion have been shown to emulate a Turing machine; i.e., this type of assembly is Turing-complete.

Arguably more interesting is the fact that the arrangement at the bottom of Figure 12-6 is an 84-tile array, yet only requires 23 unique tiles. Thus, in principle, a saving of a factor of 3.6 would be obtained if this assembly were to be made using this system, rather than using individual tiles, if this arrangement of tiles were a target for some reason.

A cumulative XOR by algorithmic assembly. A cumulative XOR assembly is an instructive example.[12.10] The XOR operation, shown in Figure 12-7a, is an XOR logic gate, with its truth table at the left: if the inputs are the same, 0 and 0 or 1 and 1, the output is 0, whereas if they are different, 0 and 1 or 1 and 0, the output is 1. Note that this operation is just addition that lacks a carry bit. Figure 12-7b shows cumulative XOR, just using the output from one XOR operation as one of the inputs to another XOR operation. The representation of this operation with tiles is shown in Figure 12-8. Panel a shows the TX tile that

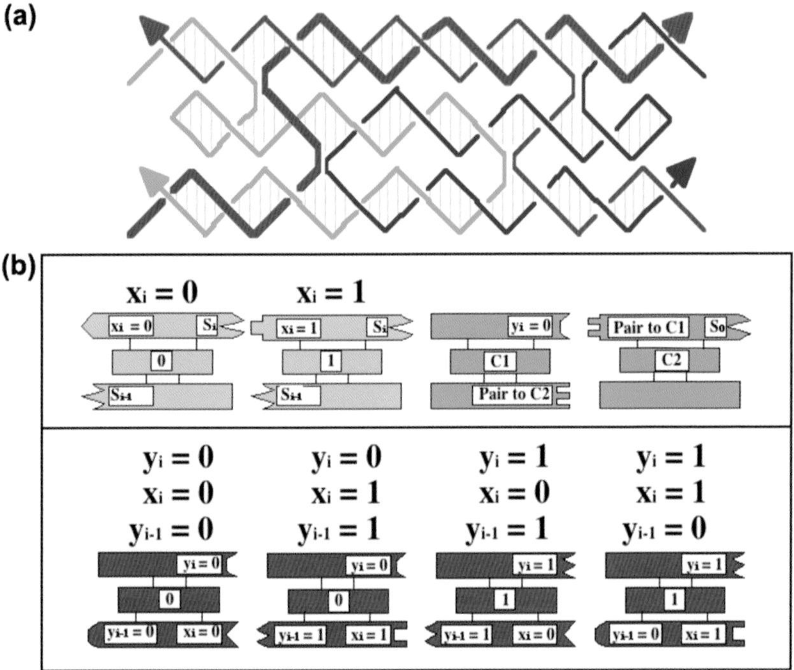

Figure 12-8 *Components of a 1D cumulative XOR operation.* (a) *The TX tile.* The tile consists of four strands. The reporter strand is drawn in a thick red line. (b) *The input (blue), calculational (red), and initiator (green) tiles.* Note that the input to the calculational tiles is on their bottom domains, and that both sides of the input must be effective for the system to work.[5.64]

is used to represent the units. The red reporter strand has been drawn with a thicker line to emphasize its importance, as will be seen below. Panel b illustrates the tiles that are used in the assembly. There are two light-blue input tiles, two green initiator tiles, and four red XOR tiles. The bottom domains of the XOR tiles represent their recognition properties: left to right, 0,0 (value 0), 1,1 (value 0), 1,0 (value 1), and 0,1 (value 1). These values are coded in their sticky ends, which are drawn as different shapes. Note that the same sticky end on the left is used for 0 in both cases, and the same sticky end on the left is used for 1 in both cases; likewise, the same sticky end on the right is used for 0 in both cases, and the same sticky end on the right is used for 1 in both cases.

Figure 12-9 shows how the calculation works for four input values. In the schematic portion in the upper left panel of the figure, the first tile ($X_1 = 1$) and the two initiator tiles (C_1 and C_2) assemble, forming a slot for the first answer

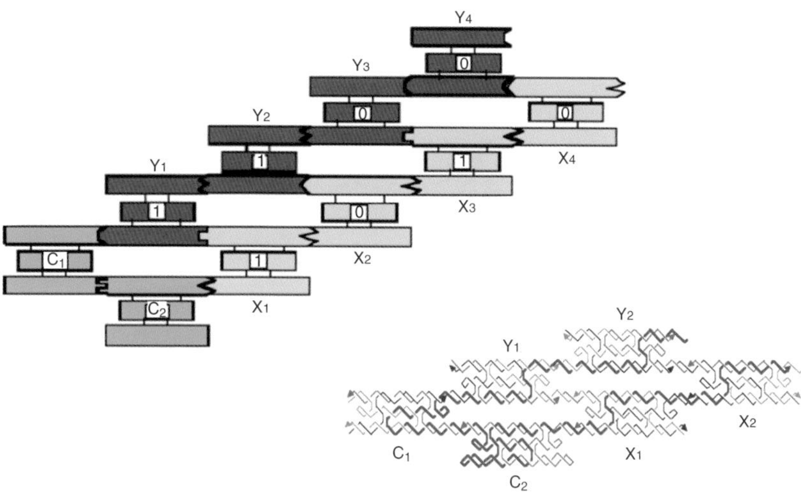

Figure 12-9 *Self-assembly of the cumulative XOR operation.* The upper left shows four input X tiles (blue), two initiator C tiles, and four calculational/output Y tiles. The region around the initiators (C1 and C2), including the X_1 and X_2 tiles on the input side and the Y_1 and Y_2 tiles on the output side, is shown in a double helical representation. The red reporter strand is emphasized.[5.64]

tile (Y_1). The value on C_1 is set to 0, so the Y_1 tile is a 1 tile. This value is reported on its upper right sticky end. When combined with X_2 (= 0), the Y_2 tile is also a 1 tile. Combining this with X_3 (= 1) binds Y_3, a 0 tile. When combined with X_4 (= 0), a 0 tile results for Y_4. This arrangement is read out through the reporter strands shown, for the region flanking the initiators, in the lower right panel of the figure. The reporter strands are ligated together to produce a long single strand connecting the input to the output. Each of the tiles contains a restriction site on its reporter strand for the two possibilities 1 and 0. Figure 12-10 shows the restriction of the reporter strand for the calculation in Figure 12-9. Inputs of X = 1,0,1,0 yield Y = 1,1,0,0. A small amount of erroneous restriction is visible near Y_3, but it is not substantial.

2D algorithmic assemblies. Other algorithmic assemblies can be produced with even fewer tiles: for example, a Sierpinski triangle (Pascal's triangle, mod 2) can be produced in principle using only seven tiles, a vertex tile, two edge tiles, and four XOR tiles, representing the four possible inputs to the XOR gate.[12.11] Likewise, Winfree and his colleagues have shown that one can count using a set of four tiles.[12.12]

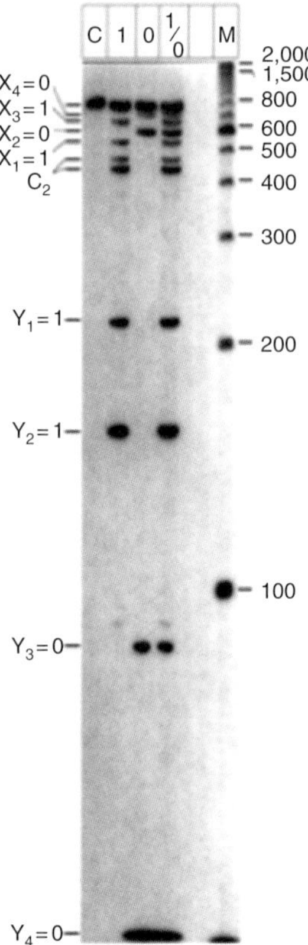

Figure 12-10 *Results of a calculation.* The input (X) of 1,0,1,0 is shown leading to the output (Y) of 1,1,0,0. Note that there is a small amount of erroneous cleavage at the Y_3 position.[5.64]

Small algorithmic assemblies vs. costs. Inspection of Figure 12-8 shows that it is more difficult to get correct assemblies from tiles representing logic gates than from conventional tiles. Thus, in Figure 7-4, a C tile, say, is competing for its slot with three other completely incorrect tiles: an A tile, a B′ tile, and a D* tile. By contrast, inspection of the four red tiles in Figure 12-8b shows that each of the XOR tiles is competing with one incorrect tile and two half-correct tiles, similar to the problem indicated in the capture experiments described in Chapter 10 (Figure 10-8). However, one must consider the alternative. If one is interested in a small algorithmic assembly (e.g., the pattern shown in Figure 12-6), how much will it cost just to do the assembly in the naïve way, with 84 unique tiles? If each tile contains 4 strands, you need 336 strands, of estimated length 36 nucleotides. The current price (summer 2014) of

DNA is $0.08 per base for 10 nanomole syntheses ($0.11 for 25 nanomoles). That's about $2.88 per strand, or ~$968. If it were to work reliably, that would be very cheap, roughly one-sixth of a month of graduate student labor (including indirect costs and tuition). In the end, this might be the best way to go for small patterns, particularly considering the success of DNA bricks and DNA origami, covered in Chapter 9. Both of these methods are an upshot of the decreasing cost of synthetic DNA. Given that the first DNA bought (admittedly in much larger quantities, for the techniques prevalent in the early 1980s) for DNA nanotechnology cost $312 per base, there does seem to be a Moore's law for DNA synthesis, just as there does for computation and for DNA sequencing, although its exponent is apparently somewhat slower, with a cost-halving time of about 30 months, compared with a roughly 15-month cost-halving time for genomic sequencing and 18–24 months for Moore's actual law of transistor density.

A successful recent implementation of DNA computation is based on cascades of "seesaw gates," built by Lulu Qian and Erik Winfree.[12.13] The seesaw gates are predicated on the use of DNA toeholds and single-stranded branch migration of DNA molecules. These gates involve single strands that have toeholds on either end, and are responsive to particular input strands that implement logical operations, such as AND or OR. They can be coupled together, leading to cascades of logical operations which basically represent complex circuitry. The DNA circuitry has been implemented so that it can extract the square roots of numbers with up to four binary bits. A detailed description of this system is beyond the scope of this book; the reader is referred to reference 12.13 for a more complete explanation, and to references 12.14–12.18 for follow-up work.

Autonomous logical devices. It is worth noting the pioneering work of Benenson, Shapiro, and their colleagues, who have developed computational systems that run autonomously,[12.19,12.20] without human intervention. They are predicated on a special class of restriction enzymes, class IIS, which cut in a place slightly removed from their recognition site. *Fok* I restriction endonuclease and DNA ligase are used as the hardware of their system, while a duplex DNA molecule containing *Fok* I recognition sites provides their input in a succession of symbols, and a series of transition molecules provide their software. The system is shown in Figure 12-11. The input molecule at the top is cleaved by *Fok* I, leaving a sticky end. Transition molecule T1 binds to the resulting sticky end through its own sticky end. Note that T1 is one of *many* transition molecules in the solution. T1 is ligated to the remainder of the input molecule, and the process repeats until it completes when all symbols on the input molecule are exhausted. At that point, one of two output reporter

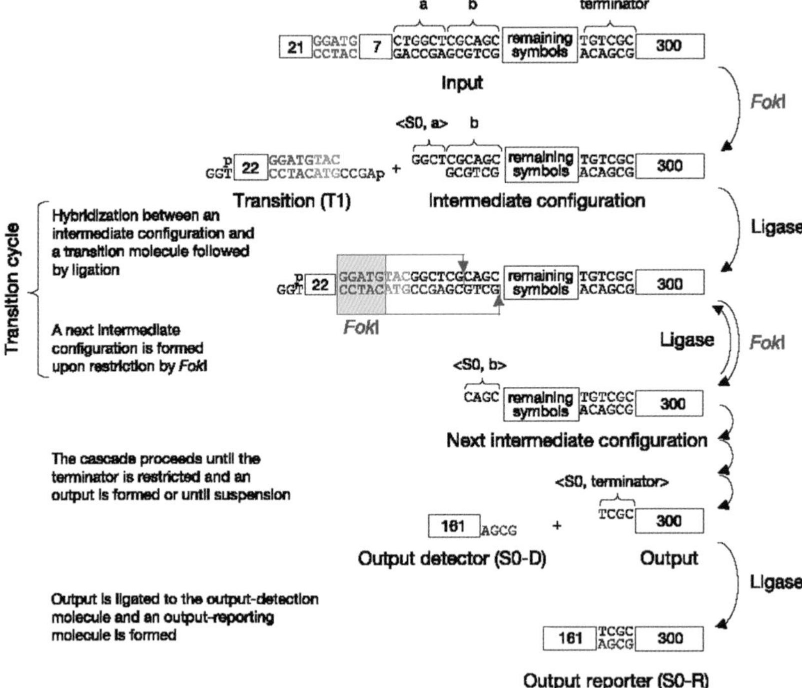

Figure 12-11 *The operation of an autonomous DNA computation system.* The input is shown at the top as a double-stranded DNA molecule encrypting a series of symbols, including a *Fok* I restriction site, shown in red. Following cleavage by the enzyme, a sticky end is exposed, which can bind to any of a variety of transition molecules; T1 is the transition molecule shown here. The transition molecule, also containing a *Fok* I site, is ligated onto the remainder of the input, and the cycle repeats until a terminator is reached. At this point, an output reporter is ligated onto the remains of the input, thereby providing the output of the calculation.

molecules is ligated to the remainder, depending on how the calculation went. Note that the system operates autonomously. An analysis of this type of computational system is provided in reference 12.21.

Final comment on DNA-based computation. It is important to recognize that this parallel field is one of the key drivers of structural DNA nanotechnology. Descriptions of all the work in this area would be out of place in this monograph. The reader is referred to reference 12.22, a multi-volume work, for a more complete introduction to this area. DNA-based computation provides novel ideas and skilled personnel for the effort to control the 3D structure of matter through information.

References

12.1 G. Church, Y. Gao, S. Kosuri, Next-Generation Digital Information Storage in DNA. *Science* **337**, 1628 (2012).

12.2 N. Goldman, P. Bertone, S. Chen, C. Dessimoz, E.M. LeProust, B. Sipos, E. Birney, Towards Practical, High-Capacity Low-Maintenance Information Storage in Synthesized DNA. *Nature* **494**, 77–80 (2013).

12.3 L.M. Adleman, Molecular Computation of Solutions to Combinatorial Problems. *Science* **266**, 1021–1024 (1994).

12.4 A.D. Ellington, J.W. Szostak, *In Vitro* Selection of RNA Molecules that Bind Specific Ligands. *Nature* **346**, 818–822 (1990).

12.5 C. Tuerk, L. Gold, Systematic Evolution of Ligands by Exponential Enrichment: RNA Ligands to Bacteriophage T4 DNA Polymerase. *Nature* **249**, 505–510 (1990).

12.6 G. Wu, N. Jonoska, N.C. Seeman, Construction of a DNA Nano-Object Directly Demonstrates Computation. *Biosystems* **98**, 80–84 (2009).

12.7 N. Jonoska, N.C. Seeman, G. Wu, On the Existence of Reporter Strands in DNA-Based Graph Structures. *Theor. Comp. Sci.* **410**, 1448–1460 (2009).

12.8 E. Winfree, Algorithmic Self-Assembly of DNA, Theoretical Motivations and 2D Assembly Experiments. In *Proc. 11th Conversation in Biomolecular Stereodynamics*, ed. R.H. Sarma, M.H. Sarma, New York, Adenine Press, pp. 263–270 (2000).

12.9 B. Grunbaum, C.G. Shephard, *Tilings and Patterns*, New York, Freeman, p. 605 (1990).

12.10 C. Mao, T. LaBean, J.H. Reif, N.C. Seeman, Logical Computation Using Algorithmic Self-Assembly of DNA Triple Crossover Molecules. *Nature* **407**, 493–496 (2000); *Erratum: Nature* 408, 750 (2000).

12.11 P.W.K. Rothemund, N. Papadakis, E. Winfree, Algorithmic Assembly of DNA Sierpinski Triangles. *PLOS Biol.* **2**, 2041–2054 (2004).

12.12 R.D. Barish, P.W.K. Rothemund, E. Winfree, Two Computational Primitives for Algorithmic Self-Assembly: Copying and Counting. *Nano Letters* **5**, 2586–2592 (2005).

12.13 L. Qian, E. Winfree, Scaling Up Digital Circuit Computation with DNA Strand Displacement Cascades. *Science* **332**, 1196–1201 (2011).

12.14 D.Y. Zhang, G. Seelig, Dynamic DNA Nanotechnology Using Strand-Displacement Reactions. *Nature Chem.* **3**, 103–113 (2011).

12.15 L. Qian, E. Winfree, J. Bruck, Neural Network Computation with DNA Strand Displacement Cascades. *Nature* **475**, 368–372 (2011).

12.16 M.R. Lakin, D. Parker, L. Cardelli, M. Kwiatkowska, A. Phillips, Design and Analysis of DNA Strand Displacement Devices Using Probabilistic Model Checking. *J. R. Soc. Interface* **9** (72), 1470–1485 (2012).

12.17 D.Y. Zhang, R.F. Hariadi, H.M.T. Choi, E. Winfree, Integrating DNA Strand-Displacement Circuitry with DNA Tile Self-Assembly. *Nature Comm.* **4**, article 1865 (2013).

12.18 W. Li, Y. Yang, H. Yan, Y. Liu, Three-Input Majority Logic Gate and Multiple Input Logic Circuit Based on DNA Strand Displacement. *Nano Letters* **13**, 2980–2988 (2013).

12.19 Y. Benenson, T. Paz-Elizur, R. Adar, E. Keinan, Z. Livneh, E. Shapiro, Programmable and Autonomous Computing Machine Made of Biomolecules. *Nature* **414**, 430–434 (2001).

12.20 Y. Benenson, R. Adar, T. Paz-Elizur, Z. Livneh, E. Shapiro, DNA Molecule Provides a Computing Machine with Both Data and Fuel. *Proc. Nat. Acad. Sci. (USA)* **100**, 2191–2196 (2003).

12.21 D. Soloveichik, E. Winfree, The Computational Power of Benenson Automata. *Theor. Comp. Sci.* **344**, 279–297 (2005).

12.22 G. Rozenberg, T. Bäck, J.N. Kok, *Handbook of Natural Computing*, Heidelberg, Springer, 4 volumes (2012).

13

Not just plain vanilla DNA nanotechnology: other pairings, other backbones

Up to this point, we have been talking largely about Watson–Crick double helical DNA. No variations in the base-pairing and no variations in the backbone. Here we are going to discuss just a little bit of the work that is going on with other DNA structures, work that entails non-DNA backbones, and other species organized by DNA nanoconstructs. This chapter is not meant to be a comprehensive review of variations on the theme of either DNA nor its interactions with other species, just a taste to stimulate the reader to pursue other materials on her own.

Paukstelis DNA structure. Perhaps the first place to start is DNA, but DNA that is not simply Watson–Crick. A robust motif discovered in a single-crystal structure[13.1] is shown in Figure 13-1. This motif contains three conventional nucleotide pairs and three parallel pairs, consisting of one A–G pair and two G–G pairs. This overall motif is shown in stereo in Figure 13-2. A salient feature of the structure is that it crystallizes in a hexagonal space group with a cavity whose volume is 300 $Å^3$, which is shown in stereo in Figure 13-3a. The robustness of the motif is demonstrated by the fact that the Watson–Crick portion of the motif can be extended by 10 nucleotide pairs, yet the space group remains the same. Although the resolution of the crystal decreases somewhat, the cavity is greatly expanded by this expansion, so its volume is now increased markedly, as shown in Figure 13-3b.

Triplex DNA. In addition to the simple double helical motif, there are other helical motifs that have been characterized. The earliest of these was the DNA triplex, wherein a pyrimidine–purine–pyrimidine structure was formed (see Figures 2-4 and 2-5).[13.2] There are a number of utilities that one can imagine for such systems in DNA constructs. Without disrupting the double helix, it is possible to address specifically designed locations within the assembly by adding a triplex-forming oligonucleotide (TFO).

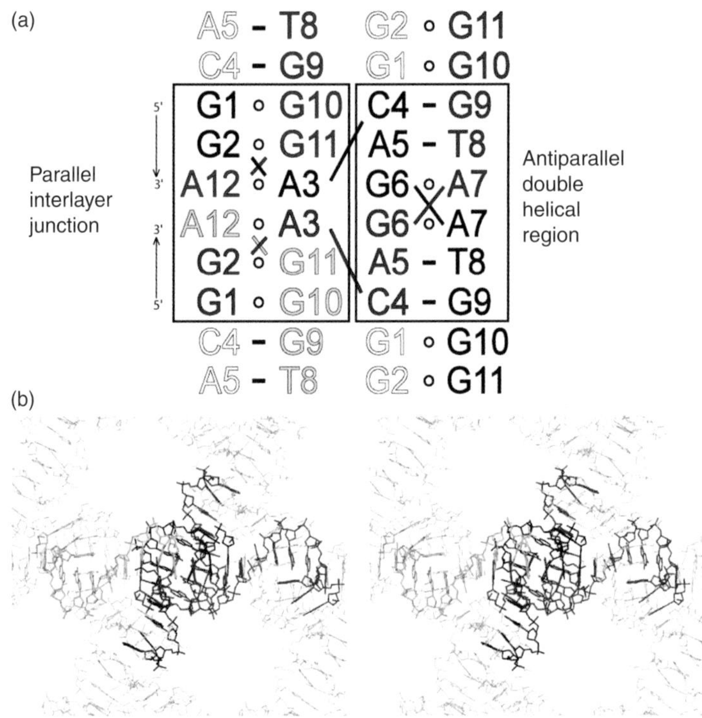

Figure 13-1 *A robust unconventional DNA motif.* (a) *A schematic diagram.* The strand connectivity is shown by color-coding. (b) *A stereo diagram of the crystal structure.* Individual strands are color-coded. Reprinted by permission from Elsevier.[13.1]

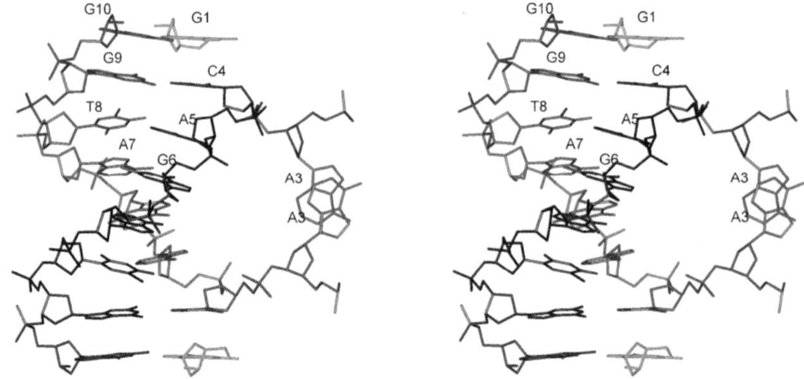

Figure 13-2 *A stereo diagram of the motif.* Note that an apparently double helical arrangement is in fact quite distorted. Reprinted by permission from Elsevier.[13.1]

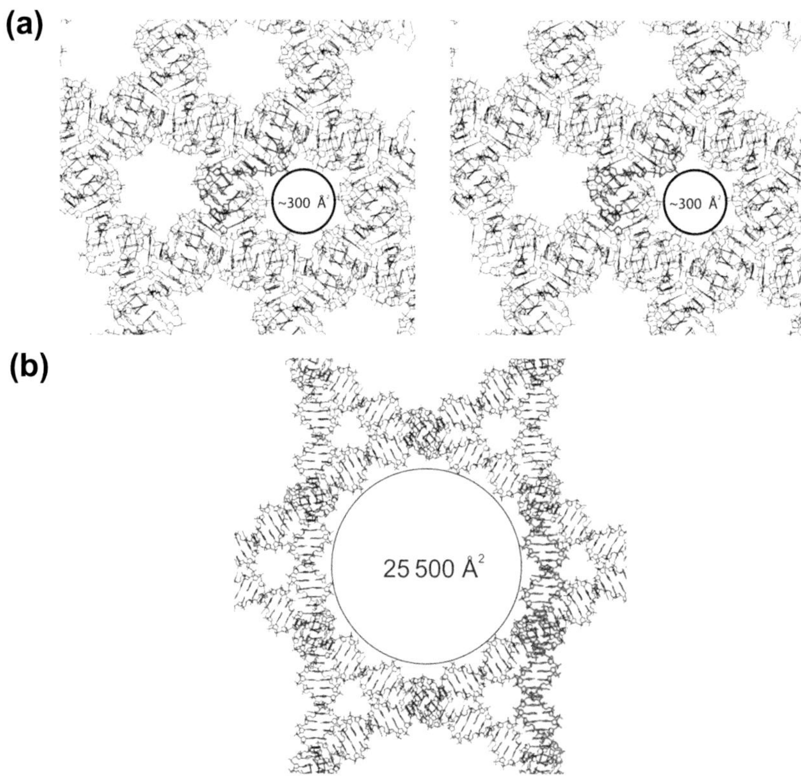

Figure 13-3 *Views down the six-fold axis of the crystal structure.* (a) *The view found in the crystal.* The cavity is indicated to have a volume of ~300 Å^3. (b) *The view down the extended molecular structure axis.* If the Watson–Crick double helical section is extended by one turn, this is the structure to be expected. Reprinted by permission from Elsevier.[13.1]

One could imagine tethering both small molecules and macromolecules to DNA constructs by the use of triplex associations. An example of triplex DNA added to DNA tensegrity-triangle crystals (see Chapter 7, especially Figure 7-23) is shown schematically in Figure 13-4; crystals to which triplex molecules containing dyes have been attached are shown in Figure 13-5.[13.3] In this illustration, panel A shows the crystal with no added TFO, and panel B shows the crystal binding a TFO containing Cy5, a turquoise dye. Panel C shows the crystal with fluorescein, a yellow dye, attached to the triangle, and panel D shows the TFO binding to the same crystal, yielding a green crystal.

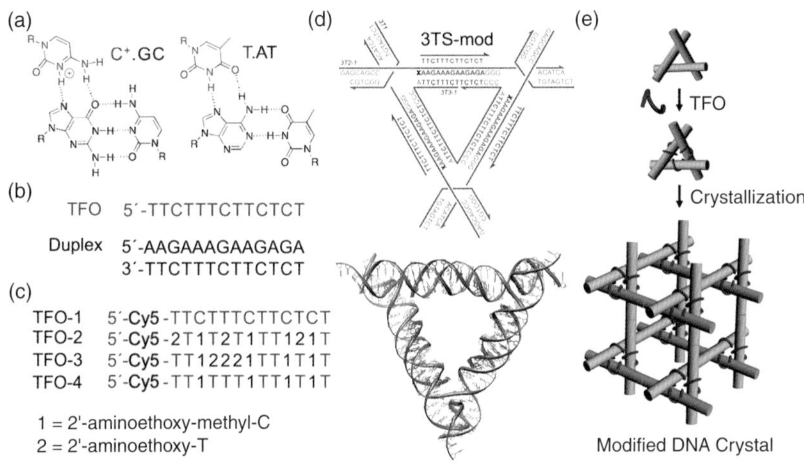

Figure 13-4 *Triplex DNA added to a 3D crystal.* This is the self-assembled crystal discussed in detail in Chapter 7. (a), (b), (c) *The nature of triplex formation, the sequences used, and the backbone modifications to encourage formation.* TFO means "triplex forming oligonucleotide." (d) *The 3-turn tensegrity triangle in schematic (top) and in double helical representation (bottom).* This is a 3-turn symmetric triangle. (e) *The protocol used to incorporate the TFO into the 3D crystal.* Reprinted by permission from John Wiley and Sons.[13.1]

3TS + Cy5-TFO	3TS-mod + Cy5-TFO	F-3TS-mod	F-3TS-mod + Cy5-TFO
Clear	Turquoise	Yellow	Green
A	B	C	D

Figure 13-5 *TFO coloration of the crystals.* Colorless crystals are shown in panel A, and crystals containing the TFO with Cy5, a turquoise dye, are shown in panel B. Crystals dyed yellow with covalently bonded fluorescein are shown in panel C, and those same crystals with Cy5 are shown as green in panel D. Reprinted by permission from John Wiley and Sons.[13.1]

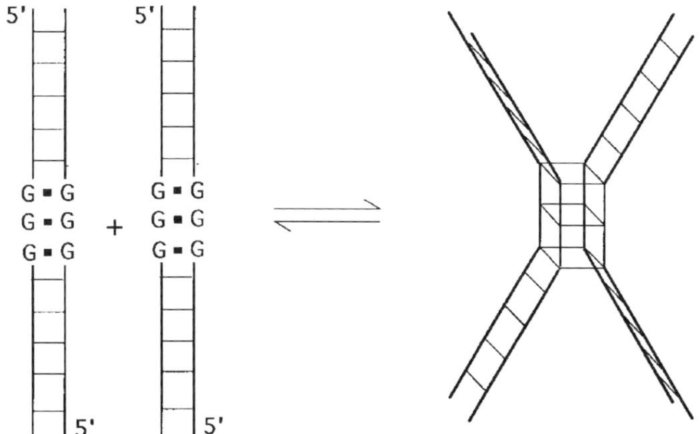

Figure 13-6 *G-quadruplex pairing of DNA duplexes.* The G4 formation at the center bonds two duplexes together. Reprinted by permission from the American Chemical Society.[13.5]

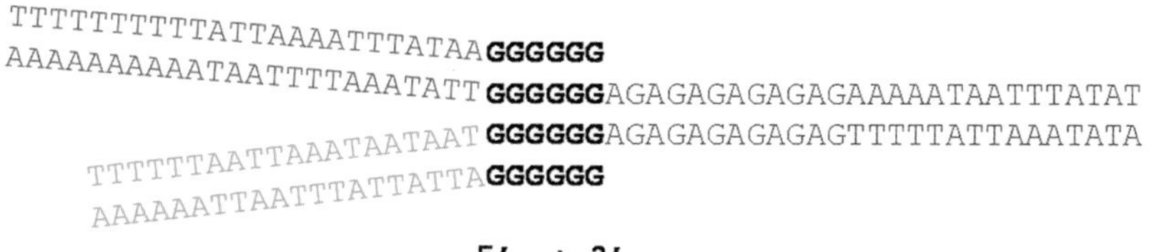

Figure 13-7 *G-quadruplex pairing to form a 3-arm junction.* A variation on the theme of Figure 13-6. Reprinted by permission from the American Chemical Society.[13.7]

G-tetrads. Triplexes are not the only unusual multi-stranded DNA structures. The G4 DNA tetrad (Figure 2-6)[13.4] is an intriguing system to be used in conjunction with Watson–Crick duplexes. The earliest work on this system with the idea of nanoconstruction dates back to Sen and Gilbert[13.5] and then Sen's later work demonstrating that the formation of G4 tetrads can be used to synapse two different DNA double helices,[13.6] as shown in Figure 13-6. This is a new way to join pairs of double helices, just as PX-DNA is a way to do that. Figure 13-7 illustrates a variation on this theme that forms a 3-arm junction with G4 at the branch point.[13.7] It is important to note that G4 can be formed from both parallel strands and a mixture of antiparallel strands, depending on the torsion angles about the glycosyl bonds. A long purely parallel structure has been assembled into a

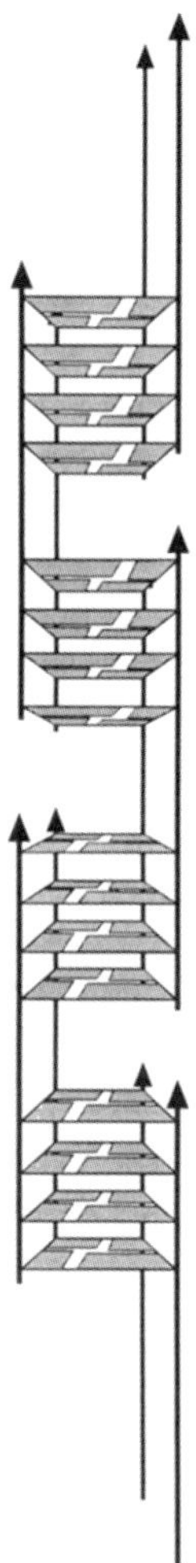

Figure 13-8 *A G-wire.* Long strands of G4 can form to produce "wires." Reprinted by permission from Oxford University Press.[13.8]

"G-wire,"[13.8] whose charge-conduction properties are currently being explored (Figure 13-8).[13.9] Kotlyar has established methods to make very long G-wires.[13.10]

The I-motif. Another non-Watson–Crick structure is the I-motif (Figure 2-7). This is the structure formed by hemi-protonated oligo-dC. The requirement of a relatively low pH to protonate the C nucleosides makes it a useful tool for estimating the pH within the cell. Krishnan[13.11] has built a number of devices containing oligo-dC in a single strand that

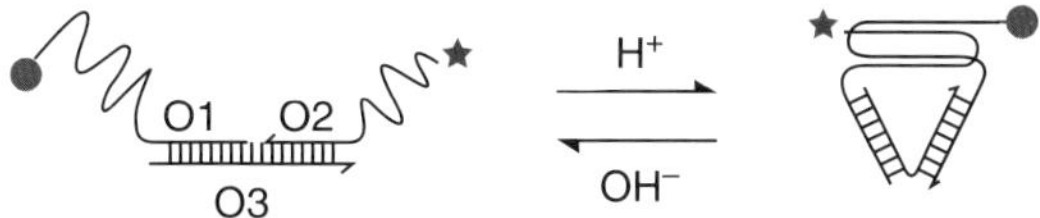

Figure 13-9 *An I-motif device.* At low pH, an I-motif forms, bringing a dye and a quencher into proximity. At high pH, the two are kept apart by duplex formation. Reprinted by permission from Macmillan Publishers Ltd.[8.6]

can fold to produce the I-motif. The device contains a FRET pair of dyes (Figure 13-9) whose average separation is dependent on the formation of the I-motif. When unfolded, as at the left of the image, they do not transfer energy. However, when folded, as at the right of the image, they produce energy transfer whose extent can be measured. The device shown is sensitive over a pH range of two units.

RNA. It is beyond the scope of this chapter to discuss further variants of DNA base-pairing within nanotechnology. To extend the system slightly, an obvious molecule that is an alternative to DNA is, of course, RNA. The first use of RNA to make a nanotechnological target dates back to 1996: the RNA trefoil knot shown in Figure 13-10.[13.12] The knot was prepared much like the DNA knots were prepared, as described in Chapter 4. A single strand was prepared containing two potentially paired domains a single turn long. On the left of Figure 13-10 the ligation takes place in the presence of a linker strand too long to accommodate the two turns, so the product is a circle; on the right, the linker is short enough that a knot is formed. The interesting thing about this construct is that it was used to search for a topoisomerase activity that works on RNA. It was found, indeed, that *E. coli* DNA topoisomerase III (but not topoisomerase I) possesses this activity.[13.13] More recently, mammalian RNA topoisomerases also have been discovered.[13.14]

It is important to recall that the RNA double helix is A-form, rather than B-form. Leontis *et al.* explored the possibilities of making PX-RNA with a 7:5 motif,[13.15] and demonstrated paranemic cohesion with it, much as discussed in Chapter 3. Thus, at least some of the unusual motifs available to DNA are available to RNA. Sometimes, but really surprisingly rarely, RNA nanotechnology uses the same sticky-ended cohesive approach as DNA nanotechnology does. However, motifs are often mined from the literature, as shown in Figure 13-11, where four molecules with a previously seen bend are combined into a square using sticky ends.[13.16] Jaeger and his colleagues have demonstrated a variety of RNA constructs based on a combination of Watson–Crick base-pairing and other motifs

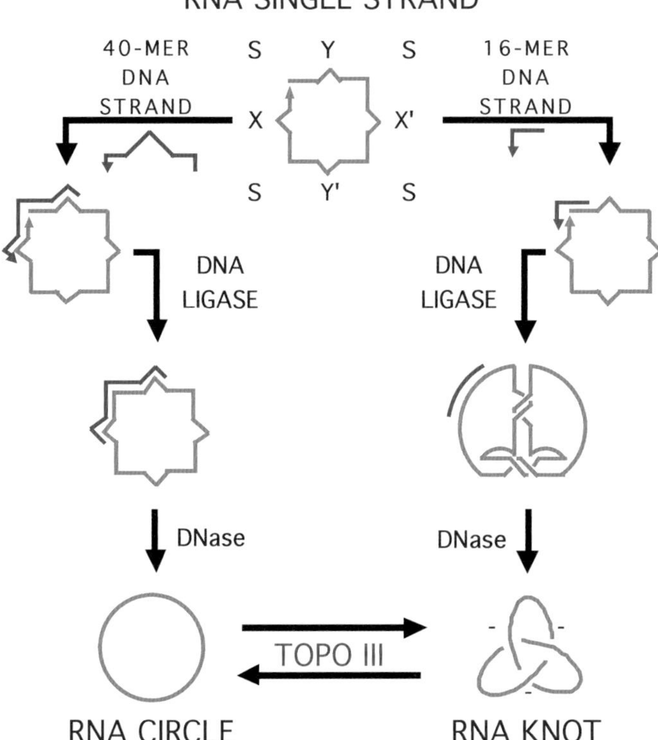

Figure 13-10 *An RNA knot.* With a short linker strand, there is enough freedom for the knotted structure to form, but that freedom is absent in the presence of a long linker tying up the strand in duplex. The circle can be interconverted to the knot by the catalysis of *E. coli* DNA topoisomerase III.

mined from the structures on naturally occurring RNA molecules. His structures include, for example, loop–loop interactions, combined with sticky ends, as shown in Figure 13-12. A number of these molecules can be combined to form 2D lattices, as illustrated in Figure 13-13.[13.17] Nevertheless, structures similar to the DNA cube (Chapter 3) can also be made from RNA, as shown in Figure 13-14.[13.18]

RNA *in vivo.* One of the intriguing things about using RNA as the basis of nanotechnological constructs is that the products can, in principle, be produced within a living cell, by the process of transcription. If transcription is under the control of a series of promoters and various transcription factors, then the structures can be generated only under certain conditions. Steps in this direction have been taken by Silver, Aldaye, and their colleagues when they

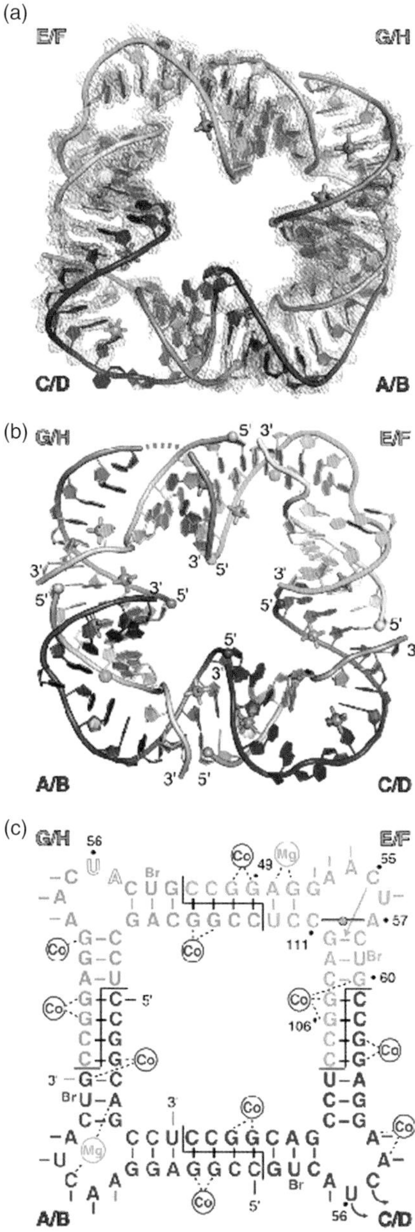

Figure 13-11 *The crystal structure of an RNA square.* Panels a and b are two views of the structure. Panel c is a schematic showing the nature of the interactions.[13.16]

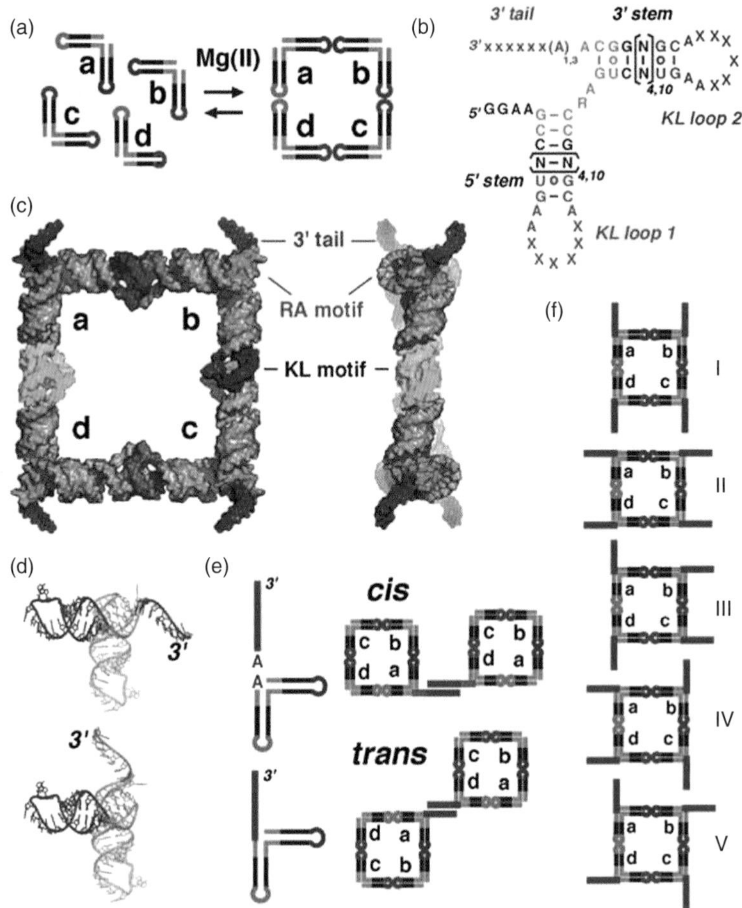

Figure 13-12 *A variety of RNA constructs based on interactions beyond Watson–Crick duplex formation.* A variety of interactions, such as loop–loop bonding, are shown. Reprinted by permission from the AAAS.[13.17]

produced a nanostructure that binds hydrogen-producing proteins in close proximity to each other, as illustrated schematically in Figure 13-15.[13.19] An elegant further step in this direction has been taken recently by Geary, Rothemund, and Andersen,[13.20] who have produced a single-stranded RNA structure that folds and combines with other molecules to produce a hexagonal lattice, as shown in Figure 13-16.

Other species. In addition to the naturally occurring nucleic acids, there are many artificial variants on this theme. One that has found popularity and

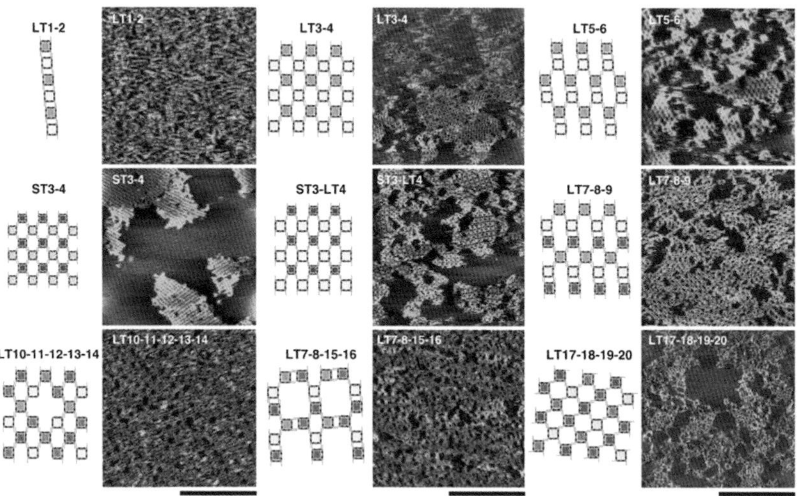

Figure 13-13 *Intermolecular RNA arrays.* 1D and 2D RNA array groupings are shown as a function of their fundamental interactions. Reprinted by permission from the AAAS.[13.17]

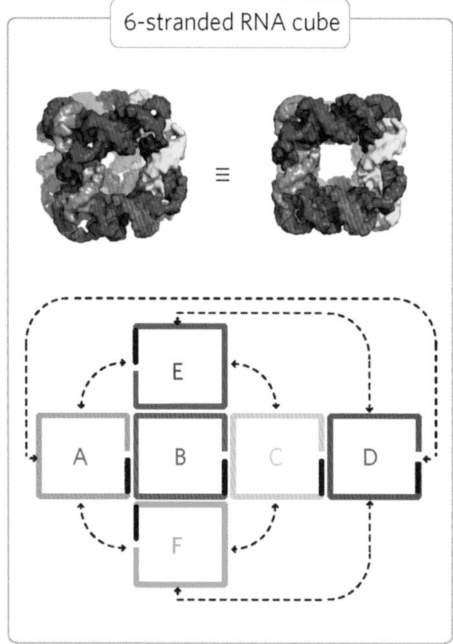

Figure 13-14 *An RNA cube-like molecule.* Several views are shown of this species, which is not topologically sealed. The schematic indicates which strands pair with one another. Reprinted by permission from Macmillan Publishers Ltd.[13.18]

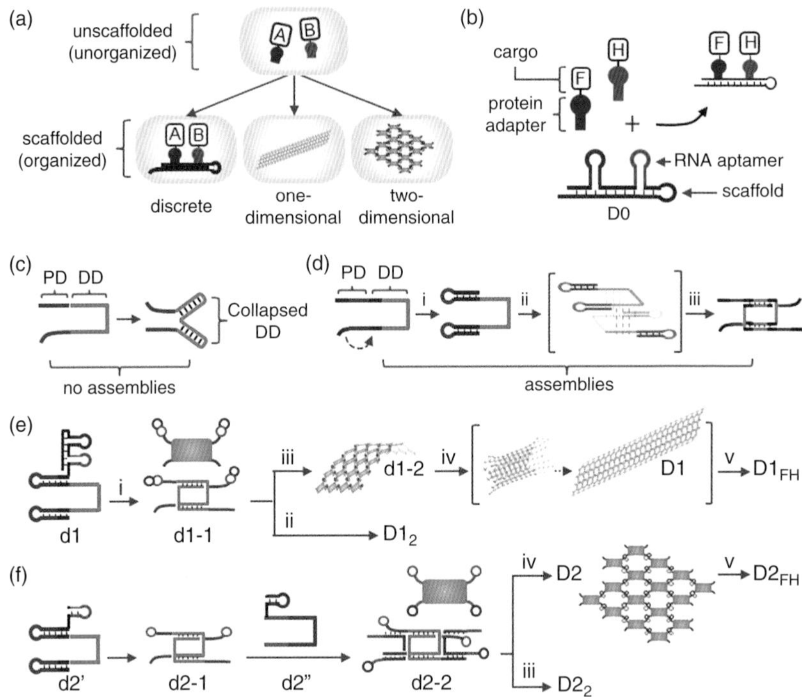

Figure 13-15 *An RNA nanostructure that binds proteins that produce hydrogen, thereby organizing them.* Reprinted by permission from the AAAS.[13.19]

interest in the nucleic acid nanotechnology community is PNA, a nucleic acid molecule with a peptide backbone. The most common variant of this structure has six bonds in its backbone (unlike the 3-bond peptide backbone found in proteins), but it has a base as a side-chain. This modification is shown in Figure 13-17.[13.21] The molecule is neutral and achiral, and it is popular in the anti-sense enterprise. The DNA-PNA double helix is not the same as the DNA-DNA double helix. One way its helicity was worked out uses the AB* system (Figure 13-18).[13.22] Only the yellow and green strands of the A tile consist of PNA. The shaded nucleotides of the B tile can be removed or extended in the experiment. By varying the lengths of those segments, different effective helicities are achieved. Figure 13-19 shows what happens when either DNA or PNA is used with variations of those shaded nucleotides. B0 corresponds to a 10.5-fold helix, B1 to 11.5, etc. If the helicity is not approximately correct, 2D AB* arrays will not form. It can be seen for the B0 tile that

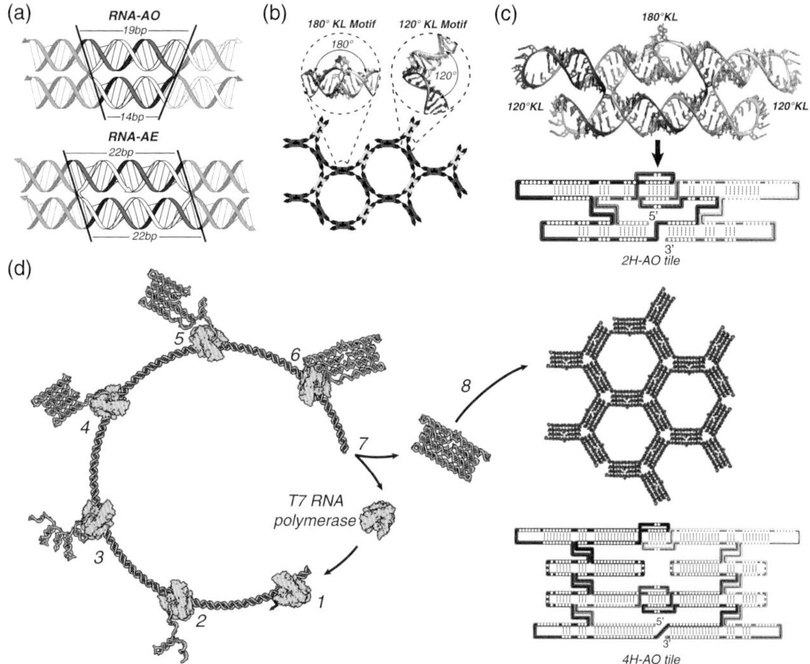

Figure 13-16 *Single-stranded RNA origami produced intracellularly.* The RNA motifs are designed to form larger arrays. Reprinted by permission from the AAAS.[13.19]

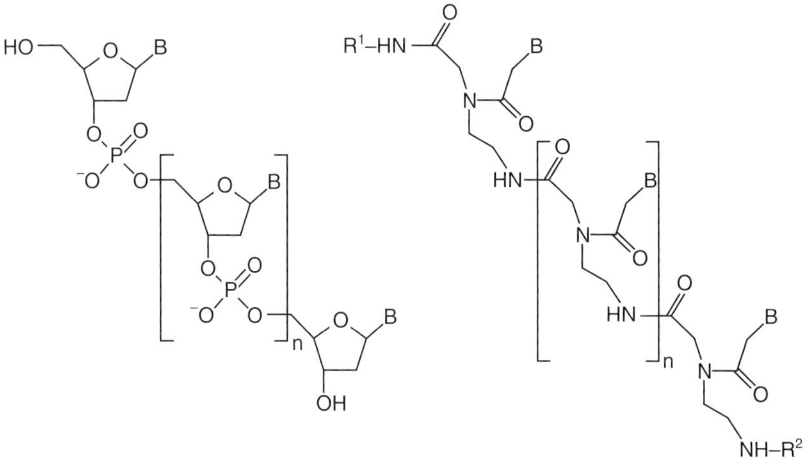

Figure 13-17 *The relationship between DNA and PNA backbones.* Reprinted by permission from the American Chemical Society.[13.21]

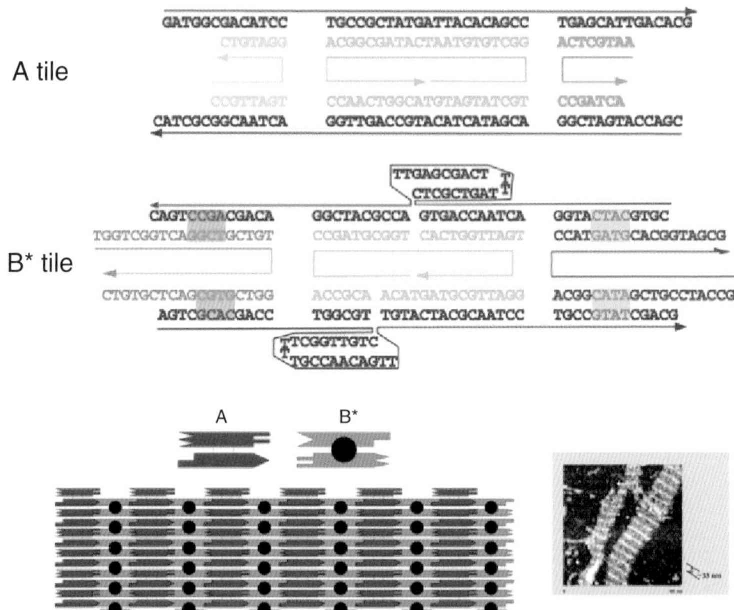

Figure 13-18 *A system to establish the helicity of PNA-DNA duplexes.* The yellow and green strands in the A tile are PNA, and the shaded parts of the B* tiles are of variable length. When good arrays are found, the corresponding helicity is taken to be correct. Reproduced by permission of the Royal Society of Chemistry.[13.22]

DNA forms a nice 2D array. By contrast, the PNA array is nonexistent with this B tile. As the number of nucleotides in the B tile is increased, the DNA 2D AB* array decreases in quality. However, as the B tile is increased in size, the PNA array starts showing 2D features. The B5 tile produces an acceptable array, suggesting that its helicity is about 15.5. Using this information, Park and his colleagues[13.23] have produced 2D arrays of hybridized DNA-PNA strands. The procedure and its results are shown in Figure 13-20. The process is assisted by taking place on a mica substrate, in a method the authors call substrate-assisted growth.

There are numerous variations on both the backbone and the bases. It is beyond the scope of this chapter to deal with them, because they have not yet been used extensively in structural DNA nanotechnology. It is my expectation that a future edition of this book will feature them prominently, and that their special properties will have had a major impact on the field.

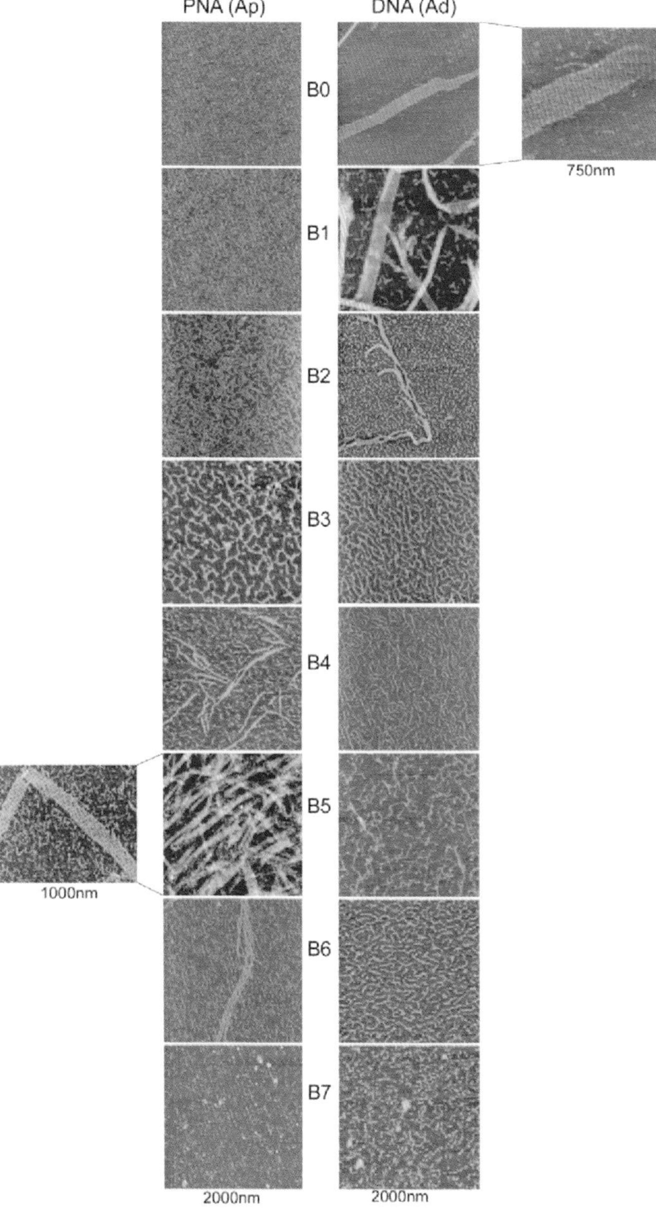

Figure 13-19 *Examples of the system in Figure 13-18.* The DNA control shows evident arrays at a helicity of 10.5, but the PNA control shows arrays at a helicity of about 15.5. Reproduced by permission of the Royal Society of Chemistry.[13.22]

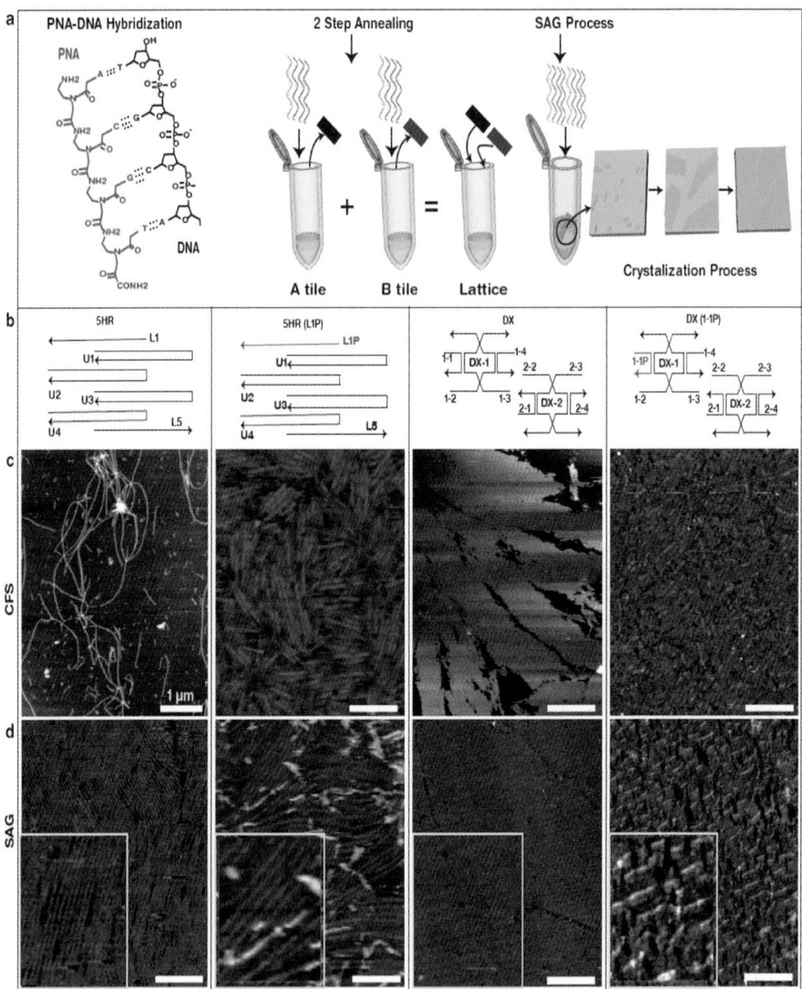

Figure 13-20 *2D arrays of DNA-PNA components.* Reproduced by permission of the Royal Society of Chemistry.[13.23]

References

13.1 P. Paukstelis, J. Nowakowski, J.J. Birktoft, N.C. Seeman, The Crystal Structure of a Continuous Three-Dimensional DNA Lattice. *Chem. Biol.* **11**, 1119–1126 (2004).

13.2 G. Felsenfeld, D.R. Davies, A. Rich, Formation of a Three-Stranded Polynucleotide Molecule. *J. Am. Chem. Soc.* **79**, 2023–2024 (1957).

13.3 D.A. Rusling, A.R. Chandrasekaran, Y. Ohayon, T. Brown, K.R. Fox, R. Sha, C. Mao, N.C. Seeman, Functionalizing Designer DNA Crystals with a Triple-Helical Veneer. *Angew. Chemie Int. Ed.* **53**, 4060–4063 (2014).

13.4 M. Gellert, M.F. Gellert, D.R. Davies, Helix Formation by Guanylic Acid. *Proc. Nat. Acad. Sci. (USA)* **48**, 2013–2018 (1962).

13.5 D. Sen, W. Gilbert, Novel DNA Superstructures Formed by Telomere-Like Oligomers. *Biochem.* **31**, 65–70 (1992).

13.6 E.A. Venczel, D. Sen, Synapsable DNA. *J. Mol. Biol.* **257**, 210–224 (1996).

13.7 L.A. Yatsunyk, O. Piétrement, D. Albrecht, P.L.T. Tran, D. Renciuk, H. Sugiyama, J.-M. Arbona, J.P. Aimé, J.-L. Mergny, Guided Assembly of Tetramolecular G-Quadruplexes. *ACSNano* **7**, 5701–5710 (2013).

13.8 T.C. Marsh, J. Vesenka, E. Henderson, A New DNA Nanostructure, the G-Wire, Imaged by Scanning Probe Microscopy. *Nucl. Acids Res.* **23**, 696–700 (1995).

13.9 E. Saphir, L. Sagiv, T. Molotsky, A.B. Kotlyar, R. Di Felice, D. Porath, Electronic Structure of G4-DNA by Scanning Tunneling Spectroscopy. *J. Phys. Chem. C* **114**, 22079–22084 (2010).

13.10 N. Borovok, N. Iram, D. Zikich, J. Ghabboun, G.I. Livshits, D. Porath, A.B. Kotlyar, Assembling of G-Strand into Novel Tetra-Molecular Parallel G4-DNA Nanostructures Using Avidin-Biotin Recognition. *Nucl. Acids Res.* **36**, 5050–5060 (2008).

13.11 S. Modi, C. Nizak, S. Surana, S. Halder, Y. Krishnan, Two DNA Machines Map pH Changes Along Intersecting Endocytic Pathways in the Same Cell. *Nature Nanotech.* **6**, 459–467 (2013).

13.12 H. Wang, R.J. Di Gate, N.C. Seeman, The Construction of an RNA Knot and Its Role in Demonstrating That E. Coli DNA Topoisomerase III is an RNA Topoisomerase. In *Structure, Motion, Interaction and Expression of Biological Macromolecules*, ed. R.H. Sarma, M.H. Sarma, New York, Adenine Press, pp. 103–116 (1998).

13.13 N.C. Seeman, M.F. Maestre, R.-I. Ma, N.R. Kallenbach, Physical Characterization of a Nucleic Acid Junction. In *Progress in Clinical and Biological Research, Vol. 172A: The Molecular Basis of Cancer*, ed. R. Rein, New York, Alan R. Liss, pp. 99–108 (1985).

13.14 D. Xu, W. Shen, R. Guo, Y. Xu, W. Peng, J. Sima, J. Yang, A. Sharov, S. Srikantan, J. Yang, D. Fox III, Y. Qian, J.L. Martindale, Y. Piao, J. Machamer, S.R. Joshi, S. Mohanty, A.C. Shaw, T.E. Lloyd, G.W. Brown, M.S.H. Ko, M. Gorospe, S. Zou, W. Wang, Top3β is an RNA Topoisomerase That Works with Fragile X Syndrome Protein to Promote Synapse Formation. *Nature Neurosci.* **16**, 1238–1250 (2013).

13.15 K.A. Afonin, D.J Cieply, N.B. Leontis, Specific RNA Assembly with Paranemic Motifs. *J. Am. Chem. Soc.* **130**, 93–102 (2008).

13.16 S.M. Dibrov, J. McLean, J. Parsons, T. Hermann, Self-Assembling RNA Square. *Proc. Nat. Acad. Sci. (USA)* **108**, 6405–6408 (2011).

13.17 A. Chworos, I. Severcan, A.Y. Koyfman, P. Weinkam, E. Oroudjem, H.G. Hansma, L. Jaeger, Building Programmable Jigsaw Puzzles with RNA. *Science* **306**, 2068–2072 (2004).

13.18 K.A. Alfonin, E. Bindewald, A.J. Yaghoubian, N. Voss, E. Jocovetty, B.A. Shapiro, L. Jaeger, *In Vitro* Assembly of Cubic RNA-Based Scaffolds Designed *in Silico*. *Nature Nanotech.* **5**, 676–682 (2010).

13.19 C.J. Delebecque, A.B. Lindner, P.A. Silver, F.A. Aldaye, Organization of Intracellular Reactions with Rationally Designed RNA Assemblies. *Science* **333**, 470–474 (2011).

13.20 C. Geary, P.W.K. Rothemund, E.S. Andersen, A Single-Stranded Architecture for Cotranscriptional Folding of RNA Nanostructures. *Science* **345**, 799–804 (2014).

13.21 M. Egholm, O. Buchardt, P.E. Nielsen, R.H. Berg, Peptide Nucleic Acids (PNA): Oligonucleotide Analogues with an Achiral Peptide Backbone. *J. Am. Chem. Soc.* **114**, 1895–1897 (1992).

13.22 P.S. Lukeman, A. Mittal, N.C. Seeman, Two Dimensional PNA/DNA Arrays: Estimating the Helicity of Unusual Nucleic Acid Polymers. *Chem. Comm.* **2004**, 1694–1695 (2004).

13.23 B. Gnapareddy, J.A. Kim, S.R. Dugasani, A. Tandon, B. Kim, S. Bashar, J.A. Choi, G.H. Joe, T. Kim, T.H. Ha, S.H. Park, Fabrication and Characterization of PNA-DNA Hybrid Nanostructure. *RSC Adv.* **4**, 35554–35558 (2014).

14

DNA nanotechnology organizing other materials

The initial goals of structural DNA nanotechnology did not stop with the organization of nucleic acid molecules into interesting and attractive shapes, or into lattices. The very first paper in the field[14.1] had the goal of organizing other molecules into 3D periodic arrays, with the hope that if they were well enough ordered those guest molecules would be susceptible to crystallographic diffraction analysis. Figure 14-1 illustrates this point, where the DNA scaffold is shown in magenta and the guest macromolecule is drawn in turquoise. This motivating goal has yet to be realized in practice, but efforts continue, nearly 35 years after its initial proposal: it is truly a holy grail of structural DNA nanotechnology.

Indeed, one of the earliest subsequent papers suggested that DNA could be used to organize the components of nanoelectronics.[14.2] Figure 14-2 shows two branched junctions forming a metallic "synapse" from two molecular wires, and Figure 14-3 shows the proposed 3D organization of a 10^7 $Å^3$ memory element. The structures of the 4-arm and 6-arm junctions illustrated there are not particularly realistic, but the notion that DNA could scaffold *the spatial organization* other species of molecules was reinforced by these suggestions. As we will see below, the organization of nanoelectronic components by DNA remains an attractive goal.

Control of polymer topology. One of the earliest attempts involving nucleic acids and heteromolecules entails the use of DNA to direct the topology of industrial polymers. The initial experiments in this program entailed hanging alternating pendent diamino and dicarboxyl groups off the 2′ position of RNA molecules (the atom furthest from the helix axis in A-form RNA)[14.3] so as to direct their topology.[14.4] Although novel topological species have not yet been produced by this approach, this system has been used to demonstrate the templated 2′, 2′ ligation of nucleic acids that produces a peptide bond.[14.5] Likewise, it has been used to generate topological targets with connectivity

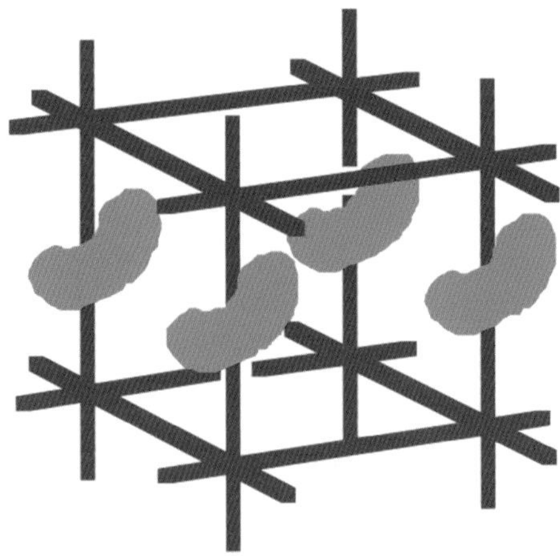

Figure 14-1 *A 3D DNA lattice organizing biological macromolecules.* This schematic depicts a cube-like lattice with sticky ends, containing blob-like macromolecules. Reprinted by permission from Elsevier.[2.12]

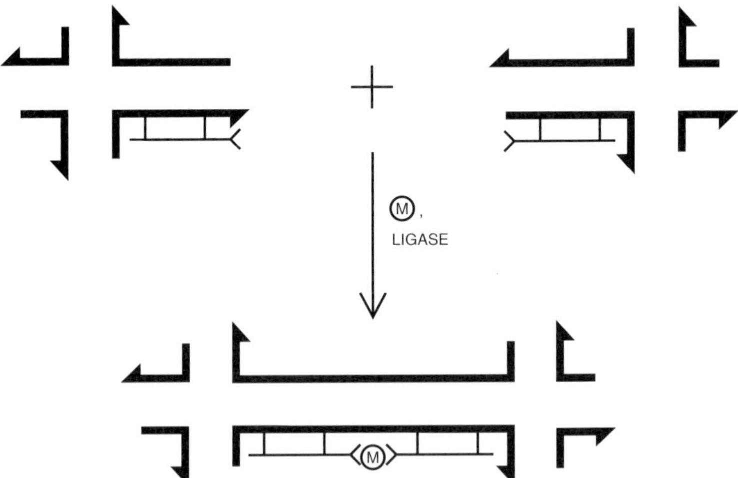

Figure 14-2 *An early proposal for using DNA lattice components to organize nanoelectronic circuitry.* The two branched junctions are programmed to associate through their sticky ends. A nanowire is pendant from the molecules, and they form a synapse through the presence of a metal. Reprinted by permission from Oxford University Press.[14.2]

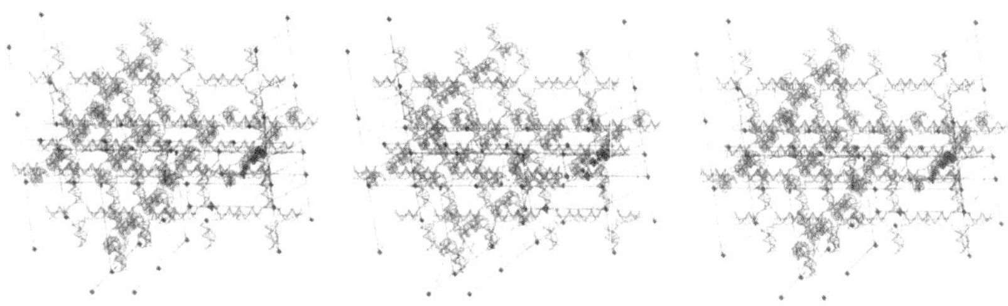

Figure 14-3 *A stereographic image of one unit cell in a molecular memory device.* Both types of stereo image are present. The DNA is shown in green, and the wires in red. The volume of this element (which contains a redox "bit") is about 10^7 Å^3. Reprinted by permission from the American Chemical Society.[14.6]

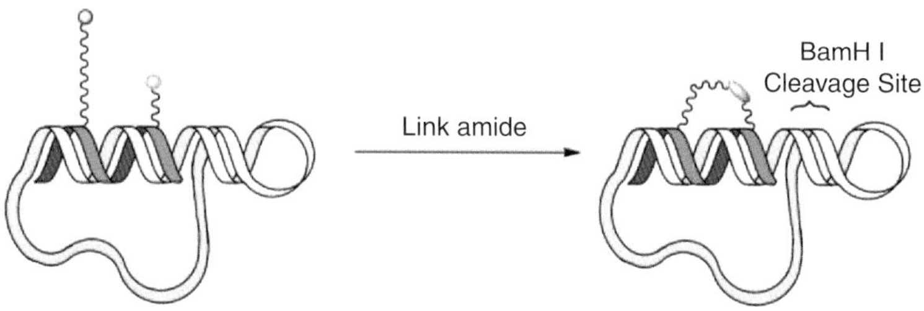

Figure 14-4 *A DNA-PEG molecule catenated to a DNA circle.* The closure of the amide bond produces the catenane. Reprinted by permission from the American Chemical Society.[14.10]

parallel to the helix axis.[14.6] The catenane produced by joining one amine with one carboxyl group is illustrated in Figure 14-4.

Metallic nanoparticles. The advent of metallic and semiconducting nanoparticles has spurred a lot of effort to organize them by DNA. Very early attempts to assemble nanoparticle clusters were performed by Alivisatos and his colleagues,[14.7] as well as by Mirkin and his collaborators.[14.8] These approaches fundamentally used DNA as "smart glue" to put DNA-coated nanoparticles together. The earliest efforts to use 2D DNA arrays to assemble nanoparticles were done by Kiehl and his co-workers,[14.9] particularly where two species of small nanoparticles were assembled in alternating rows.[14.10] This approach is shown in Figure 14-5. AFM data illustrating the success of this approach are shown in Figure 14-6.

The work above does not exploit the power of robust DNA motifs to organize large guests in 2D. The 5 nm and 10 nm particles have been organized by tensegrity triangles whose edges are 8 turns long, and whose components are DX motifs, rather than individual DNA double helices; the

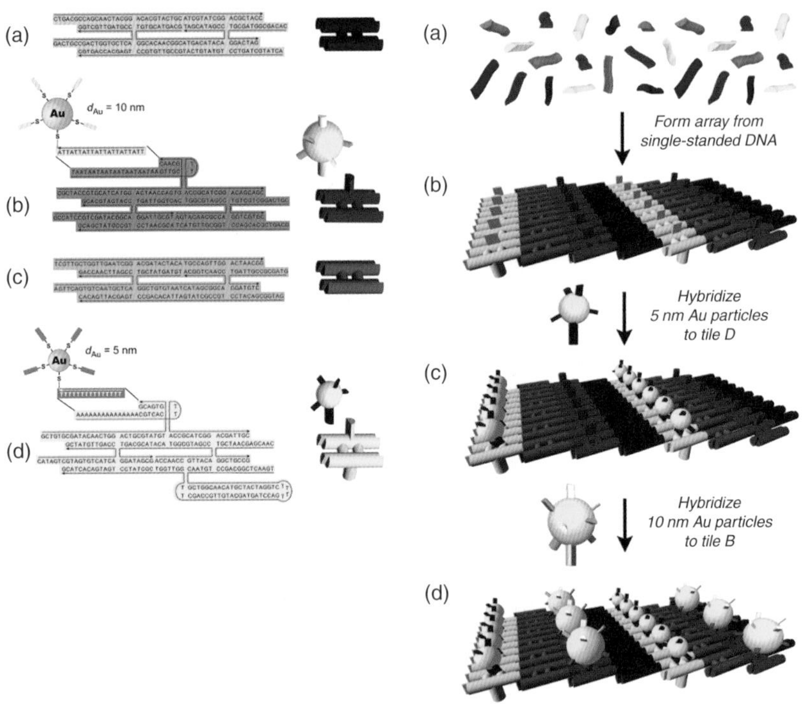

Figure 14-5 *The construction of 1D arrays of 5 nm and 10 nm gold nanoparticles.* The four components are shown in the four panels on the left. The array formation is shown at right. The stepwise formation of the 2D DNA array, the addition of the 5 nm nanoparticles, and then the addition of the 10 nm nanoparticles are indicated. Reprinted by permission from the American Chemical Society.[14.10]

motif is known as 3D-DX.[14.11] Schematic versions of this motif are illustrated in Figure 14-7. Figure 14-8a shows a schematic version of four of these motifs put together. Note that the differential coloration indicates that there are two different species of 3D-DX molecules. Figure 14-8b shows three different examples of arrays formed by attaching nanoparticles to the 3D-DX molecules: 5 nm particles attached to only one molecule, 5 nm particles attached to both molecules, and 5 nm particles attached to one molecule and 10 nm particles attached to the other one. The success of this approach depends heavily on chemistry developed in the Alivisatos laboratory that allows particles derivatized by a single DNA molecule to be isolated from particles attached to more DNA molecules.[14.12] Only the single DNA molecules are used, thereby eliminating the possibility of multiple triangles attaching to a single particle. The strand attached to the particle is simultaneously a key

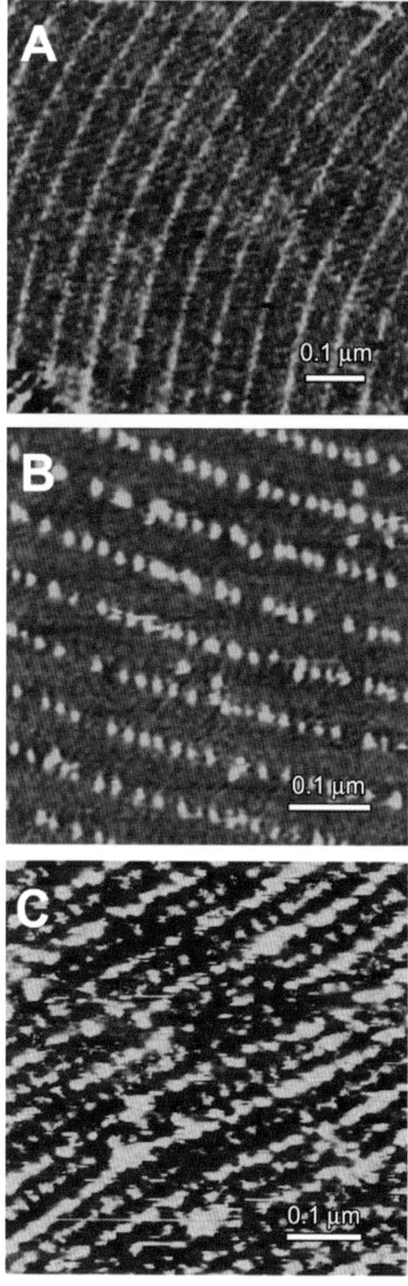

Figure 14-6 *AFM images of the system shown in Figure 14-5*. The individual nanoparticles are shown in panels A and B. The pair of them together are shown in panel C. Reprinted by permission from the American Chemical Society.[14.11]

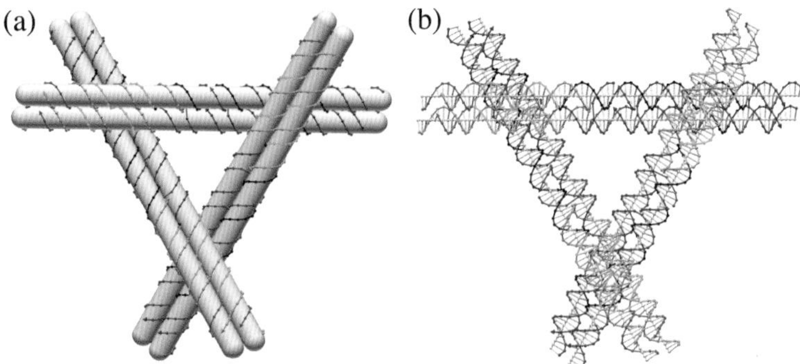

Figure 14-7 *The 3D-DX motif.* This is a tensegrity triangle whose domains are DX molecules. It is shown in two representations, with the DNA details illustrated in the molecular diagram at the right. Reprinted by permission from the American Chemical Society.[14.11]

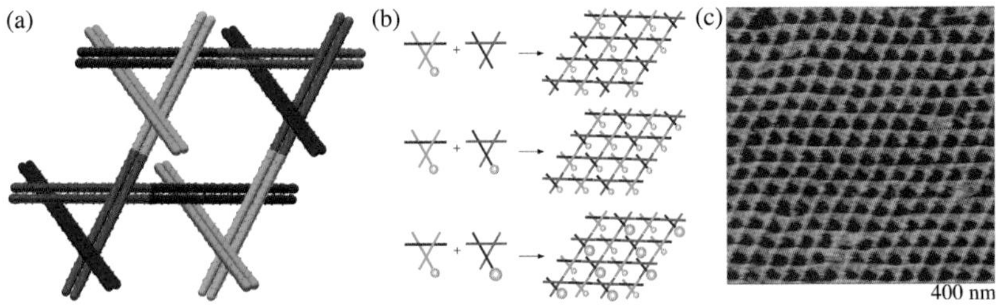

Figure 14-8 *The formation of a 2D 3D-DX array.* (a) *An array with two components.* (b) *Schematics of arrays decorated with gold nanoparticles.* From top to bottom: one component has a 5 nm particle, two components have 5 nm particles, and the two components alternate 5 nm and 10 nm particles. (c) *A 2D array of 3D-DX motifs visualized by AFM.* Reprinted by permission from the American Chemical Society.[14.11]

strand of the triangle, thereby ensuring that the particle and the triangle are combined. Figure 14-8c is an atomic force micrograph showing how 3D-DX molecules lacking particles can form a large 2D array if there are sticky ends on two of the DX domains. The three panels of Figure 14-9 display transmission electron micrographs corresponding to the three cases illustrated schematically in Figure 14-8b: only one 3D-DX triangle contains a 5 nm particle (panel a); both 3D-DX triangles contain 5 nm particles (panel b); one 3D-DX triangle contains a 5 nm particle and the other contains a 10 nm particle, leading to a checkerboard pattern (panel c).

A variety of arrangements and particulate species have been organized by DNA motifs of one sort or another. For example, Sleiman and her colleagues

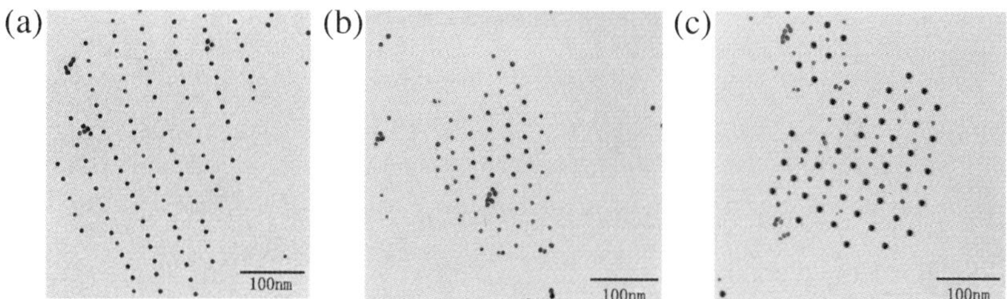

Figure 14-9 *TEM images of the arrays shown in Figure 14-8b.* (a) *The array with every other component containing a 5 nm gold nanoparticle.* (b) *The array with each component containing a 5 nm gold nanoparticle.* (c) *The array where components with 5 nm and 10 nm gold nanoparticles alternate.* Reprinted by permission from the American Chemical Society.[14.11]

have used repetition of a triangular motif in one dimension as an organizing method for nanoparticles (Figure 14-10).[14.13] Yan and his colleagues used a finite 2D motif made from individual tiles to make "nanotubes" of gold particles, as shown in Figure 14-11. Examples of cryo-EM images of these circular and helical arrangements are shown in Figures 14-12 and 14-13.[14.14] Figure 14-14 illustrates the steps that have been used to organize quantum dots on DNA origami structures.[14.15] Molecules of the protein streptavidin have been incorporated into a 4×4 DNA array by including a biotin in the array near the crossing point, as shown in Figure 14-15.[14.16] Note that the streptavidin molecules are not tightly oriented by this method. DNA origami is so large that it is convenient to put multiple species on it. In a striking example of this, a gold nanoparticle and an MS2 virus have been positioned on a single origami, as illustrated in Figure 14-16.[14.17]

Metal coordination complexes. In addition to using DNA to organize various nanoscopic species, it is also possible to incorporate metals in the DNA structures. This has been reviewed by Sleiman and colleagues.[14.18] This topic includes both the addition to specific sites between the base pairs, as done by Shionoya and his colleagues,[14.19] and also the inclusion of various coordination complexes along the backbone. Figure 14-17 from the Sleiman group illustrates this notion. The idea behind doing this is that the fixed angles associated with metallic coordination would provide fixed angles for the DNA helix axes. However, this notion has not been borne out yet, because there are a lot of single bonds attached to the coordination complexes.

Biologically oriented robotics. In a particularly elegant example of multiple species being organized by DNA, Church and his colleagues have made a

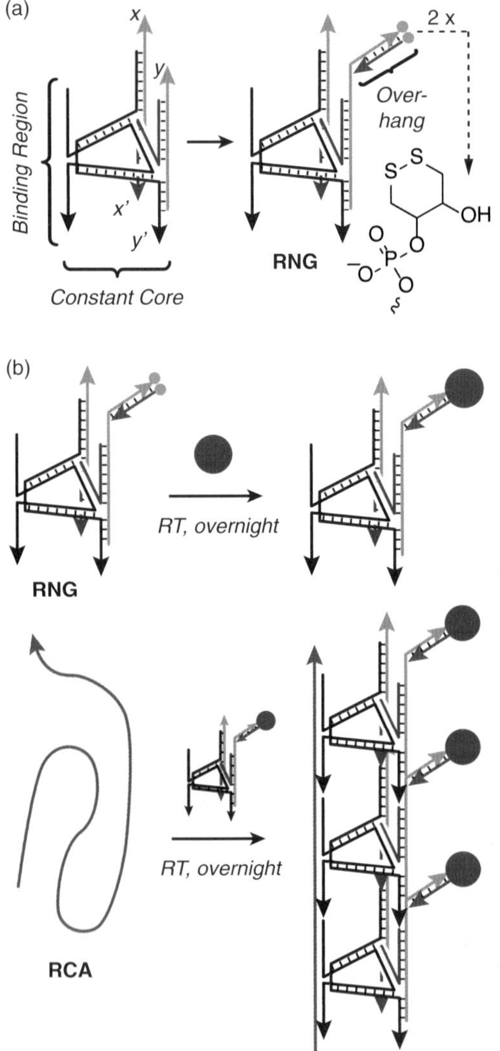

Figure 14-10 *Triangular motifs organize nanoparticles in 1D arrays.*[14.13]

triggerable device that can be programmed to deliver a variety of payloads.[14.20] Figure 14-18a illustrates the basic design of the system when it is closed. A variety of payloads are shown in pink on the inside. Figure 14-18b shows the opening mechanism for one of the two locks. An aptamer (blue) is partially unavailable (paired with the orange strand), keeping the device closed. However, the aptamer will preferentially pair with an antibody (red), thereby

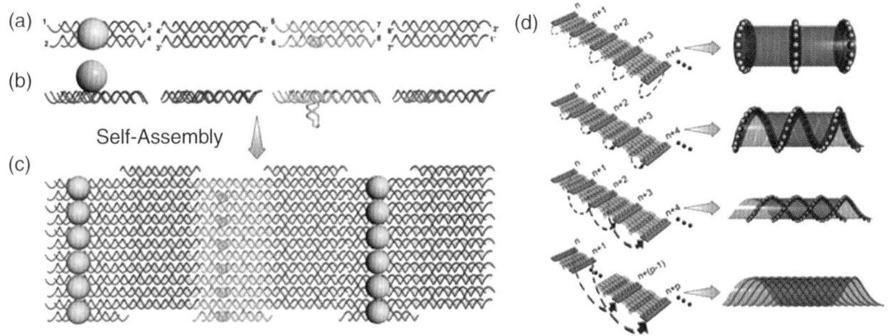

Figure 14-11 *Formation of DNA nanotubes with cylindrical and helical arrangements of nanoparticles.* Reprinted by permission from the AAAS.[14.14]

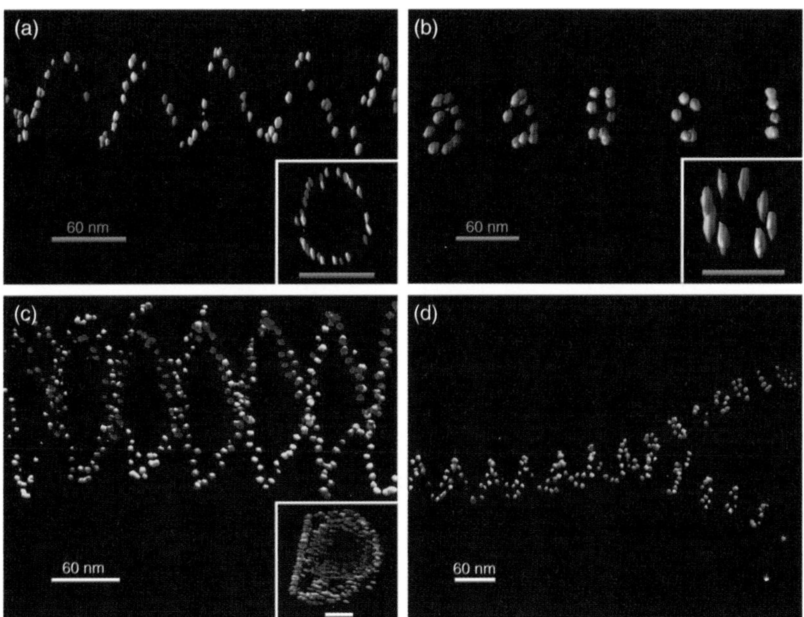

Figure 14-12 *Helical arrangements of nanoparticles.* Reprinted by permission from the AAAS.[14.14]

unlocking the device and making the payload available (Figure 14-18c). Figures 14-18d and 14-18e show aspects of the opening and closing mechanism, while Figure 14-18f displays a variety of cryo-EM images.

Organizing nanoscale carbon. When one thinks of the nanoscale, in addition to DNA many people tend to think of carbon molecules, either

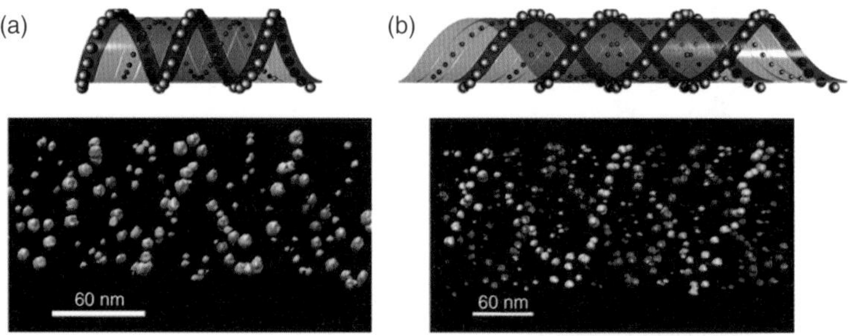

Figure 14-13 *Double helical arrangements of nanoparticles.* Reprinted by permission from the AAAS.[14.14]

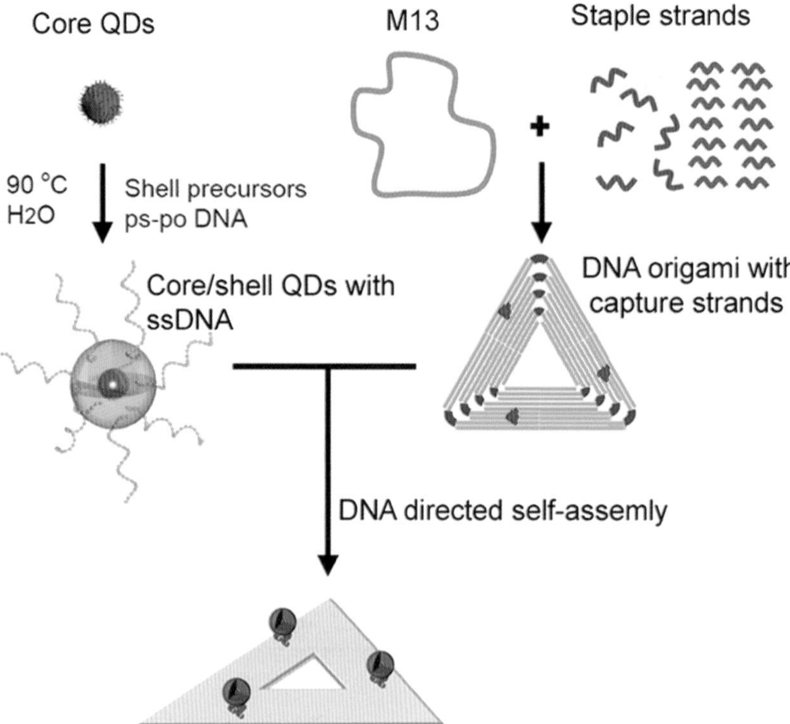

Figure 14-14 *Quantum dots captured on triangular DNA origami arrays.* Reprinted by permission from the American Chemical Society.[14.15]

buckyballs[14.21] or carbon nanotubes.[14.22] The organization of carbon nanotubes remains a challenge, but Goddard and his colleagues[14.23] have made a good start on it. Figure 14-19 illustrates their approach to the problem. The best way to dissolve carbon nanotubes in aqueous solution is by wrapping them in DNA.

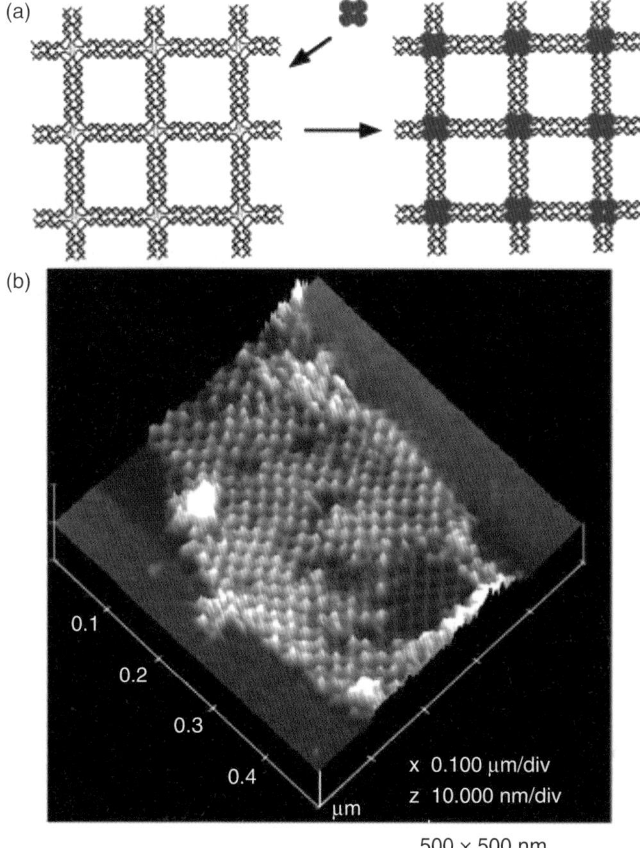

Figure 14-15 *Streptavidin organized into a 4 × 4 DNA array.* Reprinted by permission from the AAAS.[14.16]

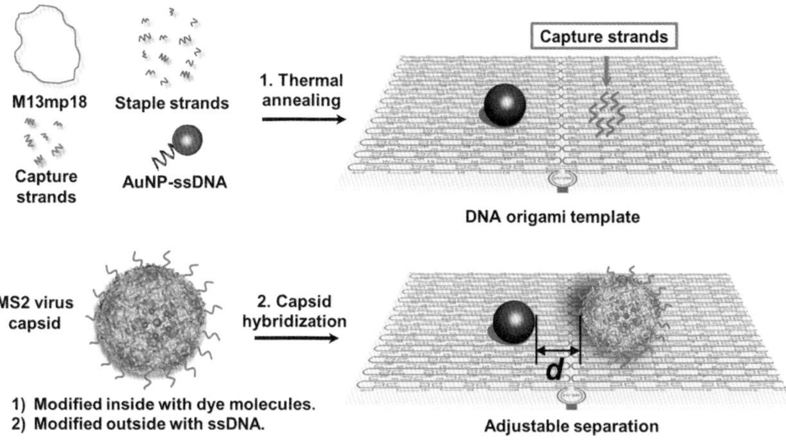

Figure 14-16 *Joint capture of a virus and a nanoparticle on a DNA origami array.* Reprinted by permission from the American Chemical Society.[14.17]

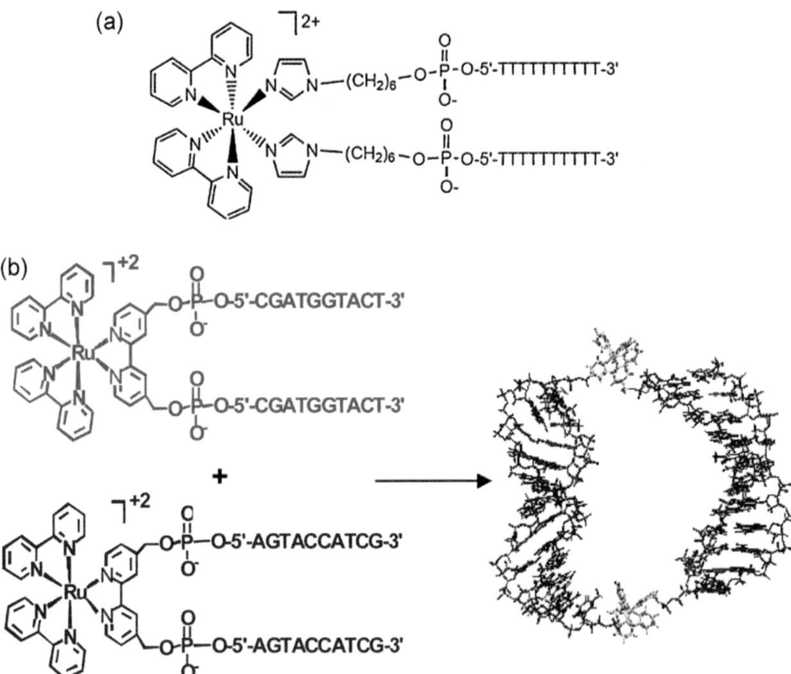

Figure 14-17 *DNA molecules joined by coordination compounds.* Reprinted by permission from Elsevier.[14.18]

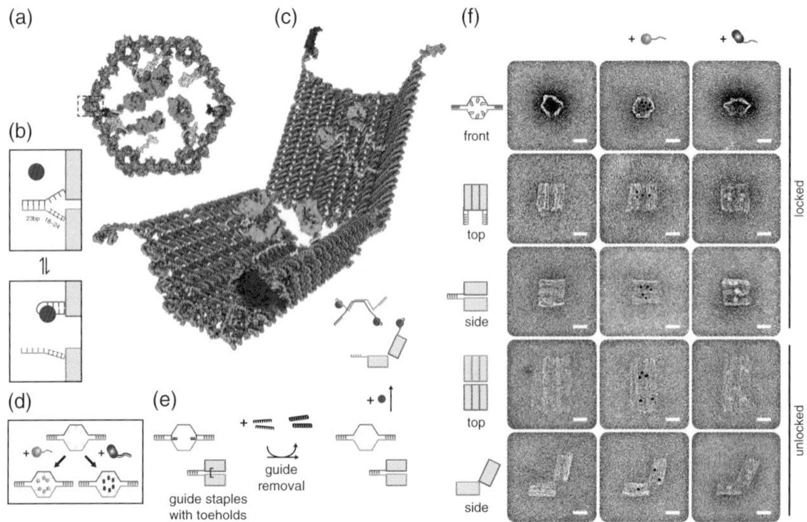

Figure 14-18 *A DNA origami device that opens in response to external signals, thereby exposing payloads.* Reprinted by permission from the AAAS.[10.5]

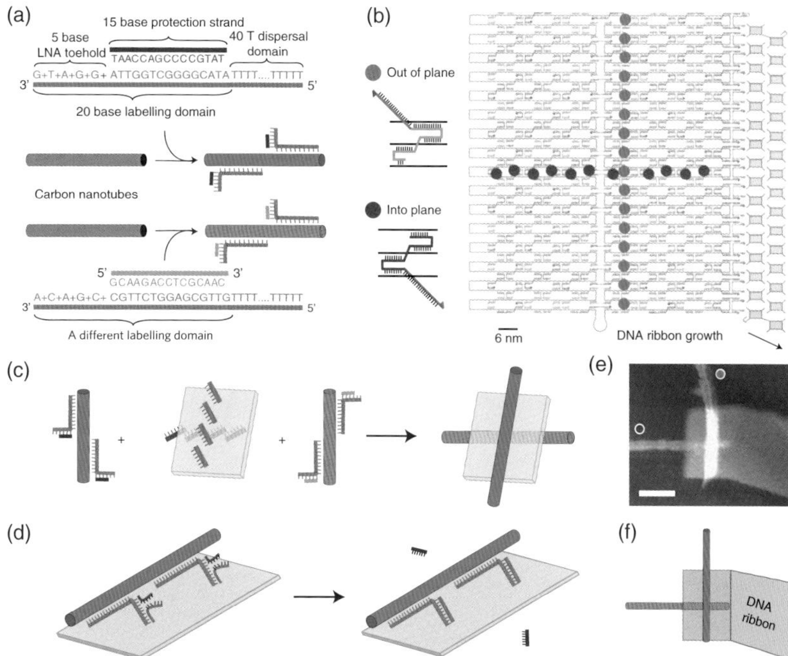

Figure 14-19 *Organization of carbon nanotubes by DNA origami.* The orientations of two different nanotubes are established by the origami. Reprinted by permission from Macmillan Publishers Ltd.[14.23]

These investigators left a little duplex DNA on the ends of the wrapping strands, and those could then be attached to a DNA origami construct. Two different nanotubes were attached on the two different sides of the origami. The duplex sections had toeholds on them so they could attach to specific sites on the origami in a crossbar arrangement. When the DNA was vaporized, a crossbar was found.

Multi-dimensional organization of metal particles by DNA. As noted above, about 15 years after my epiphany in the bar, a second pathway of DNA nanotechnology was initiated by Mirkin, Alivisatos, and their colleagues. They coated colloidal nanoparticles in DNA strands, conferring upon them an identity. Particles coated in complementary strands would cohere.[14.7,14.8] There has been a huge amount of work in this direction, which is beyond the scope of this book. However, I should mention the work from the Mirkin[14.24] and Gang[14.25] groups that has led to the formation of 3D crystals by self-assembling these particles into lattices. Figure 14-20 shows how this group was able to program both face-centered

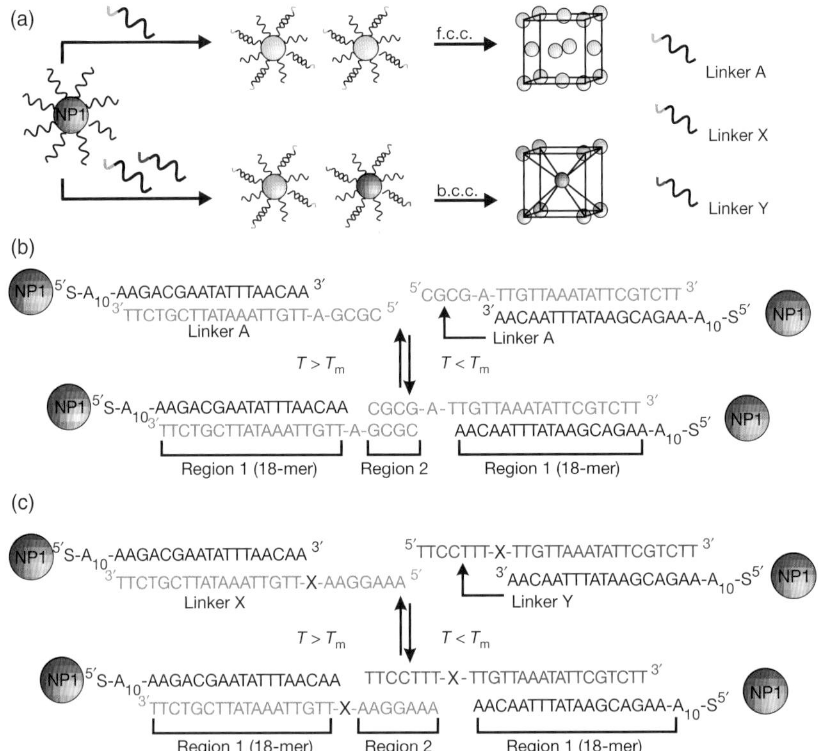

Figure 14-20 *Formation of specific nanoparticle lattices using DNA links between particles.* Reprinted by permission from Macmillan Publishers Ltd.[14.24]

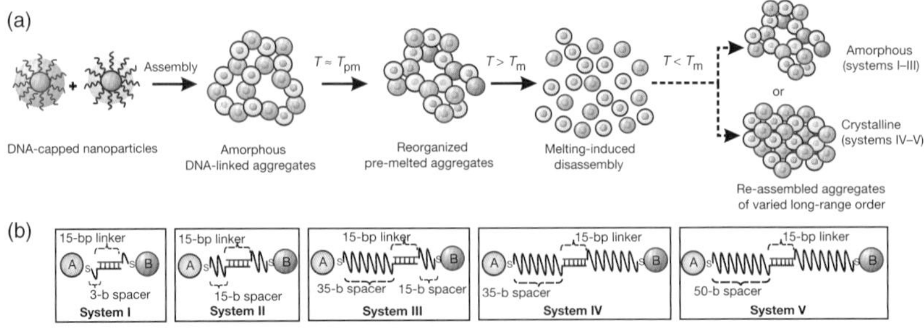

Figure 14-21 *Nanoparticle arrays formed by specific linkers between particles.* Reprinted by permission from Macmillan Publishers Ltd.[14.25]

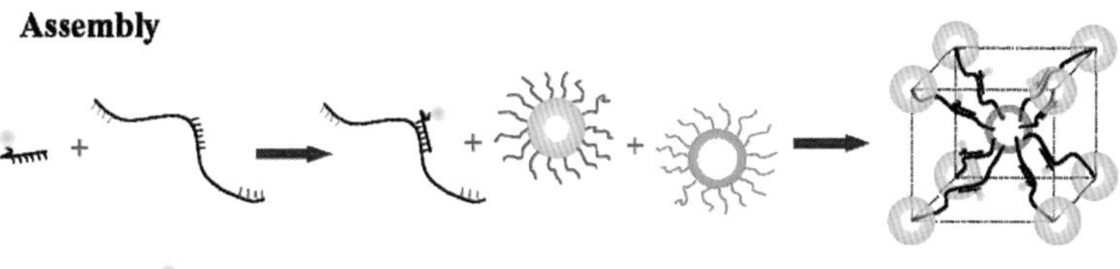

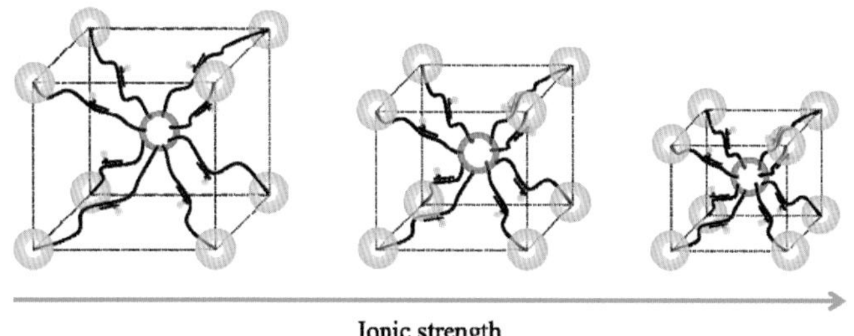

Figure 14-22 *Tuning the lattice parameters of a DNA-connected nanoparticle array.* Reprinted by permission from Macmillan Publishers Ltd.[14.26]

cubic and body-centered cubic lattices from the nanoparticles. These crystals were not large, about 1 micron or so in size, but they were an important step in the control of structure in 3D. The Gang group was able to do something quite similar, again with particles coated by DNA.[14.25] Remarkably, they were able to get similar results to the Mirkin group using a variety of spacers in their system, as shown in Figure 14-21. Figure 14-22 shows the system that Gang and his collaborators set up to alter lattice parameters as a function of salt concentration.[14.26] It is clear that we are just at the beginning of the era when the information in DNA is capable of organizing other materials on the finest scale, but such a goal is certainly within range.

References

14.1 N.C. Seeman, Nucleic Acid Junctions and Lattices. *J. Theor. Biol.* **99**, 237–247 (1982).

14.2 B.H. Robinson, N.C. Seeman, The Design of a Biochip: A Self-Assembling Molecular-Scale Memory Device. *Protein Eng.* **1**, 295–300 (1987).

14.3 J.M Rosenberg, N.C. Seeman, R.O. Day, A. Rich, RNA Double Helices Derived from Studies of Small Fragments. *Biochem. Biophys. Res. Comm.* **69**, 979–987 (1976).

14.4 L. Zhu, P.S. Lukeman, J. Canary, N.C. Seeman, Nylon/DNA: Single-Stranded DNA with Covalently Stitched Nylon Lining. *J. Am. Chem. Soc.* **125**, 10178–10179 (2003).

14.5 Y. Liu, R. Sha, R. Wang, L. Ding, J.W. Canary, N.C. Seeman, 2′, 2′-Ligation Demonstrates the Thermal Dependence of DNA-Directed Positional Control. *Tetrahedron* **64**, 9417–8422 (2008).

14.6 Y. Liu, A. Kuzuya, R. Sha, J. Guillaume, R. Wang, J.W. Canary, N.C. Seeman, Coupling Across a DNA Helical Turn Yields a Hybrid DNA/Organic Catenane Doubly Tailed with Functional Termini. *J. Am. Chem. Soc.* **130**, 10882–10883 (2008).

14.7 A.P. Alivisatos, K.P. Johnsson, X.G. Peng, T.E. Wilson, C.J. Loweth, M.P. Bruchez, P.G. Schultz, Organization of "Nanocrystal Molecules" using DNA. *Nature* **382**, 609–611 (1996).

14.8 C.A. Mirkin, R.L. Letsinger, R.C. Mucic, J.J. Storhoff, A DNA-Based Method for Rationally Assembling Nanoparticles into Macroscopic Materials. *Nature* **382**, 607–609 (1996).

14.9 S. Xiao, F. Liu, A. Rosen, J.F. Hainfeld, N.C. Seeman, K.M. Musier-Forsyth, R.A. Kiehl, Self-Assembly of Nanoparticle Arrays by DNA Scaffolding. *J. Nanoparticle Res.* **4**, 313–317 (2002).

14.10 Y.Y. Pinto, J.D. Le, N.C. Seeman, K. Musier-Forsyth, T.A. Taton, R. A. Kiehl, Sequence-Encoded Self-Assembly of Multiple-Nanocomponent Arrays by 2D DNA Scaffolding. *Nano Letters* **5**, 2399–2402 (2005).

14.11 J. Zheng, P.E. Constantinou, C. Micheel, A.P. Alivisatos, R.A. Kiehl, N.C. Seeman, 2D Nanoparticle Arrays Show the Organizational Power of Robust DNA Motifs. *Nano Letters* **6**, 1502–1504 (2006).

14.12 D. Zanchet, C.M. Micheel, W.J. Parak, D. Gerion, A.P. Alivisatos, Electrophoretic Isolation of Discrete Nanocrystal/DNA Conjugates. *Nano Letters* **1**, 32–35 (2001).

14.13 K.L. Lau, G.D. Hamblin, H.F. Sleiman, Gold-Nanoparticle 3D Building Blocks: High Purity Preparation and Use for Modular Access to Nanoparticle Assemblies. *Small* **10**, 660–666 (2014).

14.14 J. Sharma, R. Chhabra, A. Cheng, J. Brownell, Y. Liu, H. Yan, Control of Self-Assembly of DNA Tubules Through Integration of Gold Nanoparticles. *Science* **323**, 112–116 (2009).

14.15 Z. Deng, A. Samanta, J. Nangreave, H. Yan, Y. Liu, Robust DNA-Functionalized Core/Shell Quantum Dots with Fluorescent Emission Spanning from UV-Vis to Near-IR and Compatible with DNA-Directed Self-Assembly. *J. Am. Chem. Soc.* **134**, 17424–17427 (2012).

14.16 H. Yan, S.H. Park, G. Finkelstein, J.H. Reif, T.H. LaBean, DNA-Templated Self-Assembly of Protein Arrays and Highly Conductive Nanowires. *Science* **301**, 1882–1884 (2003).

14.17 D. Wang, S.L. Capehart, S. Pal, M. Liu, L. Zhang, P.J. Shuck, Y. Liu, H. Yan, M.B. Francis, J.J. De Yoreo, *et al.*, Hierchical Assembly of Plasmonic Nanostructures Using Virus Capsid Scaffolds on DNA Origami Templates. *ACS Nano* **8**, 7896–7904 (2014).

14.18 H. Yang, K.L. Metera, H.F. Sleiman, DNA Modified with Metal Complexes: Applications in the Construction of Higher-Order Metal DNA-Nanostructures. *Coord. Chem. Revs*. **254**, 2403–2415 (2010).

14.19 Y. Takezawa, M. Shionoya, Metal-Mediated DNA Base Pairing: Alternatives to Hydrogen-Bonded Watson–Crick Base Pairs. *Acc. Chem. Res*. **45**, 2066–2076 (2012).

14.20 S.M. Douglas, I. Bachelet, G.M. Church, A Logic-Gated Nanorobot for Targeted Transport of Molecular Payloads. *Science* **335**, 831–834 (2012).

14.21 R.F. Curl, R.E. Smalley, Probing C_{60}. *Science* **242**, 1017–1022 (1988).

14.22 P.M. Ajayan, S. Iijima, Smallest Carbon Nanotube. *Nature* **358**, 23 (1992).

14.23 H.T. Maune, S.P. Han, R.D. Barish, M. Bockrath, W.A. Goddard, P.W.K. Rothemund, E. Winfree, Self-Assembly of Carbon Nanotubes into Two-Dimensional Geometries Using DNA Origami Templates. *Nature Nanotech*. **5**, 61–66 (2010).

14.24 S.Y. Park, A. K.-R. Lytton-Jean, B. Lee, S. Weigand, G.C. Schatz, C.A. Mirkin, DNA-Programmable Nanoparticle Crystallization. *Nature* **451**, 553–556 (2008).

14.25 D. Nykypanchuk, M.M. Maye, D. van der Lelie, O. Gang, DNA-Guided Crystallization of Colloidal Nanoparticles. *Nature* **451**, 549–552 (2008).

14.26 M.M. Maye, M.T. Kumara, D. Nykypanchuk, W.B. Sherman, O. Gang, Switching Binary States of Nanoparticle Superlattices and Dimer Clusters by DNA Strands. *Nature Nanotech*. **5**, 116–120 (2010).

Afterword

So where's it all going? People are trying to make better bricks and bigger origami constructs, and put more automation into the field. Everybody who works in the area knows we need better molecular modeling, particularly better physical models. Everybody wants to know what's next. Figures A-1, A-2, and A-3 are how I end my seminar these days. A-1 asks what's next, A-2 says it's no longer my responsibility, and A-3 reinforces that notion.

Nevertheless, although there are far too many people in the field for it to be my own enclave, that doesn't meant that I don't have my own ideas of what should be pursued. My top priority is to *increase the control* that we already have. We have improved the 2-turn tensegrity triangle crystals to the point where they diffract to a touch better than 3 Å resolution. We got there by tweaking the sticky ends, and that seems not to be the correct route for the larger triangles that can actually host macromolecules of significant dimensions. Yossi Weizmann's recent progress with knots has encouraged me to see whether synthetic knot topoisomers can be distinguished. Current work with Henry Chapman suggests that there is a major role for DNA nanotechnology to play in the establishment of new scattering methods.

I look forward to extending the role of DNA to control the structure of matter on larger scales, such as the micron scale, as well as on the nanoscale. By contrast, I foresee the assembly line as the first of its kind, but certainly not the last. With luck, we'll be able to drop the scale on it, so that real chemical assembly can be programmed at the level of bonds. This will certainly be a challenge, but that's what Science is all about. Nature does it, meaning that sooner or later, we can too. It's just a matter of time.

Figure A-1 *René Magritte, Self Portrait* (© ADAGP, Paris and DACS, London 2015). I have hijacked Magritte's painting to ask what's coming next in structural DNA nanotechnology.

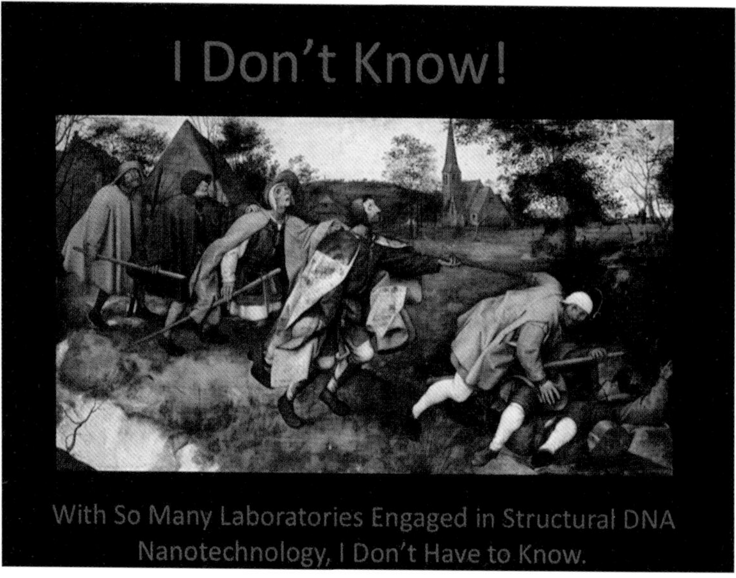

Figure A-2 *Peter Breugel the Elder, The Parable of the Blind Leading the Blind.* I have discovered that there is no way I can predict the next developments in structural DNA nanotechnology, and that's probably a good thing.

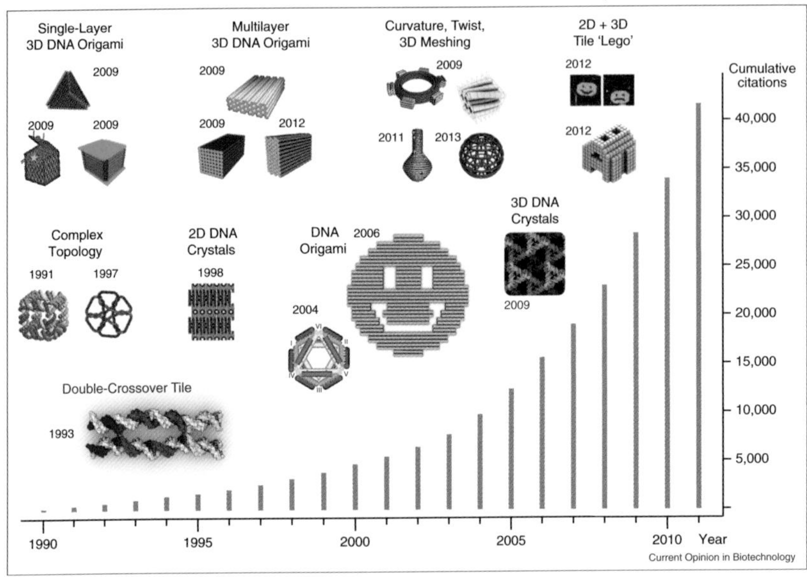

Figure A-3 *The development of structural DNA nanotechnology.* This slide from 2013 shows the remarkable growth and impact of the field in recent years. Reprinted courtesy of Hendrik Dietz.

Index

Figures and tables are noted in bold typeface.

1D arrays
  and 6-helix bundle, 113
  as analytical tool, 110, **111**, **112**, **113**
  bar-code system, **155**
  characterization, 77–79, **79**
  DX molecules, 100, **102**
  metal particles, **234**, **235**
  self-assembly, **188**, **189**, **190**, **191**, **193**, **194**
  and triangular motifs, **238**
2D arrays
  bulged junctions, 110–112, **115**
  characterization, 82–83, **83**
  and denaturing gels, 95
  DNA brick, **168**, **169**
  of DNA-PNA strands, 224–226, **226**, **227**, **228**
  and Holliday junctions, **109**, 109–110
  origami, **157**
  PX-JX cassette, 172–173, **173**, **174**, **175**, **176**
  tensegrity triangle, 113–115, **117**
2D arrays from DX molecules, **36**, **73**, **79**, **83**, 99–104, **100**, **103**, **104**, **105**
2D arrays of TX molecules
  skewed, 115–116, **119**
  structure of, **36**, 104–107, **105**, **106**, **107**, **108**
3-arm junctions, 88–92, **89**, **90**, **91**, **92**
3-colorability problem, 202–204, **203**
3D arrays
  6-helix bundle, **157**, **159**
  characterization, 82–83
  DNA brick, 166–167, **169**
  origami, **157**, **159**
3D crystals
  and DNA organization of metal particles, 243–245, **244**, **245**
  isomorphous replacement method, 119–127, **120**, **121**, **122**, **123**, **124**, **125**, **126**, **127**
3D structures
  biological organization, **232**
  circular origami, 163–164, **166**, **167**, **168**
  and X-ray crystallography, 117–119
3D-DX triangle motif
  and double cohesion, 113–115, **117**
  G-tetrad, 233–237, **236**, **237**
4-arm junction, 106–108, **107**, **108**
6-helix bundle (6HB)
  and DNA origami, 154, **157**, **159**, 160–161, **161**
  and double cohesion, 112–113, **116**
  relaxed, 161–162, **162**, **163**

Adleman experiment, 198–202, **199**, **200**, **201**
affinity interaction of sticky ends, **8**, 8–10, **9**
algorithmic assembly
  2D, 207
  and cost, 208–209
  cumulative XOR operation, **205**, 205–207, **206**, **207**, **208**
  principles of, **204**, 204–205
antijunctions
  in parallelogram structures, 110, **111**, **113**
  in single strand DNA design, **53**, 54–57, **55**, **56**, **57**, **58**
aperiodic arrays characterization, **83**
art and DNA, 4–5, 160–161, **161**, **249**

atomic force microscopy (AFM)
  bar-code system, **155**
  bowtie junctions, 110, **112**, **113**, **114**
  capture system, 177–178, **178**
  characterization, 82–83, **83**
  DNA origami self-assembly, **195**
  instead of denaturing gels, 95
  PX-JX_2 139, **139**
  TX molecules, **106**
atomic force microscopy of 1D arrays, 101, **102**, **190**, **193**, **194**, **235**
atomic force microscopy of 2D arrays, **105**, **108**
atomic force microscopy of 3D arrays, 117–119
autonomous devices, **141**, 141–142, 209–210, **210**
axis
  branched, 2–4, **4**, **5**, 77, 131–133, **132**
  and sticky end cohesion, **8**, 8–10, **9**

bar-code system in DNA origami, 154, **155**
base stacking
  in classical DNA sequences, **14**, 14–15
  in J1 junction, **29**, 31–32
B-DNA
  right-handed nature of, 47, **47**
  shape-shifters, 131–133, **132**
  and Z-DNA, 48–52, **49**, **50**, **51**, **52**
biotin, **146**
Borromean rings
  characterization, 80, **81**
  and Z-DNA, 48–51, **50**
bowtie junctions, 110, **111**, **112**, **113**, **114**
branch migration
  in non-Watson–Crick sequences, **22**, 22–24
  seesaw gates, 209
  shape-shifter, 131, **132**
branched DNA molecules
  hydroxyl radical autofootprinting analysis, 77
  shape-shifters, 131–133, **132**
  as a traffic intersection, 2–4, **4**, **5**
BTX motif, 187–188, **188**, **189**
bulged junctions
  in 2D arrays, 110–112, **115**
  3-arm, 88–92, **89**, **90**, **91**, **92**

capture system in DNA origami, 173–179, **177**, **178**, **180**
carbon and DNA organization, 239–243, **243**
cascade walkers, 145–148, **146**, **147**
catenanes, 79–80
Chapman, Henry, 248
characterization
  of arrays, 82–83, **83**
  of devices, 80–82, **81**
  of ligation targets, 77–80, **78**, **79**
  motif, 69–76, **70**, **72**, **73**, **74**, **75**
circular origami, 163–164, **166**, **167**, **168**
classic DNA structure
  1D emulation of, **188**, **189**, **190**, **193**, **194**
  and base stacking, **14**, 14–15
  complementary pairing, 61–63, **62**
  double helix, 1–2, **2**, **3**
  nucleotide and backbone variations, 213–226
clocked devices, 130–141, **145**
color control
  crystal, 125–126, **127**
  by restriction, 101–104
complementary pairing
  of DNA elements, 1, **2**
  and organization of metal particles, 243–245, **244**, **245**
  in single-strand DNA design, 61–63, **62**
construction methods, 64–69, **67**, **68**, **69**
cost of DNA, 208–209
Crick, F. H. C., 1, 61
crystal structure
  genetic material as, 97
  Paukstelis DNA, 213, **214**, **215**
  and periodic arrays, 88
  RNA, **221**, **222**, **223**
  triplex DNA, 213–215, **216**
cumulative XOR operation, **205**, 205–207, **206**, **207**, **208**

DAE motif
  2D arrays, 92, 95, **100**, 100–101
  characterization, 79, **79**
  as isomer of DX, 35, **35**
DAO motif
  characterization, 79, **79**, **100**
  as isomer of DX, 35, **35**
denaturing gels
  characterization, 77–80, **78**, **79**
  in robust motifs, 92–95, **93**, **94**
Depth (woodcut), 4–5, **6**
design for DNA branched sequences
  base pairing in, 11–15, **13**, **14**
  conditions, 24–26
  elements, 18–21, **20**

non-classical pairing, 15–18, **16**, **17**, **18**, **19**
symmetry minimization, 21–24, **22**, **23**
devices
autonomous, **141**, 141–142
DNA origami, **242**
multiple state, 139–141, **140**
protein-based, **142**, 142–143
PX-JX_2 136–139, **137**, **138**, **139**
robust assembly, 130
sequence-specific, 135–136
shape-shifters, 131–135, **132**, **133**, **134**
walkers, 143–148, **145**, **146**, **147**
DNA backbone
classic DNA structure, **3**, **13**, 13–21
I-motif, **19**
variations of, **216**, **224**, **225**
DNA bricks, 165–167, **168**, **169**, **170**
DNA concentration in branched systems, 25–26
DNA mispairs in non-Watson–Crick sequences, 15–18, **17**, **18**
DNA origami
basic structure of, 150–154, **151**, **152**, **153**
capture system, 173–179, **177**, **178**, **180**
combinations before, 172–173, **173**
distortion, 161–162, **162**, **163**, **164**, **165**
predecessors to, 154–156, **155**
self-assembly, 192–196, **194**, **195**, **196**, **197**
and sequence symmetry, 26
tensegrity, 160–161, **161**
DNA origami types
3D, 156–165
6-helix bundle, 154, **157**, **159**
box, 158–160, **160**
circular, 163–164, **166**, **167**, **168**
variations of, **240**, **241**, **243**
DNA synthesis
construction methods, 64–69, **67**, **68**, **69**
facile, 11
and Moore's law, 209
proof of, 51–52, **52**, **81**
DNAzymes as components of autonomous devices, **141**, 141–142
double cohesion
3D-DX triangle motif, 113–115, **117**
and 6-helix bundle, 112–113, **116**
in DX triangle structure, 110–112, **115**
in molecular assembly line, 180–182, **181**, **183**
parallelogram arrays, 115, **118**
skewed TX molecules, 115–116, **119**
double helical structure
base pairing in, 11–15, **13**, **14**
of classic DNA molecule, **2**
and non-classical pairing, 15–18, **16**, **17**, **18**, **19**
into a sticky-end lattice, 5–10, **7**, **8**, **9**
DX molecule
2D arrays from, **103**, **104**, **105**
double cohesion, 110–112, **115**
hydroxyl radical autofootprinting analysis, 77
reciprocal exchange, **35**, 35–37, **36**
in robust motifs, 91–95, **92**, **93**, **94**
shape-shifters, 131–133, **132**
tensegrity triangle, 113–115, **117**
DX molecule arrays
1D arrays, 100, **102**
2D arrays from, **36**, **73**, **79**, **83**, 99–104, **100**, **102**

enzymatic modification
and autonomous logical devices, 209–210, **210**
ligation, 66
error correction protocol, 179
Escher, M. C., 4–5
exponential growth
and DNA origami, 109–110, 194–196
prevention of, 192
and replication, 186

facile synthesis, 11
Ferguson analysis, 74–76, **75**
Fischer–Lerman denaturing conditions, 80
fluorescence resonance energy transfer (FRET)
integration host factor protein device, **142**, 142–143
nanorobotics, 80–81
sequence-specific, 135–136
shape analysis, 76
shape-shifters, **132**, 133, **133**

Gang, O., **243**, 245
gel electrophoresis, **70**, 70–74, **72**, **73**
glycosyl bonds
in classical DNA sequences, **3**, 13–14, **16**
in non-Watson–Crick pairings, 15, **17**
G-tetrad
in non-Watson–Crick sequences, 15, **18**
shape-shifters, 134, **134**
structures, **217**, 217–218, **218**

half-turn, 97–99, **98**
Hamiltonian path problem, 198–202, **199**, **200**, **201**
helicity determination, 151, **163**, 224–226, **226**, **227**
Holliday junction
  as analytical tool, **109**, 109–110
  base stacking, 32
  most popular, 4
  shape analysis, 76
homology and PX molecules, **38**, 40–41, **42**
Hoogsteen base pairing, 15, **17**
hydrogen bonding
  2D arrays, **109**, 109–110
  in classical DNA sequences, **2**, 4, 12–13, **13**
  non-Watson–Crick DNA pairing, 15, **17**, **18**
hydroxyl radical autofootprinting, 77

I-motif
  in non-Watson–Crick sequences, 15–18, **19**, 218–219, **219**
  shape-shifters, 133, **133**
in vitro synthesis, 1
in vivo synthesis, 220
information encryption
  2D algorithmic assemblies, 207
  3-colorability, 202–204
  the Adleman experiment, 198–202, **199**, **200**, **201**
  algorithmic assembly, **204**, 204–205
  autonomous logical devices, 209–210, **210**
  cumulative XOR, **205**, 205–207, **206**, **207**, **208**
  small algorithmic assembly cost, 208–209
integration host factor protein device (IHF), **142**, 142–143
isomorphous replacement method, 119–127, **120**, **121**, **122**, **123**, **124**, **125**, **126**, **127**

J1 branched junction molecule
  reciprocal exchange in, 30–32, **31**
  sequence design elements, 18–26, **20**
junction
  3-arm, 88–92, **89**, **90**, **91**, **92**
  4-arm, 106–108, **107**, **108**
  anti, **53**, 54–57, **55**, **56**, **57**, **58**, 110, **111**, **113**
  bowtie, 110, **110**, **111**, **113**, **114**
  bulged, 88–92, **89**, **90**, **91**, **92**, 110–112, **115**
  Holliday, 4, 32, 76, **109**, 109–110
  many types of DNA, 2–4, **4**, **5**
  meso, **53**, 54–57, **55**, **56**, **57**, **58**
junction analysis
  Ferguson analysis, **75**
  hydroxyl radical autofootprinting analysis, 77

knots
  antijunctions and mesojunctions in, **53**, 54–57, **55**, **56**, **57**, **58**
  characterization, 79–80
  and DNA half-turns, 52, **53**
  RNA trefoil, 219, **220**
  and single-strand motif design, **45**, 45–47, **46**
  and switchback DNA, **57**, 58–61, **59**, **60**
  topological protection of, **67**
  and Z-DNA, 47–52, **49**, **50**, **51**, **52**
Krishnan, Y., 133, **218**

lattices, **9**, 9–10
ligation
  characterization, 77–80, **78**, **79**
  and microscopy caveat, 101–104, **102**, **104**
  and phosphates, 65–66
  and solid-support methodology, 66–69, **68**, **69**
  of sticky-end cohesion, **7**

Mao, C., 26
melting behavior
  nanoparticle, **244**
  temperature and, 12, 76
mesojunctions, **53**, 54–57, **55**, **56**, **57**, **58**
microscopy caveat, 101
Mirkin, C., 233, 243, 245
mispairs in non-Watson-Crick sequences, 15–18, **17**, **18**
molecular assembly line, 165–167, 179–182, **181**, **183**, **184**
Moore's law, 209
motif characterization, 69–76, **70**, **72**, **73**, **74**, **75**
motif design
  in non-Watson–Crick sequences, **19**
  and reciprocal exchange, 28–42
motif replication, 24
multiple state devices, 139–141, **140**
MutS protein device, **142**, 143

nanoparticle
  assembly line, 181, **181**
  carbon, 239–243, **243**
  metallic, 233–237, **239**, **240**, 243–245

nanorobotics, 80–82, **81**
natural selection, 186–187
N-connected objects, **32**, 32–35, **34**, **36**
non-Watson–Crick DNA pairing
  bonds in, **14**, 15–21, **17**, **18**, **19**, **20**
  branch migration, **22**, 22–24
  I-motif, 218–219, **219**
nucleic acid base pairings
  in classical DNA sequences, **2**, 12–15, **13**, **14**
  in non-Watson–Crick sequences, **14**
  variations of, 213–226
nucleic acid junctions
  in 6-helix bundle, 112–113, **116**
  sequence descriptions, 6–7
  similar to *Depth* woodcut, 4–5, **6**

octahedral truss, 33–35, **34**
organization by DNA
  biological robotics, 237–239, **242**
  metal incorporation, 237, **242**
  metallic nanoparticles, 233–237, **234**, **235**, **236**, **241**
  multidimensional metal, 243–245, **244**, **245**
  nanoscale carbon, 239–243, **243**
  polymer topology, 231–233, **232**, **233**

palindromes, 61–63
parallelogram arrays, 115, **118**
paranemic cohesion, 41, **42**
PATX motif, **36**
Paukstelis DNA structure, 213, **214**, **215**
periodic arrays
  characterization, 82, **83**
  components of, 88
phosphorylation construction method, 65–66
PNA, 222–226, **225**, **226**, **227**
polarity, **29**, **36**, **38**
polymerase chain reaction (PCR), 65
protein-based devices, **142**, 142–143
purification methods, 64–65
purine
  in classical DNA sequences, 12–13, **13**
  and symmetry minimization, 24
PX molecules
  cohesion, 154
  Ferguson analysis, 76
  hydroxyl radical autofootprinting analysis, 77
  reciprocal exchange, **38**, 38–42, **40**, **42**
  and switchback DNA, 58–61, **59**, **60**
PX-JX cassette
  in capture system, 173–179, **177**, **178**
  characterization, 71–74, **72**
  in molecular assembly line, 179–182, **181**, **183**
  before origami, 172–173, **173**, **174**, **175**, **176**
PX-JX devices, 136–139, **137**, **138**, **139**
pyrimidine
  in classical DNA sequences, 12–13, **13**
  and symmetry minimization, 113

reciprocal exchange
  in classical double helix, 28–30, **29**
  and DX molecules, **35**, 35–37, **36**
  in J1 junction, 30–32, **31**
  and N-connected objects, **32**, 32–35, **34**, **36**
  and PX motif, **38**, 38–42, **40**, **42**
relaxed bundles, 161–162, **162**, **163**
replication
  complementary pairing, 1
  growth of, 186
  origami, 192–196
  using large molecules, 187–192
reporter strand
  3-colorability problem, 203, **203**
  3-arm, **91**
  bulged junction, **92**
  characterization, 77–79, **79**
  cumulative XOR operation, **206**, 206–207, **207**
  DAE, 95, **100**
resolution
  AFM, **108**, 117
  crossover sites, 28
  hydroxyl radical autofootprinting analysis, 77
  reciprocal exchange, **29**
  and self-assembly, 150
  X-ray crystallography, 118–121, 126–127
restriction analysis, 80, **81**
Rich, A., 47
RNA nanotechnology, 219–222, **220**, **221**, **222**
RNA origami, **225**
robotics, 237–239, **242**
robust motifs
  4-arm junction, 106–108, **107**, **108**
  devices, 130–148
  history of, 88–96

robust motifs (cont.)
 Paukstelis DNA, 213, **214**, **215**
Rothemund, P., 151, 222

seesaw gates, 209
self-assembly
 1D emulation, 187–192, **188**, **189**, **190**, **191**, **193**, **194**
 DNA origami, 192–196, **194**, **195**, **196**, **197**
 evolution and selection in, 186–187
 triplex DNA, 213–215, **216**
self-complementarity, 61–63
sequence-specific devices, 135–136
shape-shifter devices, 131–135, **132**, **133**, **134**
Shih, W. M., 154, **166**
Sierpinski triangle, 207
single-stranded DNA topology
 catenanes, **33**, 44–45, **45**
 knots, 45–63
skewed TX molecules, 115–116, **119**
Sleiman, H., 236, 237
Snelson, K., 160–161, **161**
solid-support object synthesis, 64–69, **68**, **69**
Solomon's knot, **52**
solution conditions
 controlling structure by, 26
 Z-DNA, **47**, 47–48
sticky ends
 2D arrays, 97–101
 3D crystals, 119–127, **122**, **123**, **125**
 capture system, 173–177, **177**
 in double helix structure, 5–10, **7**, **8**, **9**
stoichiometric determination, **74**
strand polarity, **29**, **36**, **38**
streptavidin
 in biotin, 137, **138**, 187–188, **189**
 metallic, 237, **241**
structural DNA nanotechnology
 development of, **250**
 facile synthesis, 11
 fundamental notion of, **9**, 9–10
 nucleic acid hybridization, 11
switchback DNA, **57**, 58–61, **59**, **60**
symmetry
 maximization, 26
 minimization, **20**, 20–24, **22**

temperature and non-classical pairing, 24–25
tensegrity triangle
 2D arrays, 113–115, **117**
 3D crystals, 119–127, **120**, **121**, **122**, **123**, **124**, **125**, **126**, **127**
 in molecular assembly line, 180–182, **181**, **183**
toehold
 origami box, 159, **160**
 PX-JX_2 136–139, **138**
 seesaw gates, 209
 sequence-specific, **135**, 135–136
topoisomerase (RNA), 219, **220**
topological protection construction method, 66, **67**
trefoil knot, 46, **46**
triplex DNA, 213–215, **216**
TX molecules
 2D arrays, **36**, 104–107, **105**, **106**, **107**, **108**
 hydroxyl radical autofootprinting analysis, 77, **78**
 and PX-JX cassette, 172–173, **173**, **174**, **175**
 reciprocal exchange, **35**, **36**, 36–37
 skewed, 115–116, **119**
 walkers, 144, **145**

vanilla DNA
 1D emulation of, **188**, **189**, **190**, **191**, **193**, **194**
 and base stacking, **14**, 14–15
 complementary pairing, 61–63, **62**
 double helix, 1–2, **2**, **3**
 nucleotide and backbone variations, 213–226

walkers
 devices, 143–148, **145**, **146**, **147**
 in molecular assembly line, 180–182, **181**, **183**
Wang tiles, **204**, 204–205
Watson, J. D., 1, 61
Watson–Crick DNA
 1D emulation of, **188**, **189**, **190**, **191**, **193**, **194**
 and base stacking, **14**, 14–15
 complementary pairing, 61–63, **62**
 double helix, 1–2, **2**, **3**

nucleotide and backbone variations, 213–226
Weizmann, Y., 248
Winfree, E., 83, 204, 207, 209

X-ray crystallography
and 3D structures, 117–119
and origami, **158**

Yan, H., 163, 237
Yin, P., 165

Z-DNA
left-handed nature of, **47**, 47–52, **49**, **50**, **51**, **52**
and symmetry minimization, 24